Advances in Mathematical Fluid Mechanics

Series Editors

Giovanni P. Galdi, Pittsburgh, USA
John G. Heywood, Vancouver, Canada
Rolf Rannacher, Heidelberg, Germany

Advances in Mathematical Fluid Mechanics is a forum for the publication of high quality monographs, or collections of works, on the mathematical theory of fluid mechanics, with special regards to the Navier-Stokes equations. Its mathematical aims and scope are similar to those of the *Journal of Mathematical Fluid Mechanics*. In particular, mathematical aspects of computational methods and of applications to science and engineering are welcome as an important part of the theory. So also are works in related areas of mathematics that have a direct bearing on fluid mechanics.

The monographs and collections of works published here may be written in a more expository style than is usual for research journals, with the intention of reaching a wide audience. Collections of review articles will also be sought from time to time.

More information about this series at http://www.springer.com/series/5032

Björn Gustafsson • Razvan Teodorescu
Alexander Vasil'ev

Classical and Stochastic
Laplacian Growth

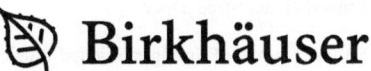 Birkhäuser

Björn Gustafsson
Department of Mathematics
KTH Royal Institute of Technology
Stockholm, Sweden

Razvan Teodorescu
Department of Mathematics
University of South Florida
Tampa, FL, USA

Alexander Vasil'ev
Department of Mathematics
University of Bergen
Bergen, Norway

ISSN 2297-0320   ISSN 2297-0339 (electronic)
ISBN 978-3-319-08286-8   ISBN 978-3-319-08287-5 (eBook)
DOI 10.1007/978-3-319-08287-5
Springer Cham Heidelberg New York Dordrecht London

Library of Congress Control Number: 2014955451

Mathematics Subject Classification (2010): 76D27, 76M40, 30C20, 30C35, 30C62, 31A05, 35Q30, 35R35

Printed on acid-free paper

Springer is part of Springer Science+Business Media (www.birkhauser-science.com)

Contents

Preface

One of the most influential works in fluid dynamics at the edge of the XIXth and XXth centuries was a series of papers, see, e.g., [265], written by *Henry Selby Hele-Shaw* between 1897 and 1899. This was a time of impetuous development of several fundamental branches of natural sciences. The following few citations related to the present monograph might be sufficient to impress an inquiring minded reader: Reynolds' description [466] of the turbulence phenomenon at higher velocities in 1873–1883; Korteweg and de Vries' description [320] of the 'solitary wave' in 1895, discovered by the Scottish engineer Russell about half a century earlier; and of course, we must mention the impressive developments in the theory of relativity and quantum physics, which used the elegant mathematical formulation of electrodynamics given by Maxwell [382], both as a benchmark for more unified theories (which emerged later in the XXth century as gauge theories), and as a guiding mathematical structure (reflected nowadays in the extensive use of principal bundles and central extensions in theoretical physics).

Hele-Shaw (1854–1941) was one of the most prominent engineering researchers of his time, a pioneer of technical education, a great organizer, president of several engineering societies, including the Royal Institution of Mechanical Engineers, Fellow of the Royal Society, and sadly, an example of one of the many undeservedly forgotten great names in Science and Engineering. In his original works, he first described his famous cell that became a subject of deep investigation only more than 50 years later. A Hele-Shaw cell is a device for investigating two-dimensional flow of a viscous fluid in a narrow gap between two parallel plates. This cell is the simplest system in which multi-dimensional convection is present. Probably the most important characteristic of flows in such a cell is that when the Reynolds number based on gap width is sufficiently small, the Navier–Stokes equations averaged over the gap reduce to a linear relation similar to Darcy's law and then to a Laplace equation for pressure. Different driving mechanisms can be considered, such as surface tension or external forces (suction, injection). Through the similarity in the governing equations, Hele-Shaw flows are particularly useful for visualization of saturated flows in porous media, assuming they are slow enough to be governed by Darcy's law. Nowadays, the principle of the Hele-Shaw cell is used as a powerful tool for modelling growth phenomena in several fields of natural sciences and engineering, in particular, condensed-matter physics, material science, crystal growth and, of course, fluid mechanics. But Hele-Shaw is known not only for his Stream-line Flow Methods (1896–1900) in which this cell plays a fundamental role. Two other of Hele-Shaw's great inventions are his Friction Clutch (1905), an early version of multi-plate wet clutch, and his Automatic Variable-Pitch Propeller (1924), jointly with T. Beacham. In fact, the full list of his inventions is much longer and comprises 82 patents. Sir George Gabriel Stokes wrote about the Hele-Shaw cell: 'Hele-Shaw's experiments afford a complete graphical solution, experimentally obtained, of a problem which from its complexity baffles mathematicians except in a few simple cases'. Stokes mentions also Hele-Shaw's experiments in his letter to Lord Kelvin from September 7, 1898: 'Hele-Shaw has some beautiful photographs, very interesting to you and me. By means of a thin stratum of viscous liquid between close glass walls, flowing past an interruption in the film, you can realise experimentally the theoretical stream lines in two dimensions in a perfect fluid flowing round a body represented in section by the obstacle' (see [532]).

Since the original works of Hele-Shaw appeared and a mathematical model of Hele-Shaw flow was formulated in the famous monograph by Lamb [341], many interesting and exciting developments have occurred. The century-long development connecting the original Hele-Shaw experiments, the conformal mapping formulation of the Hele-Shaw flow by *Pelageya Yakovlevna Polubarinova-Kochina* (1899–1999) and *Lev Aleksandrovich Galin* (1912–1981) [438, 439, 199], and the modern treatment of the Hele-Shaw evolution based on integrable systems and on the general theory of plane contour motion, was marked by several important contributions by individuals and groups.

The main idea of Polubarinova-Kochina and Galin was to apply the Riemann mapping from an appropriate canonical domain (the unit disk in most situations) onto the phase domain in order to parameterize the free boundary. The evolution equation for this map, named after its creators, allows us to construct many explicit solutions and to apply methods of conformal analysis and geometric function theory to investigate Hele-Shaw flows. In particular, solutions to this equation in the case of advancing fluid give subordination chains of simply connected domains which have been studied for a long time in the theory of univalent functions. The Löwner–Kufarev equation [328], [362] plays a central role in this study (*Charles Loewner* or *Karel Löwner* originally in Czech, 1893–1968; *Pavel Parfen'evich Kufarev*, 1909–1968). The Polubarinova–Galin and the Löwner–Kufarev equations, having some evident geometric connections, are of somewhat different nature. While the evolution of the Laplacian growth given by the Polubarinova–Galin equation is completely defined by the initial conditions, the Löwner–Kufarev evolution depends also on an arbitrary control function. The Polubarinova–Galin equation is essentially non-linear and the corresponding subordination chains are of rather complicated nature. Interestingly, it was Kufarev [330], [331] who anticipated further results in viscous fingering in 1948 by means of this equation.

Among other remarkable contributions we distinguish the discovery of the viscous fingering phenomenon by *Sir Geoffrey Ingram Taylor* (1886–1975) and

Philip Geoffrey Saffman (1931–2008) [488, 489], and the discovery, by *Stanley Richardson* (1943–2008) [470], of a complete set of integrals for the Hele-Shaw evolution, namely the harmonic moments. Contributions made by scientists from Great Britain (D. Crowdy, L.J. Cummings, C.M. Elliott, S.D. Howison, J.R. King, J.R. Ockendon, S. Richardson) are to be emphasized. They have substantially developed the complex variable approach and actually converted the Hele-Shaw problem into a modern challenging branch of applied mathematics. An even more recent mathematical physics perspective, through integrable systems in particular, allows us to look at Hele-Shaw evolution as at a general contour dynamics in the plane embedded into a dispersionless Toda hierarchy. This approach is due mainly to I. Krichever, A. Marshakov, M. Mineev-Weinstein, P. Wiegmann, A. Zabrodin among others.

The first monograph treatment [249] of Hele-Shaw appeared in 2006 and covered mostly the classical period of development of this area (a related monograph is [560]). The last decade has been marked by a burst of interest in Laplacian growth (another name for the Hele-Shaw free boundary problem used in mathematical physics literature), caused in particular by related statistical physics models, e.g., Diffusion Limited Aggregation. Several new methods, such as integrable systems and random matrices, have been employed to treat problems in Laplacian growth. Therefore, a revision of the book seemed necessary in order to give a broader and more comprehensive survey of the current status of this field, whose impetuous growth has resulted in this new text, with three authors. While Chapters 6 through 9 are entirely new (except for the first section in Chapter 7), the main ideas of [249] are also present, in particular in the first half of the introduction (Chapter 1, Sections 1–7), Sections 2, 5, and 6 of Chapter 2 (and an extended Section 1), Sections 1–2 of Chapter 3 (and partly Sections 3 and 5), Sections 1, 3, 5, and 7 of Chapter 4, and finally Sections 1–4 of Chapter 5.

In the present monograph, we aim at giving a presentation of recent and new ideas that arise from the problems of planar fluid dynamics and which are interesting from the point of view of geometric function theory, potential theory, and mathematical physics. In particular, we are concerned with geometric problems for Laplacian growth, its stochastic formulation and its treatment from the viewpoint of integrable systems and random matrices. Ultimately, we see the interaction between several branches of complex, potential analysis, mathematical physics and planar fluid mechanics.

For most parts of this book we assume the background provided by graduate courses in real and complex analysis, in particular the theory of conformal mappings, and some basic notions of fluid mechanics. We also make some historical remarks concerning the scientists who have contributed to the topic. We have tried to keep the book as self-contained as possible.

Acknowledgement. We would like to acknowledge many useful conversations with J.R. Arteaga, J. Becker, L. Cummings, P.I. Ètingof, R. Friedrich, V. Goryainov, V. Gutlyanskiĭ, Yu. Hohlov, S. Howison, D. Khavinson, J. King, K. Kornev,

M. Mineev-Weinstein, J. Ockendon, Ch. Pommerenke, D. Prokhorov, S. Rusche-weyh, H. Shahgholian, H.S. Shapiro, P. Wiegmann, A. Zabrodin. All three authors especially want to thank their wives Eva Odelman, Iuliana Teodorescu, and Irina Markina. They always inspire our work. Irina Markina is, moreover, a colleague and co-author of the third author.

The project has been supported by

- the Swedish Research Council, the Göran Gustafsson Foundation (Sweden),
- the projects of the Norwegian Research Council #177355/V30, #204726/V30, and #213440/BG,
- the Research Networking Programme 'Harmonic and Complex Analysis and its Applications' ESF,
- the EU project FP7 IRSES program STREVCOMS, grant no. PIRSES-GA-2013-612669.

Björn Gustafsson, Razvan Teodorescu, & Alexander Vasil'ev
    Stockhom-Tampa-Bergen, 2014

Chapter 1

Introduction and Background

1.1 Newtonian fluids

A *fluid* is a substance which continues to change shape as long as there are shear stresses (dependent on the velocity of deformation) present. If the force F acts over an area A, then the ratio between the tangential component of F and A gives a shear stress across the liquid. The liquid's response to this applied shear stress is to flow. In contrast, a solid body undergoes a definite displacement or breaks completely when subject to a shear stress. Viscous stresses are linked to the velocity of deformation. In the simplest model, this relation is just linear, and a fluid possessing this property is known as a *Newtonian fluid*. The constant of proportionality between the viscous stress and the deformation velocity is known as the *coefficient of viscosity* and it is an intrinsic property of a fluid.

Certain fluids undergo very little change in density despite the existence of large pressures. Such a fluid is called *incompressible*. In fluid dynamics one speaks of incompressible flows, rather than incompressible fluids. A laminar flow, that is a flow in which fluid particles move microscopically in straight parallel lines without macroscopic velocity fluctuations, satisfies Newton's Viscosity Law (or is said do be Newtonian) if the shear stress in the direction x of flow is proportional to the change of velocity V in the orthogonal direction y as

$$\sigma := \frac{dF}{dA} = \mu \frac{\partial V}{\partial y}.$$

The coefficient of proportionality μ is called the coefficient of viscosity or dynamic viscosity. Many common fluids such as water, all gases, petroleum products are Newtonian. A non-Newtonian fluid is a fluid in which shear stress is not proportional solely to the velocity gradient, perpendicular to the plane of shear. Non-Newtonian fluids do not have a well-defined viscosity coefficient. Pastes, slurries, high polymers are not Newtonian. Pressure has only a small effect on viscosity and

1

this effect is usually neglected. The kinematic viscosity is defined as the quotient

$$\nu = \frac{\mu}{\rho},$$

where ρ stands for density of the fluid. All these considerations can be made with dimensions and units taken into account, or else be made dimensionless.

1.2   The Navier–Stokes equations

Important quantities that characterize the flow of a fluid are

- m – mass;
- p – pressure;
- $\mathbf{V}$ – velocity field;
- Θ – temperature;
- ρ – density;
- μ – viscosity.

1.2.1   The transport theorem

On deriving the equations for fluid motion one often needs to take time derivatives of integrals defined over time-dependent regions. One traditional way to do this is to use the so-called *Reynolds' transport theorem* (*Osborne Reynolds* 1842–1912).

Let us consider a part of the fluid that occupies a *control region* $V(t)$ bounded by a *control surface* $S(t)$. This part is supposed to be composed of the same fluid particles all the time, hence to move with the fluid, i.e., along with the velocity field $\mathbf{V}$. Let $N(t)$ be an *extensive property* of the fluid in the region, such as mass, momentum, or energy. Let $\mathbf{x} = (x_1, x_2, x_3)$ be the spatial variable and let t be time. We denote by $\eta(\mathbf{x}, t)$ the corresponding *intensive property* which is equal to the extensive property per unit of mass, $\eta = dN/dm$,

$$N(t) = \int_{V(t)} \eta \rho \, dv, \quad dv = dx_1 dx_2 dx_3.$$

The mathematical version of Reynolds' transport theorem states that the rate of change of N for a system at time t equals the rate of change of N inside the control volume V plus the rate of flux of N across the control surface S at time t:

$$\frac{dN}{dt} = \int_{V(t)} \frac{\partial}{\partial t}(\eta \rho) \, dv + \oint_{S(t)} \eta \rho \mathbf{V} \cdot \mathbf{n} \, d\sigma. \tag{1.1}$$

Here $\mathbf{V} = (V_1, V_2, V_3)$, and $\mathbf{n}$ is the unit normal vector in the outward direction. The Gauss theorem implies

$$\frac{dN}{dt} = \int_{V(t)} \left[\frac{\partial}{\partial t}(\eta\rho) + \nabla \cdot (\eta\rho\mathbf{V}) \right] dv.$$

In this version of the transport theorem the product $\eta\rho$ remains a single quantity (as in the definition of $N(t)$).

Let us next introduce a derivative $\frac{D}{Dt}$ which is called the *convective derivative*, or Eulerian derivative, and which is defined as

$$\frac{D}{Dt} = \frac{\partial}{\partial t} + \mathbf{V} \cdot \nabla,$$

or in coordinates

$$\frac{D}{Dt} = \frac{\partial}{\partial t} + V_1 \frac{\partial}{\partial x_1} + V_2 \frac{\partial}{\partial x_2} + V_3 \frac{\partial}{\partial x_3}.$$

Then we can also write

$$\frac{dN}{dt} = \int_{V(t)} \left(\frac{D(\eta\rho)}{Dt} + \eta\rho(\nabla \cdot \mathbf{V}) \right) dv. \tag{1.2}$$

1.2.2 The continuity equation

If we choose the mass as the extensive property, then $N = m$, $\eta = 1$ and Reynolds' transport theorem (1.1) gives

$$\frac{dm}{dt} = \int_{V(t)} \frac{\partial \rho}{\partial t} dv + \oint_{S(t)} \rho\mathbf{V} \cdot \mathbf{n} \, d\sigma.$$

The physical law of conservation of mass states that $\frac{dm}{dt} = 0$. Therefore,

$$\int_{V(t)} \left(\frac{\partial \rho}{\partial t} + \nabla \cdot (\rho\mathbf{V}) \right) dv = 0.$$

Since this equation holds for any control region $V(t)$, we get

$$\frac{\partial \rho}{\partial t} + \nabla \cdot (\rho\mathbf{V}) = 0. \tag{1.3}$$

This is known as the *continuity equation*, and it can also be written as

$$\frac{D\rho}{Dt} + \rho(\nabla \cdot \mathbf{V}) = 0. \tag{1.4}$$

Using, in (1.2), the product rule for $\frac{D}{Dt}$ together with the continuity equation (1.4) the Reynold's transport theorem, for arbitrary η, simplifies to

$$\frac{dN}{dt} = \int_{V(t)} \frac{D\eta}{Dt}\,\rho\,dv = \int_{V(t)} \frac{D\eta}{Dt}\,dm. \tag{1.5}$$

This version of the transport theorem has some physical significance since its derivation is based on the principle of conservation of mass, see, e.g., [99]. Note also that η and ρ have different roles in this version.

On choosing $\eta = 1/\rho$ in (1.1), $N(t)$ becomes the volume of $V(t)$. A fluid is called *incompressible* if the volume of any control region is constant in time. By (1.2) this means that

$$\nabla \cdot \mathbf{V} = 0, \tag{1.6}$$

or equivalently (in view of (1.4)), $\frac{D\rho}{Dt} = 0$. This does not automatically imply that ρ is constant, but if the fluid is homogenous, meaning that ρ is constant in space, then incompressibility enforces ρ to be constant also in time (by (1.3) and (1.6)).

1.2.3   The Euler equation

Let us now consider only homogenous incompressible fluids. Linear momentum of an element of mass dm is a vector quantity defined as $d\mathbf{P} = \mathbf{V}\,dm$, or for the whole control volume,

$$\mathbf{P} = \int_{V(t)} \rho\mathbf{V}\,dv.$$

Applying Reynolds' transport theorem in the form (1.5) we get

$$\frac{d\mathbf{P}}{dt} = \int_{V(t)} \rho\frac{D\mathbf{V}}{Dt}\,dv = \int_{V(t)} \frac{D\mathbf{V}}{Dt}\,dm,$$

which infinitesimally is $\frac{D\mathbf{V}}{Dt}\,dm$, i.e., just the product of the mass element and acceleration.

Newton's second law for an inertial reference frame states that the rate of change of the momentum $\mathbf{P}$ equals the force exerted on the fluid in $V(t)$:

$$d\mathbf{F} = \frac{D\mathbf{V}}{Dt}\,dm = \left(\frac{\partial}{\partial t}\mathbf{V} + (\mathbf{V}\cdot\nabla)\mathbf{V}\right)dm, \tag{1.7}$$

where $\mathbf{F}$ is the vector resultant of forces. Suppose for a moment that there are no shear stresses (inviscid fluid). If the surface forces $\mathbf{F}_s$ on a fluid element are due only to pressure p and the body forces are due to gravity in the x_3-direction, then we have $d\mathbf{F} = d\mathbf{F}_s + d\mathbf{F}_b$, or

$$d\mathbf{F} = -(\nabla p)\,dv - g\nabla x_3(\rho\,dv), \tag{1.8}$$

where $\mathbf{F}_b$ is the gravity force per unit of mass and g is the gravity constant. Substituting (1.8) into (1.7) we obtain

$$-\frac{1}{\rho}\nabla p - g\nabla x_3 = \frac{\partial \mathbf{V}}{\partial t} + (\mathbf{V} \cdot \nabla)\mathbf{V},$$

or

$$-\nabla p - \rho g\nabla x_3 = \rho\frac{D\,\mathbf{V}}{D\,t}. \tag{1.9}$$

Equation (1.9) is known as the *Euler equation*, valid for inviscid fluids.

In terms of control volume we have

$$\frac{d}{dt}\int_{V(t)} \rho\mathbf{V}\,dv = -\int_{V(t)} (\nabla p + \rho g\nabla x_3)dv,$$

or, in more general notation,

$$\frac{d}{dt}\int_{V(t)} \rho\mathbf{V}\,dv = \oint_{S(t)} \sigma\cdot\mathbf{n}\,dA - \int_{V(t)} \rho g\nabla x_3 dv, \tag{1.10}$$

where $\sigma = (\sigma_{ij})_{i,j=1}^3$, $\sigma_{jj} = -p$, $\sigma_{ij} = 0$, $i \neq j$, is the stress tensor. In general, the *stress tensor* $(\sigma_{ij})_{i,j=1}^3$ is defined by the relationship $dF_i = \sum_{j=1}^3 \sigma_{ij}n_j\,dA$ between the surface force $d\mathbf{F}$ on an infinitesimal area element dA and the normal vector $\mathbf{n}$ of it ($\mathbf{F} = (F_1, F_2, F_3)$, $\mathbf{n} = (n_1, n_2, n_3)$).

1.2.4 The Navier–Stokes equation

The first term in the right-hand side of the Euler equation (1.10) is due to the surface forces and the second one is due to the body forces (or forces per unit mass in (1.9)). Let us now allow the fluid to be viscous and consider the shear and normal stresses σ_{ij} on the surface of an infinitesimal control region $dv = dx_1 dx_2 dx_3$ in form of a parallelepiped with principal diagonal joining the points $\mathbf{x} = (x_1, x_2, x_3)$ and $\mathbf{x} + d\mathbf{x} = (x_1 + dx_1, x_2 + dx_2, x_3 + dx_3)$. We denote by x_i-*surface*, that surface with one of its vertices at the point $\mathbf{x}$ and with the normal vector parallel to the x_i axis. The *surface parallel to the x_i-surface* is the one with a vertex at $\mathbf{x} + d\mathbf{x}$. We denote by σ_{jj} the normal stress on the x_j-surface in the outward direction. The normal stress on the parallel surface is $\sigma_{jj} + \frac{\partial \sigma_{jj}}{\partial x_j}dx_j$. By σ_{ij}, $i \neq j$, we denote the shear stress on the x_i-surface in the direction x_j and similarly for the parallel surface. The shear and normal stresses are given by a stress-velocity relation which is known as Stokes' viscosity law for incompressible fluids. It states that the stress tensor $(\sigma_{ij})_{i,j=1}^3$ is given by

$$\sigma_{ii} = -p + 2\mu\frac{\partial V_i}{\partial x_i}, \quad \sigma_{ij} = \mu\left(\frac{\partial V_i}{\partial x_j} + \frac{\partial V_j}{\partial x_i}\right), \quad i \neq j,$$

where μ is the viscosity coefficient. The *Navier–Stokes equation* is just a generalization of the Euler equation when allowing both normal and shear stresses for surface forces. Replacing the stress tensor in (1.10) by the above expression we obtain the Navier–Stokes equation. The Gauss theorem leads to a pointwise equation in vector form for a Newtonian incompressible fluid with constant viscosity

$$\frac{D\mathbf{V}}{Dt} = \mathbf{F}_b + \frac{1}{\rho}(-\nabla p + \mu \Delta \mathbf{V}). \tag{1.11}$$

If body forces are negligible, then we can put $\mathbf{F}_b = 0$. The equations (1.6) and (1.11) are called the Navier–Stokes equations for incompressible fluids.

1.2.5   Dynamical similarity and the Reynolds number

Letting L be a representative scale (that can be thought of as the distance between enclosing boundaries), U be a representative velocity (that can be thought of as the steady speed of a rigid boundary), we make the substitutions

$$\mathbf{x} \to L\mathbf{x}, \quad \mathbf{V} \to U\mathbf{V}, \quad t \to \frac{L}{U}t,$$

where hence the new variables $\mathbf{x}$, $\mathbf{V}$, t (on the right) are dimensionless. The pressure is similarly scaled by

$$p \to \rho U^2 p.$$

Substituting these new variables into the Navier–Stokes equation (with $\mathbf{F}_b = 0$) we have

$$\frac{D\mathbf{V}}{Dt} = -\nabla p + \frac{1}{R}\Delta \mathbf{V}, \tag{1.12}$$

where $R = \rho UL/\mu$ is the *Reynolds number*. This equation is just the Navier–Stokes equation in dimensionless variables. In terms of standard units we have

$$\rho = \frac{kg}{m^3}, \quad U = \frac{m}{s}, \quad L = m, \quad \mu = \frac{kg}{ms},$$

from which we see directly that the Reynolds number is a non-dimensional number.

Nondimensionalization, being a seemingly superficial step, becomes important when considering different flows with the same Reynolds number. A three-parameter family of solutions for a specific flow may for example be equivalent to just a one-parameter family for some modelling flow. Two flows with the same Reynolds number and the same geometry are called *dynamically similar*.

One may distinguish two different types of real fluid flow: laminar and turbulent. A well-ordered flow, free of macroscopic velocity fluctuations, is said to be *laminar*. Fluid layers are assumed to slide over one another without fluid being exchanged between the layers. In *turbulent flow*, secondary random motions are superimposed on the principal flow and there is an exchange of fluid from one adjacent segment to another. More important, there is an exchange of momentum such that slowly moving fluid particles speed up and fast moving particles give up their momentum to the slower moving particles and slow down themselves.

In an experiment in 1883, Reynolds demonstrated that, under certain circumstances, the flow in a tube changes from laminar to turbulent over a given region of the tube. He used a large water tank that had a long tube outlet with a tap at the end of the tube to control the flow speed. The tube went smoothly into the tank. A thin filament of coloured fluid was injected into the flow at the mouth, as is shown in Figure 1.1. When the speed of the water flowing through the

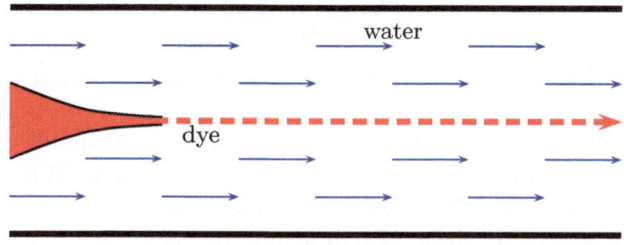

Fig. 1.1: Reynolds' experiment

tube was low, the filament of colored fluid maintained its identity for the entire length of the tube. However, when the flow speed was high, the filament broke up into the turbulent flow that existed throughout the cross section. Thus, laminar flow occurs when the Reynolds number R is not too large. When R is sufficiently large, then turbulence comes into consideration. It is observed empirically that the flow becomes turbulent whenever the Reynolds number exceeds a certain value R* which is critical. The Landau theory of the transition from steady laminar flow to turbulence suggests another limiting critical number R** > R*. Passing R* the flow becomes unstable and bifurcations occur until it arrives at turbulence passing R**. For the water flow R* = 2,300 and R** = 40,000 in Reynolds' experiment.

1.2.6 Vorticity, two-dimensional flows

When the Reynolds number is rather large, the distribution of vorticity proves to be an important entity to be taken into account. Let us consider two-dimensional flow with the velocity field $\mathbf{V} = (V_1, V_2, 0)$, subject to the restriction of incompressibility $\nabla \cdot \mathbf{V} = 0$, from which it follows that $V_1 dx_2 - V_2 dx_1$ is (locally) an exact differential $d\psi$. Then $V_1 = \partial \psi / \partial x_2$ and $V_2 = -\partial \psi / \partial x_1$ or $\mathbf{V} = \nabla \times (\psi \nabla x_3)$. If γ is a curve in the (x_1, x_2)-plane with the rightward normal vector $\mathbf{n} = (n_1, n_2, 0)$, then

$$\int_\gamma d\psi = \int_\gamma V_1 dx_2 - V_2 dx_1 = \int_\gamma \mathbf{V} \cdot \mathbf{n}\, ds.$$

Hence the flux of volume across any curve joining two points is equal to the difference between the values of ψ at these points, the function ψ is constant along a streamline, and it is called the *stream function*. The curl $\nabla \times \mathbf{V} = \boldsymbol{\omega}$ is called the *vorticity* of the fluid. In terms of the stream function, $\boldsymbol{\omega} = -\nabla x_3 \Delta\psi$. Taking the curl of Navier–Stokes equation (1.12) the term ∇p disappears and one gets an equation in $\boldsymbol{\omega}$ alone

$$\frac{D\omega}{Dt} = \frac{1}{R}\Delta\omega,$$

or for the stream function

$$\frac{D\,\Delta\psi}{D\,t} = \frac{1}{R}\Delta(\Delta\psi). \tag{1.13}$$

Equation (1.13) has several benefits. For example, it is a scalar equation rather than a vector one.

As we have remarked, the flow is laminar until the Reynolds number reaches its first critical value, or it can be thought of as a 'slow' flow. When the Reynolds number passes its second critical value the flow becomes turbulent and it can be either steady or unsteady. Even though it may be generated by a globally steady process, such as a steady volume flow through a pipe, turbulent flow is never a locally steady flow. We can see that $\mathbf{V}$ can be considered to be the sum of a time-averaged value $\tilde{\mathbf{V}}$ and a time-variable increment $\mathbf{V}'$ that is usually significantly smaller than the time-averaged value: $\mathbf{V} = \tilde{\mathbf{V}} + \mathbf{V}'$,

$$\tilde{\mathbf{V}} = \frac{1}{T}\int\limits_{t}^{t+T} \mathbf{V}(\mathbf{x},\tau)d\tau.$$

Note that the time-averaged value of $\mathbf{V}'$ is automatically zero. The random component $\mathbf{V}'$ of the velocity has some of the characteristics of random noise signals, such as electrical noise in electronic circuits. Obviously, there are small amplitude, high frequency, random motion involved in turbulent flow, the details of which are very difficult to calculate or to predict.

Adding the so-called Reynolds turbulent stress into the Navier–Stokes equation gives the equation of turbulent flow

$$\frac{D\,\mathbf{V}}{D\,t} = -\nabla p + \frac{1}{R}\Delta\mathbf{V} - (\mathbf{V}'\cdot\nabla)\mathbf{V}'.$$

An external force $\mathbf{F}_{ext}$ added to equation (1.12) or (1.13) in different forms can generate interesting flows. For example, *Andrei Nikolaevich Kolmogorov* (1903–1987) presented in 1959 a seminar in which he suggested a toy problem with which theorists might explore the transition to fluid turbulence in two dimensions. The flow is conceptually simple, and exhibits several shear instabilities before becoming fully turbulent. This flow is governed by the incompressible Navier–Stokes equation (1.12) in two dimensions with a forcing term that is periodic in one

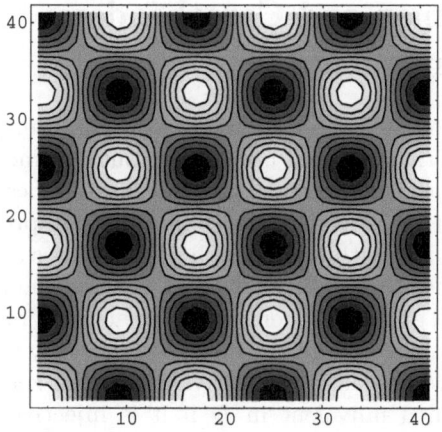

Fig. 1.2: Kolmogorov's flow

spatial direction and steady in time: $\mathbf{F}_{ext} = F_0 \sin(2\pi x_2)\nabla x_1$. Periodic boundary conditions are assumed in both directions of the a rectangular box $[0,1] \times [0,1]$. Equation (1.13) with the term corresponding to F_{ext} added becomes

$$\frac{D\,\Delta\psi}{D\,t} = \frac{1}{R}\Delta(\Delta\psi) + F_0\frac{8\pi^3}{R}\cos(2\pi x_2).$$

The stationary solution is just $\psi_{st} = -\frac{1}{2\pi}\cos(2\pi x_2)$. For small values of the forcing parameter F_0 the fluid develops a steady state spatial profile corresponding to the spatial profile of the forcing. This flow was named *Kolmogorov flow*. Above a critical value of the forcing parameter F_0, the flow becomes unstable to small velocity perturbations perpendicular to the direction of forcing. The resulting flow is a steady cellular pattern of vorticies. More generally, the external force can be chosen to be

$$\mathbf{F} = F_0 \begin{pmatrix} \sin(2\pi n x_1)\,\cos(2\pi m x_2) \\ -\cos(2\pi n x_1)\,\sin(2\pi m x_2) \end{pmatrix}.$$

For a weak forcing, i.e., for a small value of F_0, the $2n \times 2n$ array of counterrotating vortices (for the case $n = m$ see Figure 1.2) is the only time-asymptotic state. For other regimes of forcing (e.g., randomizing the phases of the individual external forces applied in each cell), a fully turbulent flow may be obtained, as numerical experiments indicate [53]. In simulations, the distribution of vorticity that were empirically observed seem to have a rather mysterious connection to the stochastic Löwner equation (discussed in the last chapters of this book).

1.3   Riemann map and Carathéodory kernel convergence

In this section we present some background on conformal maps, in particular, two basic instruments that we shall use throughout this monograph: the Riemann mapping theorem and the Carathéodory kernel convergence. A map of one domain (or surface) onto another is said to be *conformal* if it preserves angles between curves. The unit sphere S^2 without its north pole admits a stereographic projection onto the complex plane $\mathbb{C}$ which is conformal. Adding the north pole we obtain a compactification of S^2, and consequently, a compactification $\overline{\mathbb{C}}$ of $\mathbb{C}$ which is called the *Riemann sphere* or the extended complex plane. Any analytic map from $\mathbb{C}$ to $\mathbb{C}$ is conformal at a point where the derivative is non-zero. Let D be a domain in $\overline{\mathbb{C}}$. A map f is called univalent in D if it is injective (one-to-one) in D. A meromorphic function $f(\zeta)$ is univalent in D if and only if it is analytic in D except for at most one pole and $f(\zeta_1) \neq f(\zeta_2)$ whenever $\zeta_1 \neq \zeta_2$ in D. Univalence in D implies univalence in every subdomain in D. A univalent map is a conformal homeomorphism. The starting point of many considerations in this monograph is the *Riemann Mapping Theorem* (*Georg Friedrich Bernhard Riemann, 1826–1866*). Riemann had formulated his mapping theorem already in 1851 in his PhD thesis, but his proof was valid only for domains with smooth boundaries, as Weierstrass noticed. Simply connected domains with arbitrary boundaries were first treated by William Fogg Osgood (1900). Carathéodory in 1912, and Koebe (*Paul Koebe, 1882–1945*) in 1914 gave complete proofs around 1909, which were significantly simplified later by Leopold Fejér and to Frigyes Riesz.

Theorem 1.3.1. *Let Ω be a simply connected domain in $\overline{\mathbb{C}}$ whose boundary contains at least two points, and let $a \in \Omega$, $a \neq \infty$. Then there exists a real number R and a unique conformal univalent map $\zeta = f(z)$ that maps Ω onto $\mathbb{D}_R = \{\zeta : |\zeta| < R\}$ and satisfies $f(a) = 0$, $f'(a) = 1$.*

Remark 1.3.1. Generally, a domain whose universal covering surface is conformally equivalent to the unit disk is called *hyperbolic*. So the domain in the above theorem is hyperbolic.

If $f : \Omega \to \mathbb{D}_R$ is the map in Theorem 1.3.1 (or the Riemann map), then the number $R = R(\Omega, a)$ is called the *conformal radius* of the domain Ω with respect to the point a.

In the case $a = \infty$ it is more natural to let f map Ω onto the exterior of a disk $|\zeta| > R$. Then $R = R(\Omega, \infty)$ is uniquely determined by taking the expansion at infinity as $f(z) = z + a_0 + a_1/z + \cdots$.

One of the principal tools to study evolution of domains is the Carathéodory kernel convergence. *Constantin Carathéodory* (1873–1950) gave in 1912 [87] a complete characterization of convergence of univalent maps in terms of convergence of the images of a canonical domain under these maps. Its formulation is found also in [17], [156], [446].

Let $\{\Omega_n\}_{n=1}^{\infty}$ be a sequence of domains in the Riemann sphere $\overline{\mathbb{C}}$ such that a fixed point z_0 belongs to all Ω_n excluding possibly a finite number of them. A domain Ω is said to be the *kernel* of $\{\Omega_n\}_{n=1}^{\infty}$,

$$\Omega = \operatorname{Ker}_{z=z_0}\{\Omega_n\},$$

if Ω satisfies the following three conditions:

- $z_0 \in \Omega$;
- any compact set of Ω belongs to all Ω_n starting with certain number N;
- any domain $\tilde{\Omega}$ satisfying the preceding conditions is a subset of Ω.

If the point z_0 belongs to all Ω_n, starting with certain number $N(z_0)$, but there is no neighbourhood of z_0 that is contained in all Ω_n for $n > N$, then $\operatorname{Ker}_{z=z_0}\{\Omega_n\} = z_0$ and the kernel degenerates. For the kernel with respect to the origin we write simply $\Omega = \operatorname{Ker}\{\Omega_n\}$.

A sequence $\{\Omega_n\}_{n=1}^{\infty}$ is said to converge to the kernel Ω with respect to z_0 if every subsequence $\{\Omega_{n_k}\}_{k=1}^{\infty}$ has Ω as its kernel. This type of convergence is called *kernel convergence*. If Ω_n is a decreasing sequence with $0 \in \bigcap_{n=1}^{\infty}\Omega_n$ and Ω denotes the set of interior points of $\bigcap_{n=1}^{\infty}\Omega_n$, then Ω_n converges to the component of Ω that contains 0 if $0 \in \Omega$, and to $\{0\}$ if $0 \notin \Omega^0$, where Ω^0 is the interior of Ω.

Theorem 1.3.2 (Carathéodory kernel theorem). *Let the functions $f_n(\zeta)$ be analytic and univalent in $\mathbb{D} \equiv \mathbb{D}_1$, $f_n(0) = 0$, $f_n'(0) > 0$, and let $\Omega_n = f_n(\mathbb{D})$. Then the sequence f_n converges locally uniformly in $\mathbb{D}$ if and only if Ω_n converges to its kernel Ω, $\Omega \neq \mathbb{C}$, with respect to the origin. If $\operatorname{Ker}\Omega_n \neq \{0\}$, then the limiting function is a univalent map of $\mathbb{D}$ onto Ω. If $\operatorname{Ker}\Omega_n = \{0\}$, then $\lim_{n\to\infty} f_n(z) \equiv 0$.*

The kernel convergence can be generalized to continuous intervals as follows. Let $\{\Omega(t)\}$, $t \in [a, b]$ be a one-parameter family of domains in the Riemann sphere $\overline{\mathbb{C}}$ such that a fixed point z_0 belongs to all $\Omega(t)$. Consider first the case $t_0 \in [a, b]$, and let there be a neighbourhood of z_0 that belongs to all $\Omega(t)$, $t \neq t_0$. A domain Ω is said to be the *kernel* of $\{\Omega(t)\}$ with respect to z_0, if Ω satisfies the following three conditions:

- $z_0 \in \Omega$;
- for any compact set D of Ω there is a positive number ε, such that $D \subset \Omega(t)$ for all $0 < |t - t_0| < \varepsilon$;
- any domain satisfying the preceding conditions is a subset of Ω.

If there is no such neighbourhood, then we say that the kernel degenerates and $\operatorname{Ker}_{z=z_0}\{\Omega(t)\} = \{z_0\}$.

A generalized Carathéodory kernel theorem states that if the functions $f(\zeta, t)$ are analytic and univalent in $\mathbb{D}$, $f(0, t) = 0$, $f'(0, t) > 0$, $\Omega(t) = f(\mathbb{D}, t)$, then the family $f(\zeta, t)$ converges locally uniformly in $\mathbb{D}$ if and only if $\Omega(t)$ converges to its kernel Ω, $\Omega \neq \mathbb{C}$, as $t \to t_0$ with respect to the origin. If $\operatorname{Ker}\Omega(t) \neq \{0\}$, then the limiting function is a univalent map of $\mathbb{D}$ onto Ω. If $\operatorname{Ker}\Omega(t) = \{0\}$, then $\lim_{t\to t_0} f(z, t) \equiv 0$.

1.4   Hele-Shaw flows

Henry Selby Hele-Shaw (1854–1941), an English mechanical and naval engineer, published in the Nature magazine (in 1898) [265], see also [266], a short note proposing to study the following situation. For a liquid flow in a tube or in a channel with wetted sides, the velocity reaches its maximum in the middle and vanishes at the sides. Thus, the transition from laminar flow to turbulent can be observed somewhere in between. To make the separation interface visible Hele-Shaw proposed to inject a gas (an inviscid fluid) into the system. This injection can be interpreted as a suction of the original viscous fluid. To avoid gravity effect he suggested considering a flow between two parallel horizontal plates with a narrow gap between them.

Later a model with slightly different geometry appeared in [199, 438, 439, 470], see Figure 1.3. In this model the viscous fluid occupies a bounded domain with free boundary and more fluid is injected or removed through a point well. The free boundary starts moving due to injection/suction. Similar problems appear in metallurgy in the description of the motion of phase boundaries by capillarity and diffusion [396]; in the dissolution of an anode under electrolysis [190]; in the melting of a solid in a one-phase Stefan problem with zero specific heat [117], etc.

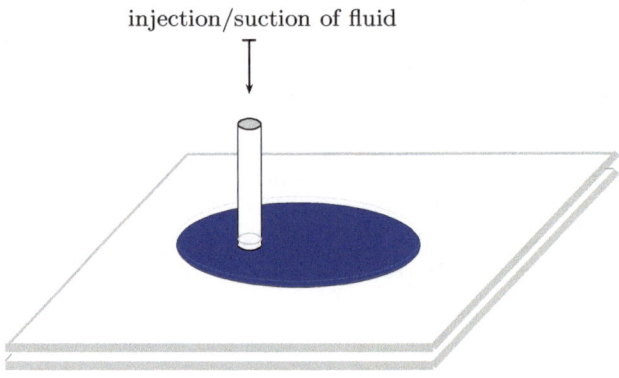

injection/suction of fluid

Fig. 1.3: A Hele-Shaw cell

This book will expose some of the developments in two-dimensional Hele-Shaw theory, or Laplacian growth, that have taken place the last few decades, including some connections to quantum physics, statistical physics, random matrix theory, and stochastic Löwner–Kufarev theory. These new areas have revived the subject considerably, and were initiated by seminal papers such as [6], on the semiclassical dynamics of an electronic droplet confined in the plane in a quantizing inhomogeneous magnetic field, and [518], on stochastic Löwner evolutions with Brownian motion driving mechanisms.

Several other models, methods, and applications exceed the scope of our work (or are discussed only briefly), like the treatment of the rectangular dam by Polubarinova-Kochina [440], who gave solutions in terms of the Riemann ℘-function [118, 282]; mathematical treatment of rotating Hele-Shaw cells [114, 176]; some nice analytical and numerical results are found in [93, 94, 95, 408]; a study of Hele-Shaw flows on hyperbolic surfaces [261, 263]; applications to electromagnetic problems [120, 190]; models of diffusion-limited aggregation [89, 583, 584]; Hele-Shaw flows with multiply connected phase domains [235, 472, 124]; development of singularities in non-smooth free boundary problems [270, 311, 312]; connections between Stokes and Hele-Shaw flows [119] (a large collection of references on Hele-Shaw and Stokes flows is found in [208]), two phase Muskat problem [4, 281, 523]; some applications of quasiconformal maps are found in [65, 378].

1.4.1 Lamb model

A mathematical treatment of the Hele-Shaw flow was suggested by *Sir Horace Lamb* (1849–1934) in [341] in 1906. Sometimes it is also referred to as the Leibenzon–Stokes model due to the model of the motion of gases in porous media developed by Leibenzon in 1921–1922. Approximately at that time the finite-source model for the Hele-Shaw cell was proposed. We consider a slow parallel flow of an incompressible fluid between two parallel flat plates which are fixed at a small distance h. The reference velocity $\mathbf{V}$ is generated by some external pumping mechanism. A vertical section is given in Figure 1.4. We assume that the flow attains its

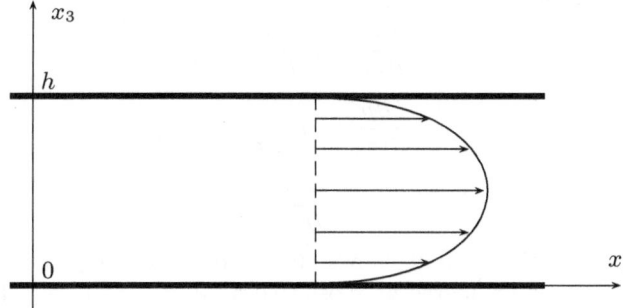

Fig. 1.4: The section of a Hele-Shaw cell in the x_1-direction

maximal velocity at the middle of the cell and the velocity vanishes at the sides. We follow Lamb's method [342] of deriving the Hele-Shaw equation starting from the Navier–Stokes equations (1.6), (1.11), which neglecting gravity become

$$\frac{\partial \mathbf{V}}{\partial t} + (\mathbf{V} \cdot \nabla)\mathbf{V} = \frac{1}{\rho}(-\nabla p + \mu \Delta \mathbf{V}), \quad \nabla \cdot \mathbf{V} = 0. \tag{1.14}$$

We assume that the injection of fluid is slow enough for the flow to be approximately steady and parallel. This means that

$$\frac{\partial \mathbf{V}}{\partial t} = 0, \quad V_3 = 0.$$

These assumptions reduce (1.14) to

$$\left(V_1 \frac{\partial}{\partial x_1} + V_2 \frac{\partial}{\partial x_2} \right) V_1 = -\frac{1}{\rho} \frac{\partial p}{\partial x_1} + \frac{\mu}{\rho} \Delta V_1,$$

$$\left(V_1 \frac{\partial}{\partial x_1} + V_2 \frac{\partial}{\partial x_2} \right) V_2 = -\frac{1}{\rho} \frac{\partial p}{\partial x_2} + \frac{\mu}{\rho} \Delta V_2,$$

$$0 = -\frac{1}{\rho} \frac{\partial p}{\partial x_3},$$

with boundary conditions

$$V_1 \bigg|_{x_3=0,h} = V_2 \bigg|_{x_3=0,h} = 0.$$

If h is sufficiently small and the flow is slow, then we can assume that the derivatives of V_1 and V_2 with respect to x_1 and x_2 are negligible compared to the derivatives with respect to x_3. Therefore, we can simplify the system by putting

$$\frac{\partial V_1}{\partial x_j} = \frac{\partial V_2}{\partial x_j} = \frac{\partial^2 V_1}{\partial x_j^2} = \frac{\partial^2 V_2}{\partial x_j^2} = 0, \quad j = 1, 2,$$

which gives the system

$$\frac{\partial p}{\partial x_1} = \mu \frac{\partial^2 V_1}{\partial x_3^2},$$

$$\frac{\partial p}{\partial x_2} = \mu \frac{\partial V_2}{\partial x_3^2},$$

$$0 = \frac{\partial p}{\partial x_3}.$$

The last equation in the system shows that p does not depend on x_3, whence V_1, V_2 are polynomials of degree at most two as functions of x_3. The boundary conditions then imply

$$V_1 = \frac{1}{2} \frac{\partial p}{\partial x_1} \left(\frac{x_3^2}{\mu} - \frac{h x_3}{\mu} \right), \quad V_2 = \frac{1}{2} \frac{\partial p}{\partial x_2} \left(\frac{x_3^2}{\mu} - \frac{h x_3}{\mu} \right).$$

The integral means $\tilde{V}_1$ and $\tilde{V}_2$ of V_1 and V_2 across the gap are

$$\tilde{V}_1 = \frac{1}{h} \int_0^h V_1 \, dx_3 = -\frac{h^2}{12\mu} \frac{\partial p}{\partial x_1}, \quad \tilde{V}_2 = \frac{1}{h} \int_0^h V_2 \, dx_3 = -\frac{h^2}{12\mu} \frac{\partial p}{\partial x_2},$$

so the integral mean $\tilde{\mathbf{V}}$ of $\mathbf{V}$ satisfies

$$\tilde{\mathbf{V}} = -\frac{h^2}{12\mu}\nabla p. \tag{1.15}$$

Here $\tilde{\mathbf{V}}$ and p depend only on x_1 and x_2, so we may consider (1.15) as a purely two-dimensional equation. Thus equation (1.15) describes a two-dimensional potential flow for which the potential function is proportional to the pressure. By incompressibility (1.6) the pressure is a harmonic function. Equation (1.15) is called the *Hele-Shaw equation*. It is of the same form as Darcy's law, which governs flows in porous media.

In the sequel we write just $\mathbf{V}$ instead of $\tilde{\mathbf{V}}$. The Lamb model suggests a point sink/source (x_1^0, x_2^0) of constant strength within the system. The rate of area (or mass) change is given as

$$\int_{\partial \mathbb{D}_\varepsilon} \rho \mathbf{V} \cdot \mathbf{n}\, ds = \text{const},$$

where $\mathbb{D}_\varepsilon = \mathbb{D}_\varepsilon(x_1^0, x_2^0)) = \{(x, y) : (x_1 - x_1^0)^2 + (x_2 - x_2^0)^2 < \varepsilon^2\}$ for ε sufficiently small. Equality (1.15) and Green's theorem imply

$$\int_{\mathbb{D}_\varepsilon} \left(-\frac{h^2 \rho}{12\mu}\right) \Delta p \, dx_1 dx_2 = \text{const},$$

for any ε. So $-\Delta p = Q\delta_{(x_1^0, x_2^0)}$ for some constant Q, where $\delta_{(x_1^0, x_2^0)}$ is Dirac's distribution, and the potential function p has a logarithmic singularity at (x_1^0, x_2^0).

On the fluid boundary the balance of forces in the three-dimensional view gives that

$$p = \text{exterior air pressure} + \text{surface tension}.$$

The air pressure can be taken to be constant while the surface tension is roughly proportional to the curvature of the boundary. If the gap h is sufficiently small, then the curvature in the x_1, x_2 plane is negligible compared to the curvature in the x_3 direction. Due to capillary forces the boundary profile in the x_3 direction will be somewhat similar to the graph in Figure 1.4 which is more or less the same everywhere. Hence, the surface tension effect to p is more or less constant (at least with respect to x_1, x_2). Finally, rescaling p we can take $p = 0$ on the boundary. By slight abuse of language we shall refer to this as the zero surface tension assumption.

1.4.2   The Polubarinova–Galin equation

Now let us study the motion of the boundary. Galin [199] and Polubarinova-Kochina [438, 439] first proposed a complex variable method by introducing the Riemann mapping from an auxiliary parametric plane (ζ) onto the phase domain in the physical (z)-plane and derived an equation for this parametric mapping. So the resulting equation is known as the *Polubarinova–Galin equation* (see, e.g., [280], [269]) (see a survey on the Polubarinova-Kochina contribution and its influence in natural sciences and industry in [414]).

We denote by $\Omega(t)$ the bounded simply connected domain in the phase z-plane occupied by the fluid at instant t, and we consider suction/injection through a single well placed at the origin as a driving mechanism (Figure 1.5). We assume

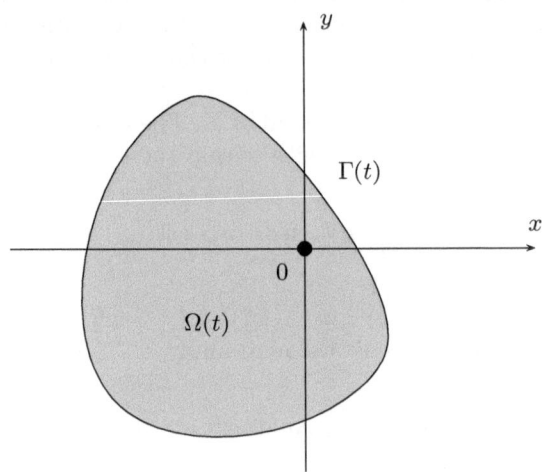

Fig. 1.5: $\Omega(t)$ is a bounded simply connected phase domain with the boundary $\Gamma(t)$ and the sink/source at the origin

the sink/source to be of constant strength Q which is positive ($Q > 0$) in the case of injection and negative ($Q < 0$) in the case of suction. The dimensionless pressure p is scaled so that $p = 0$ corresponds to the atmospheric pressure. We put $\Gamma(t) \equiv \partial\Omega(t)$ and assume that it is given by the equation $\phi(x_1, x_2, t) \equiv \phi(z, t) = 0$, where $z = x_1 + ix_2$. The initial situation is represented at the instant $t = 0$ as $\Omega(0) = \Omega_0$, and the boundary $\partial\Omega_0 = \Gamma(0) \equiv \Gamma_0$ is defined by an implicit function $\phi(x_1, x_2, 0) = 0$. The potential function p is harmonic in $\Omega(t) \setminus \{0\}$ and satisfies

$$-\Delta p(z, t) = Q\delta_0(z), \quad z = x_1 + ix_2 \in \Omega(t), \qquad (1.16)$$

where $\delta_0(z)$ is the Dirac distribution supported at the origin. The zero surface tension *dynamic* boundary condition is given by

$$p(z, t) = 0 \text{ for } z \in \Gamma(t). \qquad (1.17)$$

It follows that p is essentially the Green function of $\Omega(t)$. In general, the *Green function* $G_\Omega(z, a)$ of a domain Ω is defined by the properties

$$\begin{cases} G_\Omega(z, a) = -\frac{1}{2\pi} \log|z - a| + \text{harmonic} \quad (z \in \Omega), \\ G_\Omega(z, a) = 0 \quad (z \in \partial\Omega). \end{cases} \qquad (1.18)$$

The normalization factor for the singularity is chosen so that $-\Delta G_\Omega(\cdot, a) = \delta_a$. Thus the pressure is exactly given by $p(z,t) = Q\, G_{\Omega(t)}(z,0)$.

The resulting motion of the free boundary $\Gamma(t)$ is given by the fluid velocity $\mathbf{V}$ on $\Gamma(t)$. This means that the boundary is formed by the same set of particles all the time. The normal velocity in the outward direction is

$$V_n = \mathbf{V}\Big|_{\Gamma(t)} \cdot \mathbf{n}\,(t),$$

where $\mathbf{n}\,(t)$ is the unit outer normal vector to $\Gamma(t)$. Rewriting this law of motion in terms of the potential function and rescaling equation (1.15) to the simpler form $\mathbf{V} = -\nabla p$, we get the *kinematic* boundary condition

$$\frac{\partial p}{\partial n} = -V_n, \tag{1.19}$$

where $\frac{\partial p}{\partial n} = \mathbf{n} \cdot \nabla p$ denotes the outward normal derivative of p on $\Gamma(t)$.

Since p is harmonic (away from the origin) it is, locally, the real part of an analytic function $W = W(z,t)$. This is the *complex potential*. Writing $W = p - i\psi$, with a conventional minus sign, ψ is the previously discussed *stream function* of the flow. It is determined only up to an additive constant, and is in the present case moreover multi-valued due to the presence of a source. The real part p of W solves in $\Omega(t)$ the Dirichlet problem (1.16), (1.17). The Cauchy–Riemann system for p, ψ express exactly that the velocity field $\mathbf{V}$ is divergence free and curl free. The Cauchy–Riemann equations also give that

$$\frac{\partial W}{\partial z} = \frac{\partial p}{\partial x_1} - i\frac{\partial p}{\partial x_2}.$$

Since Green's function solves (1.16), (1.17), we have the representation

$$W(z,t) = -\frac{Q}{2\pi}\log z + W_{\mathrm{reg}}(z,t), \tag{1.20}$$

where $W_{\mathrm{reg}}(z,t)$ is a a regular analytic function in $\Omega(t)$. In terms of W the fluid velocity $\mathbf{V}$ gets represented by $-\overline{\partial W/\partial z}$.

To derive the equation for the free boundary $\Gamma(t)$ we consider an auxiliary parametric complex ζ-plane, $\zeta = \xi + i\eta$. The Riemann Mapping Theorem yields a unique normalized conformal univalent map $f(\zeta, t)$ from the unit disk $\mathbb{D} = \{\zeta : |z| < 1\}$ onto the phase domain:

$$f(\cdot, t) : \mathbb{D} \to \Omega(t), \quad f(0,t) = 0, \quad f'(0,t) > 0.$$

The function $f(\zeta, 0) = f_0(\zeta)$ parameterizes the initial boundary $\Gamma_0 = \{f_0(e^{i\theta}), \theta \in [0, 2\pi)\}$ and the moving boundary is parameterized by $\Gamma(t) = \{f(e^{i\theta}, t), \theta \in [0, 2\pi)\}$. The normal velocity V_n of $\Gamma(t)$ in the outward direction is given by (1.19).

From now on and throughout the monograph we use the notations $\dot{f} = \partial f/\partial t$, $f' = \partial f/\partial \zeta$. The normal outward vector is given by the formula

$$\mathbf{n} = \zeta \frac{f'}{|f'|}, \quad \zeta \in \partial \mathbb{D}.$$

Therefore, the normal velocity is obtained as

$$V_n = \mathbf{V} \cdot \mathbf{n} = -\mathrm{Re}\left(\frac{\partial W}{\partial z} \zeta \frac{f'}{|f'|}\right).$$

Because of the conformal invariance of Green's function we have

$$(W \circ f)(\zeta, t) = -\frac{Q}{2\pi} \log \zeta, \tag{1.21}$$

and by taking the derivative we get

$$\frac{\partial W}{\partial z} f'(\zeta, t) = -\frac{Q}{2\pi\zeta}.$$

Hence

$$V_n = \frac{Q}{2\pi|f'|}.$$

On the other hand, in general for a moving boundary we have $V_n = \mathrm{Re}\,[\dot{f}\,\overline{\zeta f'/|f'|}]$, and so we finally deduce that

$$\mathrm{Re}\left[\dot{f}(\zeta, t)\,\overline{\zeta f'(\zeta, t)}\right] = \frac{Q}{2\pi}, \quad \zeta = e^{i\theta}. \tag{1.22}$$

Galin [199] and Polubarinova-Kochina [438], [439] first derived the equation (1.22), so (1.22) is known as the *Polubarinova–Galin equation* (see, e.g., [280, 269, 414]).

The Polubarinova–Galin equation is a nonlinear partial differential equation of a rather special form. It is natural to think of (1.22) basically as an evolution equation for f and try to solve it for $\dot{f}$. This can be achieved by dividing both members of (1.22) by $|f'|^2$, so that the left member becomes the real part of a function which is analytic in $\mathbb{D}$ and whose imaginary part vanishes at the origin:

$$\mathrm{Re}\left[\frac{\dot{f}(\zeta, t)}{\zeta f'(\zeta, t)}\right] = \frac{Q}{2\pi|f'(\zeta, t)|^2}, \quad \zeta \in \partial \mathbb{D}.$$

Then in terms of the *Schwarz–Poisson integral* of the right member,

$$P(\zeta, t) = \frac{Q}{4\pi^2} \int\limits_0^{2\pi} \frac{1}{|f'(e^{i\theta}, t)|^2} \frac{e^{i\theta} + \zeta}{e^{i\theta} - \zeta} d\theta, \quad \zeta \in \mathbb{D}, \tag{1.23}$$

the equation becomes

$$\dot{f}(\zeta, t) = \zeta f'(\zeta, t) P(\zeta, t), \quad \zeta \in \mathbb{D}, \tag{1.24}$$

which is an equation of Löwner–Kufarev type. The system (1.24), (1.23) is actually equivalent to the Polubarinove–Galin equation (1.22), as can easily be seen by letting ζ tend to the unit circle in (1.24), (1.23) and running the previous steps backwards.

We call (1.24) a *Löwner–Kufarev type equation* because of the analogy with the linear partial differential equation that describes monotone deformations of simply connected domains (see, e.g., [17, 156, 446]). In the classical Löwner–Kufarev theory the function $P(\zeta, t)$ is a free function (subject to choice), having positive real part and often normalized by $P(0, t) = 1$, whereas in the Hele-Shaw case $P(\zeta, t)$ is coupled to f by the nonlinear relation (1.23). This makes it produce special subordination Löwner chains which are uniquely determined by their initial domains. The study of such Löwner chains is one of the basic objects of this book.

1.4.3 Local existence and ill/well-posedness

Under some assumptions on smoothness of $\partial\Omega(0)$ it is known that in the case of an expanding fluid ($Q > 0$) there exists a unique solution to the problem (1.16)–(1.19), or (1.22), or (1.23)–(1.24), in terms of analytic functions $f(\zeta, t)$ (strong or classical solution), locally forward in time. The first proof appeared in 1948 [569] by *Yurii P. Vinogradov* and *Pavel Parfen'evich Kufarev* (1909–1968). This proof was rather involved, and later, Gustafsson [233] gave a simpler proof in the case when a polynomial or a rational univalent function f_0 parameterizes the initial phase domain. In 1993 Reissig and von Wolfersdorf [467] made clear that this model could be interpreted as a particular case of an abstract Cauchy problem and that the strong solvability (locally in time) could be proved using a nonlinear abstract Cauchy–Kovalevskaya Theorem (see [410]). More precisely, they proved that if the initial function $f_0(z)$ is analytic and univalent in the disk $\mathbb{D}_r = \{\zeta : |\zeta| < r\}$ for some $r > 1$, then there exists $t_0 > 0$, such that the solution $f(\zeta, t)$ to the Polubarinova–Galin equation exists and is unique in some time interval $t \in [0, t_0)$. Other approaches to the existence question are given in [361] and [484]. In the multidimensional case proofs of local existence and uniqueness can be found in, e.g., [180] and [554].

We note that the problem (1.16)–(1.19) is formally time reversible by changing $Q \to -Q$, $p \to -p$, $t \to -t$. However, the cases of suction and injection differ considerably: the suction case is severely ill posed, whereas the injection case is correspondingly well behaved, even though one may need a smooth boundary to start up a classical (strong) solution. Generally speaking, in the injection case and starting with a smooth boundary, the boundary immediately becomes analytic, and the asymptotic behavior is that it approaches circular shape. It may still happen that in the course of development the solution breaks down as a classical solution (i.e., a smooth evolution for which (1.22) holds in a pointwise sense), due to topological events, essentially that two different parts of the boundary collide, or in exceptional cases that cusps develop. There however exists a good concept of weak solution (variational inequality weak solution) for the injection version of the Hele-Shaw problem. This admits changes of topology, and indeed there is a basic theorem saying that for any (bounded) initial domain there exists a unique, and global in time, weak solution. Of course, whenever the strong solution exists it agrees with the weak solution.

For the suction case the situation is completely opposite: the initial boundary needs to be analytic even for a solution to start, and after start the solution quickly deteriorates in most cases, by development of cusps and other singularities on the boundary. What one sees in experiments is typically a development of characteristic fingers. There are also examples of solutions which break down due to the fluid boundary reaching the sink, without all fluid having been sucked up. Only in the case that the initial domain is a disk centered at the origin can all fluid be sucked up. Several attempts have been made to formulate a good concept of a weak solution for the suction problem, however as far as we know none of them has as yet proved to be completely successful.

If the initial function $f_0(\zeta)$ is a polynomial of degree at least two, then the solution in the suction case always breaks down by development of cusp(s) on the boundary, and this occurs before the boundary reaches the sink. In exceptional cases the solution can be continued after a cusp has developed (this can occur only for cusps of order $(4n + 1)/2$, see [278] and [492] for complete classification), but eventually new cusps will stop the evolution.

An attempt to classify the solutions to the zero surface tension model for the Hele-Shaw flows in bounded and unbounded regions with suction has been launched by Hohlov, Howison [269] and Richardson [471]. They also described cusp formation. Another typical scenario is fingering, first described in the classical work by Saffman, Taylor [488]. Recently it has become clear that in the model with injection fingering does not occur [243].

1.4.4 Regularizations

There are several proposals for regularization of the ill-posed Hele-Shaw problem. One of them is the 'kinetic undercooling regularization' [271], where the condition (1.17) is replaced by

$$\beta \frac{\partial p}{\partial n} + p = 0, \quad \text{on} \quad \Gamma(t), \quad \beta > 0.$$

It was shown in [271, 469] that there exists a unique solution locally in time (even a strong solution) in both the suction and injection cases in a simply connected bounded domain $\Omega(t)$ with an analytic boundary. We remark that at the conference about Hele-Shaw flows, held in Oxford in 1998, V.M. Entov suggested using a nonlinear version of this condition motivated by applications. Reissig and Pleshchinskii discussed this model in [432] where local existence and uniqueness were obtained as well as some numerical results were presented.

Another proposal is to introduce surface tension as a regularization mechanism. A model with non-zero surface tension is obtained by modifying the boundary condition for the pressure p to be the product of the mean curvature κ of the boundary and surface tension $\gamma > 0$. Thus the problem (1.16)–(1.19) becomes, with this new condition:

$$-\Delta p = Q\delta_0 \quad \text{in} \quad \Omega(t), \tag{1.25}$$

$$p = \gamma\kappa \quad \text{on} \quad \Gamma(t), \tag{1.26}$$

$$V_n = -\frac{\partial p}{\partial n} \quad \text{on} \quad \Gamma(t). \tag{1.27}$$

A similar problem appears in metallurgy in the description of the motion of phase boundaries by capillarity and diffusion [396]. The condition (1.26) is found in [383] (it is known as the Gibbs–Thomson law or the Laplace–Young condition). *Pierre-Simon Laplace* (1749–1827) and *Thomas Young* (1773–1829) obtained independently this law in 1805. Later *Josiah Willard Gibbs* (1839–1903) and *William Thomson* (Lord Kelvin) (1824–1907) in the 1870s derived an analogous relation. It takes into account how the surface tension modifies the pressure through the boundary interface.

The problem of existence of a solution in the non-zero surface tension case is more difficult. Duchon and Robert [154] proved the local existence in time for weak solution for all γ. Prokert [452] obtained even global existence in time and exponential decay (in the case of flow driven by surface tension) of the solution near equilibrium for bounded domains. We refer the reader to the works by Escher and Simonett [178, 179] who proved the local existence, uniqueness and regularity of strong solutions to one- and two-phase Hele-Shaw problems with surface tension when the initial domain has a smooth boundary. The case of the initial domain bounded by a non-smooth boundary was considered in [22, 180]. The global existence in the case of the phase domain close to a disk was proved in [181]. If the domain occupied by the fluid is unbounded and its boundary extends to infinity, then the corresponding result about short-time existence and uniqueness for positive surface tension was obtained by Kimura [309] (he also shows that the problem is ill posed in the case of suction). More results on existence for general parabolic problems can be found in [182]. Most authors work with a weak formulation of the problem (see this formulation in Chapter 3 and in [139, 173, 234]). It is worth remarking that the weak solution to the problem with injection exists all the time and coincides with the strong one if the latter exists.

A good survey of models with surface tension, kinetic undercooling, and other regularization methods is given in the doctoral thesis [123] of M. Dallaston. See also [125, 126].

1.4.5   Numerical treatment

There exists a considerable amount of work dedicated to numerical study of the Hele-Shaw problem. One may distinguish several directions of the development of numerical methods which are principally divided by well/ill-posedness and by presence/absence of the surface tension effect. A complete review of numerical methods exceeds the scope of this book but we give some remarks on them. For a recent survey, see [123], and for numerical treatment of free and moving boundary problems in general, [110].

Numerical computation of receeding evolution is known to be difficult due to the ill-posedness of the zero-surface tension problem [133, 386]. Reasonably accurate computation of solutions of the ill-posed problem is possible if one works with the conformal mapping from the flow domain into the unit circle rather than the usual free boundary problem formulation. Cusp formation can be described analytically in many cases and some of these cases are used as tests for the numerical scheme, which uses a finite element/boundary integral method, see, e.g., [15]. In addition, the possibility of singularity formation demands extremely high resolution. The cusp singularities are good candidates for potential topological singularities in the presence of small surface tension but even perturbations at the low error level can lead to a rapid growth of the solution high-frequency components [48, 122]. The boundary integral method with fourth-order time integration is typically employed, see [93]–[95]. It uses the small-scale decomposition technique of Hou *et al.* [275] to remove the high-order stability constraint induced by surface tension. A good review of numerical methods may be found in, e.g., [276]. Some other relevant references are [31, 468, 579].

There are several works where the Helmholtz–Kirchhoff method is used and the free boundary problem is reformulated as an integro-differential equation. Numerical simulations are also performed and qualitative properties are demonstrated. For example, the Hele-Shaw problem is considered with the initial domain being a non-analytic perturbations of a disk [136, 137]. The authors develop the quasi-contour method in order to give a numerical simulation.

1.4.6   Stochastic homogenization in disordered 2D systems

The zero-surface limit problem derived above arises not just in the fluid dynamics context, but also from an impressive number of different physical systems, as the result of (seemingly) distinct approximations. The first instance where this model was derived arises from multiple-scales homogenization of water seepage through porous soils, derived by H. Darcy in 1856 [141], and leading to the law named after him. Later, many other systems characterized by multiple scales were considered (for viscous Newtonian flows around small obstacles, for non-homogenous granular flows, etc.) and proved to be characterized by the same limiting behavior [62, 189, 485, 543].

At the turn of XXth century, weak nonlinear corrections were considered for the Darcy law, notably by Forchheimer (see [62] for a more recent exposition), who proposed a generalization of the Darcy law in the form

$$\frac{\partial p}{\partial n} = V_n + \alpha V_n^3, \tag{1.28}$$

with α an effective nonlinearity parameter, obtained through homogenization of the multiple-scales transport problem, in the sense of the Lamb model discussed before.

Effective transport with Darcy law was also considered in the case of electric currents through disordered conductors, correlated percolation on lattices, disordered networks at large scale, and it was observed that they share the same properties and universal features as fluid-dynamics [52, 164, 187, 302].

For instance, the effective Darcy law

$$\mathbf{V} = -\lambda \nabla p \tag{1.29}$$

arises naturally by multiple-scale averaging in the study of transport through 2D disordered media, where the function $\lambda(\mathbf{x}) \geq 0$ may, e.g., describe the electrical conductivity of the medium (in which case the velocity field in (1.29) is replaced by current density $\mathbf{j}(\mathbf{x})$, and pressure p becomes electrostatic potential ϕ), or the filtration coefficient as in (1.29). This parameter is allowed to vanish on certain sets in the domain of interest, in order to capture the physics of conductive-insulating mixed media.

A seemingly modest goal of this model is computing the *effective* conductivity λ_e, defined in the following way: assume that our problem is defined on a domain D whose boundary allows a partition $\partial D = B_i \cup B_a \cup B_o$, such that the normal component of the velocity (or current density) represents *inward* transport through B_i, there is no flow through B_a (reflecting boundary), and we measure the *outward* flow of matter or current through B_o. The pressure (or potential) field p is assumed to have sources of (signed) density ρ, supported on some subset Σ:

$$\nabla \cdot \mathbf{V} = \rho(\mathbf{x})\chi_\Sigma(\mathbf{x}). \tag{1.30}$$

Moreover, we require that B_i, B_o belong to (different) levels sets of p. The effective conductivity is then defined as the conductivity of a *uniform* medium $\lambda_e = $ constant, which would carry the same amount of matter for the same potential difference:

$$\lambda_e(\lambda) = \frac{\int_{B_o} V_n d\ell - \int_{B_i} V_n d\ell}{p_{B_o} - p_{B_i}}, \tag{1.31}$$

with V_n the normal component of velocity. If the domain D is not bounded, then we define λ_e via an obvious limiting procedure. The fact that λ_e is a functional of the field $\lambda(\mathbf{x})$ was indicated explicitly. In the context of fluid dynamics, computation of the effective conductivity was carried over for specific microscopic models of transport, see [543], [189].

Introducing now the notation $\lambda(\mathbf{x}) = e^{\mu(\mathbf{x})}$, we can formally solve the problem (1.29), subject to the boundary conditions specified above, in the form

$$p = [\Delta + \nabla\mu \cdot \nabla]^{-1} \lambda^{-1} \rho, \qquad (1.32)$$

where the inverse (integral) operator satisfies the given boundary conditions.

In the most interesting physical set-up, the conductivity field λ is a random variable, taking values in $\mathbb{R}^+$. In that case, the true effective conductivity is obtained by computing the average with respect to the distribution of $\lambda(\mathbf{x})$ of the quantity defined in (1.31): $\overline{\lambda_e} = \mathbb{E}_\lambda[\lambda_e(\lambda)]$. Alternatively, we need to find the inverse of a Laplace–Beltrami operator with *stochastic* coefficients (generically non-local).

A discretized version of this problem (known in that form as the *random resistor network* model) has been studied theoretically and numerically for more than three decades [302], [187], but little attention was paid to the detailed properties of the ensemble of level lines of the potential p.

It has been known from physical considerations (but never proven rigorously) that, for a class of distributions of $\lambda(\mathbf{x})$, which we will refer to in the following as *critical*, the effective conductivity of the medium exhibits a transition from zero to positive values (alternatively, the medium changes from an insulator to a conductor). Strictly speaking, this behavior may only occur in infinite-size samples.

1.5   Harmonic moments

Consider Hele-Shaw flow with injection ($Q > 0$) and assume that a strong solution to the Polubarinova–Galin equation (1.22) exists for some time interval $[0, t_0)$. Since the velocity of the boundary in the outward normal direction is strictly positive, we have $\Omega(s) \subset \Omega(t)$ for $0 < s < t < t_0$. The *harmonic moments* of $\Omega(t)$ are defined by

$$M_n(t) = \frac{1}{\pi} \int\limits_{\Omega(t)} z^n d\sigma_z = \frac{1}{2\pi i} \int\limits_{\partial\Omega(t)} z^n \bar{z} dz, \qquad (1.33)$$

where $d\sigma_z = dxdy$ denotes the area elements in the z-plane and where the factor $\frac{1}{\pi}$ is introduced for later convenience. The zeroth-order moment $M_0(t)$ is just the normalized area of the domain, which increases linearly in time. However, as S. Richardson discovered in his seminal paper [470], all other moments are constants of motion for (1.22). Thus

$$\begin{cases} M_0(t) & = M_0(0) + \frac{Qt}{\pi}, \\ M_n(t) & = M_n(0) \quad \text{for } n \geq 1. \end{cases} \qquad (1.34)$$

For further consideration, and for explanations of the above, it is convenient to study the time evolution of more general integrals. If $h = h(z)$ is any function which is harmonic in a neighborhood of $\overline{\Omega(t)}$, the Reynolds transport theorem together with Green's formula gives

$$\frac{d}{dt} \int_{\Omega(t)} h \, d\sigma = \int_{\Gamma(t)} h \, V_n ds = - \int_{\Gamma(t)} h \frac{\partial p}{\partial n} \, ds$$

$$= - \int_{\Gamma(t)} p \frac{\partial h}{\partial n} \, ds - \int_{\Omega(t)} h \, \Delta p \, d\sigma = Qh(0).$$

Integrating we obtain

$$\int_{\Omega(t)} h d\sigma = \int_{\Omega(0)} h d\sigma + Qth(0), \tag{1.35}$$

for all $0 \le t < t_0$. The function h is allowed to be complex-valued, in particular analytic, and choosing $h(z) = z^n$ we obtain (1.34).

It is easy to see that one can run the above arguments backward (see, e.g., [234]) to show that a smooth family $\Omega(t)$ of simply connected domains is a strong solution to the Hele-Shaw problem if and only if the equality (1.35) holds for every function h which is harmonic and integrable in all the domains $\Omega(t)$. The conservation law (1.35) is closely related to the notion of a weak solution (discussed in Chapter 3), which actually is just a strengthened form of (1.35) for subharmonic test functions h.

Let

$$f(\zeta, t) = \sum_{n=1}^{\infty} a_n(t)\zeta^n \tag{1.36}$$

be the power series expansion of the normalized conformal map $f : \mathbb{D} \to \Omega$. When pulled back to the unit disk by f the formulas (1.33) for the moments become (suppressing the dependence on t)

$$M_n = \frac{1}{\pi} \int_{\mathbb{D}} f(\zeta)^n |f'(\zeta)|^2 d\sigma_\zeta = \frac{1}{2\pi i} \int_{\partial \mathbb{D}} f(\zeta)^n \overline{f(\zeta)} f'(\zeta) d\zeta.$$

By inserting (1.36) into the boundary integral, replacing $\overline{f(\zeta)}$ by $\overline{f(1/\bar\zeta)}$ and then using the residue theorem one arrives, with Richardson [470], at the following explicit formula for harmonic moments in terms of the coefficients a_n:

$$M_n = \sum_{(i_1,\ldots,i_{n+1})} i_1 a_{i_1} \cdots a_{i_{n+1}} \overline{a_{i_1+\cdots+i_{n+1}}}. \tag{1.37}$$

Here $n = 0, 1, 2, \ldots$ and the summation runs over all $(n + 1)$-tuples $(i_1, \ldots, i_{n+1})$ of integers with $i_j \ge 1$.

1.6 The Polubarinova–Galin equation in terms of Poisson brackets

Writing $\zeta = e^{i\theta}$ on $\partial\mathbb{D}$ we have $\frac{\partial f}{\partial\theta} = i\zeta\frac{\partial f}{\partial\zeta}$. Therefore the Polubarinova–Galin equation (1.22) can be written as

$$\mathrm{Im}\left(\frac{\partial f}{\partial\theta}\overline{\frac{\partial f}{\partial t}}\right) = \frac{Q}{2\pi}. \tag{1.38}$$

Decomposing f into real and imaginary parts, $f = u + iv$, the equation becomes

$$\frac{\partial(u,v)}{\partial(\theta,t)} = -\frac{Q}{2\pi}, \tag{1.39}$$

where

$$\frac{\partial(u,v)}{\partial(\theta,t)} = \frac{\partial u}{\partial\theta}\frac{\partial v}{\partial t} - \frac{\partial v}{\partial\theta}\frac{\partial u}{\partial t}$$

is the Jacobi determinant of the map $(\theta,t) \mapsto (u,v)$.

Equation (1.39) can be regarded as a differential equation for the two real-valued functions u and v defined on the circle. As such it expresses that the map $(\theta,t) \mapsto (u,v)$ shall be area preserving up to a constant factor. The two functions u and v in (1.39) are, however, not independent of each other, but are linked via the condition that, as a function of $e^{i\theta}$, $u + iv$ has an analytic continuation to all of $\mathbb{D}$. In other words, v is to be the Hilbert transform of u. The interpretation is still nice: (1.39) says that $f(\zeta,t)$ maps any cylinder $\partial\mathbb{D} \times [t_1, t_2]$ onto the part of $\mathbb{C}$ swept out by $\partial\Omega(t)$, $t_1 \le t \le t_2$, in an area preserving fashion (up to a constant factor).

Moreover, if we interpret the area-preserving diffeomorphism condition (choosing $Q = -2\pi$ in (1.39)) as the fact that the 1-form $\Lambda_1 \equiv u\,dv - \theta\,dt$ is closed, $d\Lambda_1 = 0$ for u, v, θ taken along a Laplacian growth trajectory, then we arrive at a one-dimensional Hamiltonian system by identifying θ with its Hamilton function, t with time, and u, v with momentum and coordinate, respectively. Of course, these quantities depend smoothly on a (continuous) index giving the position along the boundary, so we are in fact dealing with a Hamiltonian 1D (classical) field theory. This turns out to be the essential link between Laplacian growth and other integrable subordinations processes, as discussed in Chapter 7.

Notice that the boundary integral version in (1.33) gives a meaning also to the moments M_n for n a negative integer. These negative moments play an important role in some developments initiated by M. Mineev-Weinstein, P. Wiegmann, A. Zabrodin and others (see, e.g., [6, 321, 370, 580]). In this context one reinterprets the above Jacobi determinant as a Poisson bracket defined for functions $f(\zeta; M_0, M_1, \ldots)$, $g(\zeta; M_0, M_1, \ldots)$ analytic with respect to ζ in a neighborhood of $\partial\mathbb{D}$ and smooth with respect to the moments. Defining

$$\{f, g\} = \zeta\frac{\partial f}{\partial\zeta}\frac{\partial g}{\partial M_0} - \zeta\frac{\partial g}{\partial\zeta}\frac{\partial f}{\partial M_0} \tag{1.40}$$

and setting, in general,

$$f^*(\zeta) = \overline{f(1/\bar{\zeta})} \tag{1.41}$$

we see that

$$\{f, f^*\} = 2\mathrm{Im}\ \left(\frac{\partial f}{\partial \theta}\frac{\overline{\partial f}}{\partial t}\right)$$

on $\partial \mathbb{D}$, taking into account that $\frac{\partial}{\partial t} = \frac{Q}{\pi}\frac{\partial}{\partial M_0}$ by (1.34). Hence the Polubarinova–Galin equation (1.38) takes the form

$$\{f, f^*\} = 1. \tag{1.42}$$

It holds on $\partial \mathbb{D}$, but since the left member now is analytic in a neighborhood of the circle, the equality automatically extends to this neighborhood. The equation (1.42) is called the *string equation*, which by definition is the equation of type $[L, A] = 1$ for the coefficients of two linear ordinary differential operators L and A, which appeared in physics literature in the 90s, see, e.g., [73, 146, 228]. One then thinks of the evolution of a one-dimensional object (the string), which in the present case is $\Gamma(t) = \partial \Omega(t)$.

1.7 The Schwarz function

1.7.1 Definition and relation to moving boundaries

The function now called the Schwarz function appeared explicitly in a paper by Grave [224] in 1895, and was later employed by *Gustav Herglotz* in 1914 [268]. In the works by *Hermann Amandus Schwarz* (1843–1921) it does not seem to appear explicitly, whereas this designation (due to Philip Davis) is now immutably connected with his name. Let Γ be a non-singular analytic Jordan curve in $\mathbb{C}$, i.e., Γ possesses a real-analytic bijective parametrization with non-vanishing derivative. Then there is a neighbourhood U of Γ and a uniquely determined analytic function $S(z)$, $z \in U$, such that $S(z) = \bar{z}$ for $z \in U$. This is the *Schwarz function*. The map $z \mapsto \overline{S(z)}$ is an anti-conformal reflection in Γ, which agrees with the classical Schwarz reflection in case Γ is a straight line or a circular arc. Clearly the consistency relation $\overline{S(\overline{S(z)})} = z$ holds identically in U, saying that the reflection is involutive. Thorough treatments of the Schwarz function can be found in [132, 517].

One way to construct $S(z)$ is to write the equation for Γ in implicit form as $\phi(x_1, x_2) = 0$ with ϕ real-analytic, and having non-vanishing gradient on Γ, then substituting $x_1 = (z + \bar{z})/2$, $x_2 = (z - \bar{z})/2i$ to obtain $\psi(z, \bar{z}) = 0$ (say), and finally solving this equation for $\bar{z}$, which is always possible because the non-vanishing of the gradient means that $\partial \psi / \partial \bar{z} \neq 0$ on Γ. This gives $S(z)$, satisfying $\psi(z, S(z)) = 0$ identically. For moving curves $\Gamma(t)$, like the fluid fronts in the Hele-Shaw problem, the Schwarz function will of course depend on time t, i.e., we have

$$\bar{z} = S(z, t) \quad \text{for } z \in \Gamma(t), \tag{1.43}$$

with $S(z,t)$ defined and analytic in a neighbourhood of $\Gamma(t)$. Differentiating (1.43) with respect to an arc length parameter s on $\partial\Omega(t)$ for fixed t gives the expression

$$\frac{dz}{ds} = \frac{1}{\sqrt{S'(z,t)}}, \tag{1.44}$$

for the unit tangent vector on $\partial\Omega(t)$. The outward normal vector $\mathbf{n}$ is represented (as a complex number) by the same expression multiplied by $-i$.

Since the map $z \mapsto \overline{S(z,t)}$ is the anti-conformal reflection in $\Gamma(t)$ it follows that, for fixed z, the point $\overline{S(z,t)}$ moves with the double speed of $\Gamma(t)$. So if $\Gamma(t)$ moves with the normal velocity V_n, then the point $\overline{S(z,t)}$ moves with speed $|\dot{\bar{S}}| = 2V_n$. This formula holds on $\Gamma(t)$. Taking the direction into account this gives the general formula

$$V_n = \frac{\dot{S}(z,t)}{2i\sqrt{S'(z,t)}} \quad (z \in \Gamma(t)),$$

for the choice of square root for which dz/ds in (1.44) is the positively oriented tangent vector (cf. [280, 305]).

In the Hele-Shaw case, with $\Gamma(t) = \partial\Omega(t)$ and $\Omega(t)$ the fluid domain, the velocity vector $\frac{1}{2}\dot{\bar{S}}$ equals $-\overline{\partial W/\partial z}$, where $W(z,t)$ is the complex potential (1.20), hence the Hele-Shaw equation (or Darcy law) becomes

$$W'(z,t) = -\frac{1}{2}\dot{S}(z,t). \tag{1.45}$$

Initially this equation holds for $z \in \Gamma(t)$, but since both members are analytic functions of z it remains valid as an identity throughout the region of analyticity. In our particular case this means that (1.45) holds in all $\Omega(t)$, and on using $W(z,t) = -\frac{Q}{2\pi}\log\zeta$, by (1.21), it can also be written

$$\frac{\partial}{\partial t}S(z,t) = \frac{Q}{\pi}\frac{\partial}{\partial z}\log\zeta,$$

where $\zeta = f^{-1}(z,t)$.

1.7.2   Relation to Cauchy transform and moments

Let $\Gamma = \partial\Omega$, with Ω a bounded simply connected domain containing the origin. A somewhat indirect way to construct the Schwarz function for Γ is to consider the Cauchy integral

$$g(z) = -\frac{1}{2\pi i}\int_{\Gamma} \frac{\bar{\zeta}d\zeta}{\zeta - z}.$$

It defines one analytic function, $g_e(z)$, for $z \in \mathbb{C}\setminus\overline{\Omega}$ of Γ and one, $g_i(z)$, for $z \in \Omega$. On Γ the jump condition

$$g_e(z) - g_i(z) = \bar{z}, \quad z \in \Gamma \tag{1.46}$$

holds for the boundary values. When Γ is analytic both g_i and g_e extend analytically across the boundary so that $g_i(z) - g_e(z)$ is analytic in a full neighbourhood of Γ. Hence the left member of (1.46) must be the Schwarz function of Γ:

$$S(z) = g_e(z) - g_i(z). \tag{1.47}$$

In view of (1.47) it is natural to define $S_-(z) = g_e(z)$, $S_+(z) = -g_i(z)$, so that $S_-(z)$ is holomorphic in $\mathbb{C} \setminus \overline{\Omega}$, $S_+(z)$ is holomorphic in Ω, both extend analytically across $\partial\Omega$, and (1.47) becomes

$$S(z) = S_-(z) + S_+(z). \tag{1.48}$$

The *Cauchy transform* of Ω (or, more precisely, of the measure $\chi_\Omega d\sigma$) is defined in all $\mathbb{C}$ by

$$C_\Omega(z) = -\frac{1}{\pi} \int_\Omega \frac{d\sigma_\zeta}{\zeta - z} \quad (z \in \mathbb{C}). \tag{1.49}$$

For $z \in \mathbb{C} \setminus \overline{\Omega}$ this equals $g(z)$, hence

$$S_-(z) = C_\Omega \quad \text{for } z \in \mathbb{C} \setminus \overline{\Omega}, \tag{1.50}$$

and $S_-(z)$ also represents the analytic continuation of this Cauchy transform across $\partial\Omega$. Similarly, $S_+(z)$ equals a renormalized version of the Cauchy transform of $\mathbb{C} \setminus \Omega$ for $z \in \Omega$, and its analytic continuation across $\partial\Omega$.

The Laurent expansion of $S_-(z)$ at infinity is

$$S_-(z) = \sum_{k=0}^{\infty} \frac{M_k}{z^{k+1}},$$

hence has the (positive) harmonic moments as coefficients. The negative moments appear as coefficients in the expansion of $S_+(z)$ at the origin:

$$S_+(z) = \sum_{k=1}^{\infty} M_{-k} z^{k-1}.$$

The radii of convergence of the above two series need not overlap, but one can still insert them into (1.47) to get at least a formal Laurent series for the Schwarz function:

$$S(z) = \sum_{k=-\infty}^{\infty} \frac{M_k}{z^{k+1}}. \tag{1.51}$$

In some good cases the series does converge on $\partial\Omega$, and in general the coefficients at least contain complete information of $S(z)$ and $\partial\Omega$ since the positive and negative parts give the germs of S_- and S_+ at infinity and the origin, respectively.

1.7.3   Hydrodynamic interpretation of the Schwarz function

A major case of interest is when the Schwarz function for $\Gamma = \partial\Omega$ is analytic in all Ω except for singularities at isolated points and along arcs. It is then useful when computing averages of integrable analytic functions $h(z)$ over the domain Ω:

$$\frac{1}{\pi}\int_\Omega h(z)dxdy = \sum_{k=1}^{m}\sum_{j=0}^{n_k-1} c_{kj}h^{(j)}(z_k) + \sum_{r=1}^{\ell}\int_{\gamma_r}\rho_r(z)h(z)dz, \qquad (1.52)$$

if the Schwarz function $S(z)$ has poles of order n_k at $z = z_k$ and branch cuts γ_r, with $\rho_r(z)$ representing the jump of $S(z)$ across γ_r. Applying formula (1.52) for the constant function $h(z) = 1$ we obtain

$$M_0 = \sum_{k=1}^{m}\operatorname{Res} S(z_k) + \sum_{r=1}^{\ell}\int_{\gamma_r}\rho_r(z)dz,$$

which shows that the singularity data of the Schwarz function in Ω can be interpreted as giving the location and strength of fluid sources (isolated or line-distributed) [470]. In the case when the Schwarz function is meromorphic in Ω, (1.52) becomes

$$\frac{1}{\pi}\int_\Omega h(z)dxdy = \sum_{k=1}^{m}\sum_{j=0}^{n_k-1} c_{kj}h^{(j)}(z_k), \qquad (1.53)$$

and the domain is called a *quadrature domain*, or *algebraic domain*. When the Schwarz function in addition has branch cuts in Ω, so that a quadrature formula of type (1.52) holds, Ω is called a *generalized quadrature domain* (*Abelian domain* if all the ρ_r are constant). This is the typical scenario for our problem. The rigorous theory of quadrature domains is outlined in Section 3.3.1.

The hydrodynamic interpretation of the Schwarz function arises from the Darcy law (1.45), which we rewrite here with W rescaled by a factor:

$$\partial_t S = \partial_z W. \qquad (1.54)$$

Let C be some closed time-independent contour, boundary of a domain B, and integrate equation (1.54) over it. We obtain, as a general consequence of (1.54),

$$\partial_t \oint_C S(z,t)dz = \int_B \omega\, dxdy - i\int_B \nabla\cdot\mathbf{V}\, dxdy, \qquad (1.55)$$

where $\omega = \partial_y V_x - \partial_x V_y$ is the vorticity field, and $\nabla\cdot\mathbf{V} = \partial_x V_x + \partial_y V_y$ is the divergence of the velocity field. The real part of this identity shows that the zero vorticity of the flow implies that

$$\operatorname{Re}\,\partial_t \oint S(z,t)dz = 0.$$

The imaginary part of (1.55) illustrates again the interpretation of the singularity set of $S(z,t)$ as sources: assume that the contour C in (1.55) encircles the droplet without crossing any other branch cuts, then the contour integral may be performed using Cauchy's theorem, giving the total flux Q:

$$\int_B \nabla \cdot \mathbf{V}\, dxdy = Q.$$

1.8 Other geometries and other kinds of sources

Further interesting considerations can be obtained from decomposing the power series expansion of $S(z)$ as

$$S(z) = \sum_{k=1}^{\infty} M_{-k} z^{k-1} + \frac{M_0}{z} + \sum_{k=1}^{\infty} \frac{M_k}{z^{k+1}}.$$

Notice that the moment M_0 is real (in fact positive), and it has a special role. It turns out that the moments $M_0, M_1, M_2, \ldots$ are enough to determine the domain Ω completely, as long as only domains close enough to a given simply connected domain with analytic boundary are considered. Similarly, the moments $M_0, M_{-1}, M_{-2}, \ldots$, which can be viewed as the moments for the complementary set $\mathbb{C}\backslash\Omega$, also characterize Ω locally. Thus, as for local variations, we can set the domain Ω in one-to-one correspondence with either set of moments $(M_0, M_1, M_2, \ldots)$ or $(M_0, M_{-1}, M_{-2}, \ldots)$. The discoveries by Richardson discussed in the previous subsection mean that Hele-Shaw evolution in the coordinates $(M_0, M_1, M_2, \ldots)$ just amounts to $M_1, M_2, \ldots$ remaining constant and M_0 developing linearly in time. The negative moments $M_{-1}, M_{-2}, \ldots$, however, turn out to develop in a rather complicated manner.

But we can turn everything around and consider another type of evolution, for which the negative moments $M_{-1}, M_{-2}, \ldots$ remain constant, with M_0 still being a linear (or affine) function of time. This turns out to be exactly the same as considering a Hele-Shaw evolution for which the fluid occupies the exterior domain and there is a source (or sink) at infinity. The mathematical meaning is that the boundary $\partial\Omega$ moves with velocity proportional to the gradient of the Green function of the exterior domain and with the pole at infinity. In other words, the speed in the normal direction is proportional to the density of the *equilibrium measure* on $\partial\Omega$. This geometry for Hele-Shaw flow is natural in many applications, and a considerable amount of the literature actually refers to this case. The complement of the fluid domain, which in this geometry is a compact set, turns out to have several physical interpretations, for example as a cluster of charge particles (electrons) in statistical models such as Coulomb gases, which mathematically relates to normal random matrix ensembles, crystal aggregates in Brownian motion processes related to freezing (DLA patterns), etc. In some applications this complement is referred to as a "droplet" (or several droplets, if Ω is not simply connected). See Section 6.2.3.

As for the mathematical treatment the difference between the exterior and interior case is relatively minor. To give a few details, let $\Omega(t)$ be the now unbounded fluid domain, so that $\infty \in \Omega(t) \subset \overline{\mathbb{C}}$. It is advantageous to let f be the normalized conformal map from the exterior of the unit disk, $\mathbb{D}^* = \{|\zeta| > 1\} \cup \{\infty\}$, to the fluid domain: $f : \mathbb{D}^* \to \Omega(t)$, normalized according to

$$f(\zeta,t) = b(t)\zeta + b_0(t) + \sum_{j=1}^{\infty} \frac{b_j(t)}{\zeta^j}, \quad b(t) > 0. \tag{1.56}$$

Apart from a sign change the Polubarinova–Galin equation looks the same as before:

$$\mathrm{Re}\left[\dot{f}(\zeta,t)\overline{\zeta f'(\zeta,t)}\right] = -\frac{Q}{2\pi}, \quad \zeta = e^{i\theta}. \tag{1.57}$$

Since either set of moments, $(M_0, M_1, M_2, \dots)$ or $(M_0, M_{-1}, M_{-2}, \dots)$, determines the domain (locally) it is clear that the negative moments may be considered as functions of the positive ones, or vice versa. In the work of M. Mineev-Weinstein, P. Wiegmann, A. Zabrodin previously mentioned, and in several subsequent papers, for example [370, 327], remarkable integrability properties among the moment sequences were shown, for example

$$\frac{1}{k}\frac{\partial M_{-k}}{\partial M_j} = \frac{1}{j}\frac{\partial M_{-j}}{\partial M_k} \quad (k,j \geq 1).$$

This can be explained in terms of the presence of a certain prepotential $\mathcal{E}(\Omega)$ such that

$$\frac{1}{k}M_{-k}(\Omega) = \frac{\partial \mathcal{E}(\Omega)}{\partial M_k}.$$

In fact, this $\mathcal{E}(\Omega)$ simply is the renormalized energy (unique up to an additive constant) of the external domain:

$$\mathcal{E}(\Omega) = \frac{1}{4\pi} \int_{\mathbb{D}_R \backslash \Omega} \int_{\mathbb{D}_R \backslash \Omega} \log|z - \zeta| d\sigma_z d\sigma_\zeta,$$

where R is chosen so that $\mathbb{D}_R$ contains Ω. The exponential of $\mathcal{E}(\Omega)$ can be identified with a τ-function which appears as a partition function in mathematical models in statistical mechanics, see [321]. It will be discussed from different points of view in Chapters 6–8 below.

1.9 Historical remarks

This chapter is a shorter version of a historical overview [566] of the development of the topic now commonly known as Laplacian Growth, from the original Hele-Shaw experiment to the modern treatment based on integrable systems. Some of information and the picture were taken from from [253].

Hele-Shaw (1854–1941) was one of the most prominent engineering researchers at the edge of the XIXth and XXth centuries, a pioneer of technical education, and a great organizer. The son of a successful solicitor Mr Shaw, he was a very religious person, influenced by his mother from whom he adopted her family name 'Hele' in his early twenties. Hele–Shaw was born on 29 July 1854 at Billericay (Essex). At the age of 17, he finished a private education and was apprenticed at the Mardyke Engineering Works, Messr Roach & Leaker in Bristol.

His brother Philip E. Shaw (Lecturer and then Professor in Physics, University College Nottingham) testifies: '... Hele's life from 17 to 24 was a sustained epic: 10 hrs practical work by day followed by night classes'.

Hele-Shaw applied for a 3 year Whitworth Scholarship in Bristol and he was a leading candidate in the list before an exam, when the congestion of lungs happened and the effort and exposure would be dangerous. Nevertheless, he went by cab to the examination and again headed the list and got the highest award of £740. It is interesting that later in 1923 he founded the Whitworth Society.

Fig. 1.6: H.S. Hele-Shaw, from [253]

1.9.1   1876–1885

In 1876 he entered the University College Bristol (founded in 1872) and in 1878 he was offered a position of Lecturer in Mathematics and Engineering under Professor J.F. Main. In 1880 he got a Miller Scholarship from the Institution of Civil Engineers for a paper on *Small motive power*.

In 1882 Main left the College and Hele-Shaw was appointed as Professor of Engineering while the Chair in Mathematics was dropped. At that time he organized his first Department of Engineering at the age of 27 and became its first professor.

1.9.2   1885–1904

In 1885 Hele-Shaw was invited to organize the Department of Engineering at the University College Liverpool (founded in 1881), his second department, where he served as a Professor of Engineering until 1904 when he moved to South Africa. During this period Hele-Shaw carried out his seminal experiments at University College Liverpool, designing the cell that bears his name.

1.9.3   1904–1906

In 1904 Hele-Shaw became the first Professor of Civil, Mechanical and Electrical Engineering of the Transvaal Technical Institute (founded in 1903) which then gave rise to the University of Johannesburg and the University of Pretoria.

It became his third department. In 1905 he was appointed as a Principal of the Institute and an organizer of Technical Education in the Transvaal. Hele-Shaw thus became one of the pioneers of technical education not only in the metropolitan area but also in the colonies.

Moreover, he was an exceptional teacher and his freehand drawing always attracted special attention. He always tried to present difficult experiments in an easier way, creating new devices in order to visualize certain phenomena.

An interesting fact is that after the paper [566] had appeared, the last author was contacted by prof. Jane Carruthers, a historian of sciences, and she published [91] an extended account of Hele-Shaw's Transvaal period.

1.9.4   1906–1941

Upon returning from South Africa, Hele-Shaw abandoned academic life, setting up as a consulting engineer in Westminster, concerning with development and exploitation of his own inventions. In 1920 Hele-Shaw became the Chairman of the Educational Committee of the Institution of Mechanical Engineers, the British engineering society, founded in 1847 by the Railway 'father' George Stephenson. In 1922 Hele-Shaw became the President of the Institution of Mechanical Engineers.

Hele-Shaw took a very active part in the professional and technical life of the Great Britain. He was:

- President of the Liverpool Engineering Society (1894);
- President of the Institution of Automobile Engineers (1909);
- President of the Association of Engineers in Charge (1912);
- President of Section G of the British Association for the Advancement of Science (1915);
- President of the Institution of Mechanical Engineers (1922);
- Fellow of the Royal Society (1899).

One of his greatest contributions to Technical Education was the foundation of 'National Certificates' in Mechanical Engineering. He was joint Chairman of the corresponding Committee (1920–1937).

Hele-Shaw was a man of great mental and physical alertness, of great energy and of great courage. He was a self-made person and was successful and recognized during his professional life. He possessed a great sense of humor, was a good conversationalist (testimonies of his brother Philip, colleagues), loved companies.

He married Miss Ella Rathbone, a member of a prominent Liverpool family. They had 2 children, the son was killed in combat during the First World War, the daughter was married to Mr Harry Hall.

He retired at the age 85 from his office in London and died one and a half year later on 30 January 1941.

1.9.5   Hele-Shaw's inventions

Two of Hele-Shaw's greatest inventions are his Stream-line Flow Methods (1896–1900), and Automatic Variable-Pitch Propeller (1924), jointly with T. Beacham. However, the full list of his inventions is much longer.

His earliest original work was registered in 1881 on the measurement of wind velocity. At that time many engineers tried to model the Tay Bridge disaster (28 December 1879). Designed by Thomas Bouch, the bridge over the river Tay (near Dundee) was one of the most remarkable engineering constructions of that time taking six years to build, and costing £300 000, as well as a lot of constructive means and human resources.

During a stormy night on 28 December 1879 the central sections of the bridge collapsed taking with them a train of 6 carriages and 75 passengers, all of whom perished. While modelling possible reasons for this disaster Hele-Shaw invented a new integrating anemometer (a device to measure wind speed).

Continuing the list of his inventions let us mention a special stream-line filter to purify water from oil pollution and the Hele-Shaw Friction Clutch (the first of its kind; 1905) for cars, patent #GB795974. At a notable Paris Motor Show (1907) about 80% of exhibited cars had the Hele-Shaw clutch.

His other inventions include the Hele-Shaw hydraulic transmission gear (1912), Hele-Shaw pump (1923), and many others; a total of 82 patents.

Let us return to the most important inventions mentioned at the beginning of this section. H.S. Hele-Shaw and T.E. Beacham patented the first constant speed, variable pitch propeller in 1924, patent #GB250292.

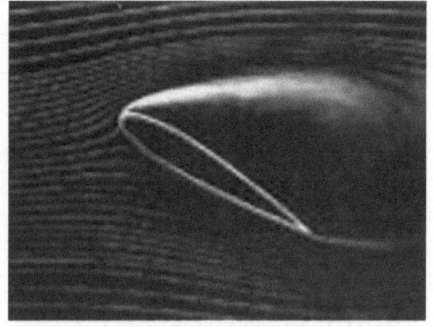

Later in 1929 Fairey and Reed in the UK and Curtiss in the USA improved it and in 1932 variable pitch propellers were intro-

Fig. 1.7: Stream line method, from Hele-Shaw 1898 [265]

duced into air force service in both countries. Further developments of the Hele-Shaw variable pitch propeller include:

- 1929 Adjustable pitch propeller drive, patent #GB1723617;

- 1931 Control system for propeller with controllable pitch, patent #GB1829930.

- 1932 Hele-Shaw and Beacham invented 'Exactor Control', a remote mechanism to reproduce the control movements in aircrafts. Hele-Shaw was then 78 years old!

The most notable result of Hele-Shaw's scientific research came from his desire to exhibit on a large screen the character of the flow past an object contained in a lantern slide for students in Liverpool. He proposed a device consisting of two parallel glass plates fixed at a small distance sandwiching a viscous fluid. Hele-Shaw wanted to visualize stream lines of the flow. He tried to inject colouring liquid (but it turned to be unsuitable, it immediately mixed with the rest of fluid), and then sand (but it formed eddies, and then, it modified the flow).

He finally achieved his objective quite by accident. Apparently the glass developed a small accidental leak providing small air bubbles acting as continuous tracers (1897). In 1897 Hele-Shaw presented his method at the Royal Institution of Naval Architects. Later in 1898, Osborne Reynolds (1842–1912) criticized the experiments by Hele-Shaw expecting turbulence at higher velocities. Indeed, O. Reynolds (1873) revealed the turbulence phenomenon under higher velocities (his experiment is shown in Figure 1.8 with sketches taken from the original Reynolds' paper [466]).

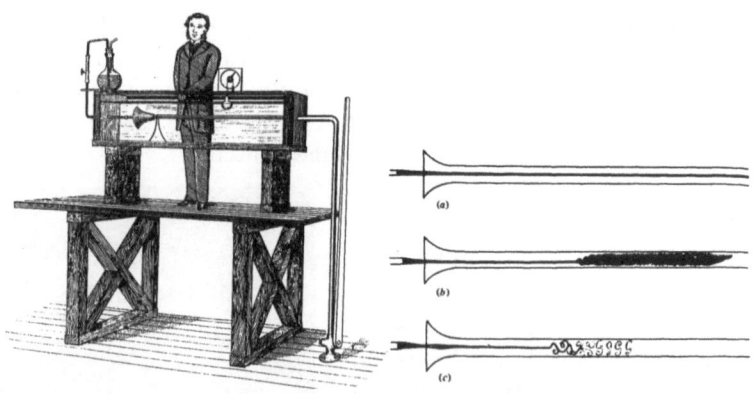

Fig. 1.8: Reynolds' experiment, from Reynolds 1833 [466]

Hele-Shaw's greatest discovery in this context was that if the glass plates on the lantern slide are mounted sufficiently close (0.02 inch) to each other, then the flow is laminar at all velocities! He got the Gold Medal from the Royal Institution of Naval Architects in 1898 for his stream line method.

Sir George Gabriel Stokes wrote: 'Hele-Shaw's experiments afford a complete graphical solution, experimentally obtained, of a problem which from its complexity baffles mathematicians except in a few simple cases'. Stokes mentions also

Hele-Shaw's experiments in his letter to Lord Kelvin from September 7, 1898: 'Hele-Shaw has some beautiful photographs, very interesting to you and me. By means of a thin stratum of viscous liquid between close glass walls, flowing past an interruption in the film, you can realise experimentally the theoretical stream lines in two dimensions in a perfect fluid flowing round a body represented in section by the obstacle' (see, [532]).

Hele-Shaw published several papers on this method, the most known of which is [265]. They are:

- Experiments on the flow of water. *Trans. Liverpool Engn. Soc.*, 1897;
- Investigation of the nature of surface resistance of water and of stream line motion under certain experimental conditions, *Trans. Inst. Nav. Archit.*, 1898 [Gold Medal];
- Experimental investigation of the motion of a thin film of viscous fluid, *Rep. Brit. Assoc.*, 1898 [Appendix by G. Stokes]
- Experiments on the character of fluid motion, *Trans. Liverpool Engn. Soc.*, 1898;
- The flow of water, *Nature* **58** (1898), 33–36;
- Flow of water, *Nature* **59** (1899), 222–223;
- The motion of a perfect fluid, *Not. Proc. Roy. Inst.*, **16** (1899), 49–64.

To conclude this section let us mention that Universities of Bristol and Liverpool perpetuated Hele-Shaw's contributions and established prizes in his memory:

- Hele-Shaw Prize (University of Bristol) to the students in their Final Year in any Department with a good academic or social record not otherwise covered;
- Hele-Shaw Prize (University of Liverpool) for a candidate who has specially distinguished himself in the Year 2 examination for the degree of Bachelor or master of Engineering.

Fig. 1.9: Hele-Shaw cell

1.9.6   Sir Horace Lamb

The first written mathematical treatment of the phenomenon discovered by Hele-Shaw appeared in the famous monograph by Lamb [341]. Sir Horace Lamb (1849–1934) created one of the most famous texts in fluid dynamics known as 'Hydrodynamics' to most applied mathematicians. The first edition of this work appeared in 1879 [340] as 'Treatise on the Motion of Fluids', a time at which the subject was actively developing, with intriguing experiments and new theories. When Lamb was 17 years old he won a scholarship to Queen's College, Cambridge (to read classics) but soon he turned to mathematics and entered Trinity College Cambridge.

He was taught by Stokes and Maxwell and graduated in 1872 with the second rank in the list of those students awarded the First Class degree. In 1875 Lamb moved to Australia where he was appointed as a chair in mathematics at Adelaide. He remained there with his family for 10 years before taking an appointment as a chair at the Victoria University in England (now the University of Manchester). In 1920 he retired and returned to Cambridge where he was awarded an honorary Rayleigh lectureship, specially created for him. He was elected as Fellow of the Royal Society in 1884. A part of his life was dedicated to improvement and completion of his famous book which ran six editions from the first [340] 258 page text to the sixth 738 page edition [342] then reprinted several times, e.g.,

Fig. 1.10: Sir Horace Lamb, photo by Maull & Fox (1885), see [347]

1945, 1993. Chronologically they were 1879 1st, 1895 2nd, 1906 3rd, 1916 4th, 1924 5th, and 1932 6th. One sees that the book matured simultaneously with the subject.

Lamb's mathematical model for the Hele-Shaw cell was based on the fact that a Newtonian fluid sandwiched between two closely situated parallel plates moves according to the parabolic distribution of the velocity vectors. 'Folklore' opinion suggests that Stokes knew the Hele-Shaw equation already at the time of the experiments, but we found no written evidence of this.

1.9.7   Darcy's law

The Hele-Shaw equation reminds us of Darcy's law in higher dimensions for flow of a viscous fluid through a porous medium, where the flow is studied in a macroscopic scale although the microscopic behaviour is rather complex.

Averaging the slow flow equations through the complicated microscopic random geometry leads to a very simple equation

$$\mathbf{V} = -\frac{k}{\mu}\nabla p,$$

where k is permeability and μ continues to stand for viscosity. This equation was obtained in 1855 [141] experimentally by a French scientist and engineer Henry Philibert Gaspard Darcy (1803–1858).

At that time he became sick and retired from his position of Chief Director for Water and Pavements, Paris, and returned back to his hometown Dijon where he conducted research on viscous fluid filtration through sand, which led him to the law called after him. He died in 1858 soon after his discovery.

However it took a long time to derive Darcy's law from the Navier–Stokes equations via homogenization. Hubbert [283] and Hall [255] were the first who tried to derive Darcy's law by integrating the Navier–Stokes equation over a representative volume element. However, these works were in a mechanistic fashion and were not sufficiently rigorous. Following these first attempts many authors improved them (e.g., Poreh, Elata, Whitaker, Ahmed, Sunada, etc.). We do not include all references to these works in order to keep the reference list reasonable for the book; however, the reader can find them, e.g., in [407]. Perhaps the most rigorous derivation appeared only in 1977 by Neuman [407]. Darcy's law became one of the most important tools in Hydrodynamics and Engineering possessing numerous applications, e.g., in oil recovery.

Fig. 1.11: Protrait of Darcy by F. Perrodin from the collection of the Bibliothèque Municipale de Dijon, from [75] (left), and L.S. Leibenzon, from [225, p. 380] (right)

The next step was made by the Soviet scientist Leonid Samuilovich (Leib Shmulevich) Leibenzon (1879–1951). One of the most prominent Soviet mechanists of that time, he was born in Kharkov (Ukraine) in the family of a doctor. In 1897 he entered Moscow University where he studied under N.E. Joukowski's supervision. During a difficult time in Russia Leibenzon traveled a lot working in different institutions (in Tbilisi, Tartu, Moscow). In 1925 he organized the first laboratory of the oil industry in the USSR. One of the most important researches of Leibenzon was dedicated to Gas Dynamics. The theory of the motion of gases in porous media was developed by Leibenzon in 1921–1922 and was published in a series of papers [355] in 1923, eight years earlier than Muskat [400], see later complete publications in [356, 401]. Approximately at that time the finite-source model for the Hele-Shaw cell was proposed.

1.9.8   Conformal mapping

The finite sink/source model was first proposed because of its applicability in the oil industry (Polubarinova-Kochina, Galin 1945), and then to the problem of injection moulding (Richardson, 1972). Molten polymer is forced into a mould of appropriate shape through a strategically placed hole [470]. The step from the local behaviour of flow to the large scale evolution yields many interesting consequences, the most important of which is the Free Boundary formulation of the problem.

Polubarinova-Kochina and Galin were the first to reformulate the free boundary problem for the pressure as a boundary-value problem for a conformal map from some canonical domain (the unit disk in our case) to the phase domain $\Omega(t)$. Let us mention that this approach became one of the most widely used in field theory, where such a map is called dynamical variables.

Pelageya Yakovlevna Polubarinova-Kochina (1899–1999) died two months before her centenary. A conference in 'Modern approaches to flow in porous media' in her honor in Moscow in 1999 became the conference in her memory. Polubarinova-Kochina was one of the most prominent modern Russian-Soviet mathematicians and mechanists, a member of the Academy of Sciences, a laureate of many academic and governmental distinctions, a Hero of Socialistic Labour (the highest award in Soviet Union), and a very kind person. It is impossible to overestimate Polubarinova-Kochina's influence in Mathematics and Industry. Her published works span the period from 1924 to 1999.

Pelageya Polubarinova was born in Astrakhan, a city situated in the delta of the Volga River, 100 km from the Caspian Sea. Her father Yakov Stepanovich

Fig. 1.12: N.Ye. Kochin and P.Ya. Polubarinova, from [566]

Polubarinov, an accountant, discovered Pelageya's particular interest in science and decided to go to St. Petersburg where she graduated from Pokrovskii Women's Gymnasium. In 1918, after her father's death, Pelageya Polubarinova accepted a job at the Main Geophysical Laboratory to bring in enough money to allow her to continue her education. She worked under the supervision of Aleksandr Aleksandrovich Friedmann (1888–1925). In 1921 she obtained a degree in pure mathematics. In 1921–23 she met Nikolai Yevgrafovich Kochin (1901–1944) who graduated from the Leningrad State University.

They married in 1925 and had two daughters Ira and Nina. In 1934 she returned to a full-time position being appointed as a professor at Leningrad University. In the following year her husband N.Ye. Kochin was appointed to Moscow University and the family moved to Moscow. In 1939 Kochin became a Head of the Mechanics Institute of the USSR Academy of Sciences, and a member of the USSR Academy of Sciences. Pelageya worked at the same institute.

During the World War II Polubarinova-Kochina and her two daughters were evacuated to Kazan in 1941 when Germans approached Moscow. However, N.Ye. Kochin remained in Moscow carrying out military research. In 1943 she returned to Moscow but Kochin became ill and died soon after. He had been in the middle of lecture courses and his wife took over the courses and completed their delivery. His research was on meteorology, gas dynamics and shock waves in compressible fluids.

Fig. 1.13: L.A. Galin, from [566]

In 1958 P.Ya. Polubarinova-Kochina was elected a member of the USSR Academy of Sciences, and moved to Novosibirsk to help build the Siberian Branch of the Academy of Sciences. For the next 12 years she worked in Novosibirsk where she was Director at the Hydrodynamics Institute and also Head of the Department of Theoretical Mechanics at the University of Novosibirsk. In 1970 she returned to Moscow and became the Director in the Mathematical Methods of Mechanics Section of the USSR Academy of Sciences.

One of her major contributions is the complete solution of the problem of water filtration from one reservoir to another through a rectangular dam. In this work she established connections with the Riemann P-function, Hilbert problems and Fuchsian equations [440].

Lev Alexandrovich Galin (1912–1981) was born in Bogorodsk (Gor'kii region), graduated from the Technology Institute of Light Industry in 1939 and started to work at the Mechanics Institute led by N.Ye. Kochin. He became a professor at the Moscow State University from 1956 and Correspondent Member of the Soviet Academy of Sciences from 1953.

P.Ya. Polubarinova-Kochina [438], [439] and L.A. Galin [199] independently gave a conformal formulation of the Hele-Shaw problem in 1945. Polubarinova and Galin proposed the first non-trivial solution to the free boundary problem with suction taking a polynomial Ansatz $f(\zeta, t) = a_1(t)\zeta + a_2(t)\zeta^2$. This allowed them to reduce the problem to a simple system of ODEs for the coefficients a_1 and a_2, see the next chapter.

1.9.9   Kufarev and the existence of a solution

Taking different Ansatzs one can obtain many explicit solution (e.g., polynomial, rational, logarithmic). However, the problem becomes much more complex in its general formulation: *given an initial domain, and therefore, an initial function find the solution to the Polubarinova–Galin equation.*

The problem of existence is very difficult even for an advancing fluid. The Polubarinova–Galin equation can be easily reformulated in the unit disk making use of the Cauchy–Schwarz representation. The corresponding equation is a first-order integro-PDE of Löwner type. The Löwner equation is linear and the solution obviously exists. In contrast, the Polubarinova–Galin equation contains feedback in the driving integral term. Kufarev with his student Vinogradov were the first to address the problem of existence [569] in 1948.

Pavel Parfenievich Kufarev (1909–1968) was born in Tomsk on 18 March, 1909. His life was always linked with the Tomsk State University where he studied (1927–1932), was appointed as a docent (1935), and finally as a professor (1944). He got the State Honor in Sciences (1968) just before his death. His main achievements are in the theory of Univalent Functions where he general-

Fig. 1.14: P.P. Kufarev, from [568]

ized in several ways the famous Löwner parametric method. But the first works were in Elasticity Theory and Mechanics.

Kufarev was greatly influenced by Fritz Noether (Erlangen 1884–Orel 1941), the brother of Emmy Noether, and by Stefan Bergman (1895–1977), who immigrated from nazi Germany (under anti-Jewish repressions) to Tomsk (1934). Bergman fortunately moved to Paris in 1937. Noether's life turned out to be more tragic. He was arrested during the Great Purge, and sentenced to a 25-year imprisonment for being a 'German spy'. While in prison, he was accused of 'anti-Soviet propaganda', sentenced to death, and shot in the city of Orel (Orlovskii Zentral concentration camp) in 1941. He was shot together with Sergei Efron (Marina Tsvetaeva's husband), Maria Spiridonova, the left SRs (Socialist Revolutionary Party) leader, and many other political prisoners gathered in the infamous 'Orlovskii Zentral' because of panic due to Germany's immediate occupation of Orel. There was no chance of evacuation, so the NKVD preferred to kill them rather than to leave the prison in the hands of the Germans.

The results of Vinogradov and Kufarev's 1948 paper and Kufarev's other results in exact solutions (including logarithmic ones) [570, 328, 330, 331] had remained unknown until Vladimir Gutlyanskiĭ and Yuri Hohlov brought them to a Western audience in the early 90's. The result of existence was reproved in modern language as a particular case of the nonlinear abstract Cauchy–Kovalevskaya Theorem in 1993 by Reissig and von Wolfersdorf [467]. Kufarev gave many exact solutions: when the initial domain is a strip or a half-plane; when the initial domain is a disk with a non-centered sink; rational exact solutions; the case of several sinks/sources, etc. It is interesting that the review in Mathematical Reviews (MR0097227, 20# 3697) on the now famous 1958 paper by Saffman and Taylor [488] says '... the authors do not seem to be aware of the fact that there exists a vast amount of literature concerning viscous fluid flow into porous (homogeneous and non-homogeneous) media in Russian and Romanian', referring first of all to Kufarev's works.

1.9.10   Saffman and Taylor fingering

Of course, one of the most important steps in the early treatment of the Hele-Shaw experiment was the phenomenon of fingering described by Saffman and Taylor in 1958 [488]. It is the most common instability type occurring for a receding viscous fluid (ill-posed problem). It never occurs for an advancing fluid. At the same time an analytic travelling wave solution, known now as the Saffman–Taylor solution, became an important example of an exact solution which exists for all time for the ill-posed problem. The fingering phenomenon and its prediction is especially

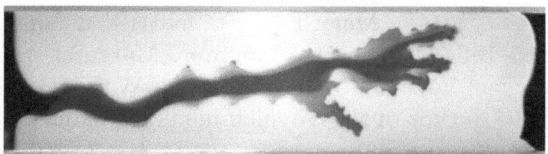

Fig. 1.15: Instability of an interface moving towards a more viscous fluid, from [289] with permission from © 2008 The Amernican Physical Society.

important in industrial applications, e.g, oil recovery. During primary recovery, the natural pressure of the reservoir, combined with pumping equipment, brings oil to the surface. Primary recovery is the easiest and cheapest way to extract oil from the ground. But this method of production typically produces only about 10 percent of a reservoir's original oil in place reserve. Getting as much oil as possible out of a reservoir has always been the industry's prime goal. Where pressure has

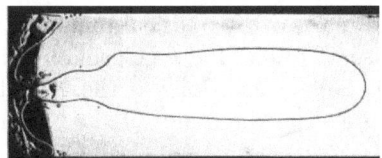

Fig. 1.16: Growth of a single long
finger, from [566]

dipped, this often involves using water as a means of flushing the oil out, in the
secondary phase. However, due to lower viscosity, water forms long fingers in the
sand-oil medium apparently reaching the pumping tube. This ends the secondary
phase and generally results in the recovery of 20 to 40 percent of the original oil
in place. The tertiary phase may use thermal or chemical recovery.

Fig. 1.17: Sir G.I. Taylor,
from [566]

Sir Geoffrey Ingram Taylor (1886, London–
1975, Cambridge) is cited as one of the greatest
physical scientists of the 20th century. He was
born in St. John's Wood, London, his father was
an artist, and his mother, Margaret Boole, came
from a family of mathematicians (his aunt was Alicia Boole Stott and grandfather George Boole).
One can observe some interesting connections with
other Boole family members. Margaret's grandfather Colonel Sir George Everest (1790–1866)
was a British surveyor, geographer and Surveyor-
General of India from 1830 to 1843. After the
Great Trigonometric Survey of India along the
meridian arc, the famous (and the highest on the
planet) mount was named in his honor. His niece,
Mary Everest, married a famous mathematician
George Boole (1815–1864). Their 5th daughter, Margaret's sister, Ethel Lilian
married Wilfrid Michael Voynich (Michał Habdank-Wojnicz), 1865–1930, a polish
revolutionary and the author of the Voynich medieval manuscript.

Ethel Lilian became a writer and she is widely known as author of *The Gadfly*,
1897. Other sisters entered the academic career as well: Alicia was a Lecturer in
Geometry, Lucy became professor in Chemistry, and Mary Ellen was a teacher.

In 1910 Taylor was elected to a Fellowship at Trinity College, and the following year he was appointed to a meteorology post, becoming Reader in Dynamical
Meteorology. His work on turbulence in the atmosphere led to the publication of
'Turbulent motion in fluids', which won him the Adams Prize in 1915. During
World War II Taylor again worked on applications of his expertise to military
problems such as the propagation of blast waves, studying both waves in air and
underwater explosions.

These skills were put to the service of scientists at Los Alamos when Taylor was sent to the United States as part of the British delegation to the Manhattan project between 1944 and 1945. In 1944 he also received his knighthood and the Copley Medal from the Royal Society. His final research paper was published in 1969, when he was 83.

Philip Geoffrey Saffman (1931–2008) was a student of George Keith Batchelor (1920–2000), FRS–1957, an Australian scientist, a student of Taylor, and a founder of the Journal of Fluid Mechanics (1956). Saffman was a Theodore von Kármán Professor at the California Institute of Technology, USA.

The Saffman–Taylor exact solution in its conformal formulation depend on the relative channel width λ. But in experiments λ was found to be close to $1/2$ except in some very special cases (very slow flow, Saffman's unsteady solution). Why is $\lambda = 1/2$ selected? This is a famous *selection problem* which Saffman and Taylor posed in 1958. They also proposed to use a small surface tension as a selection mechanism. Despite many attempts both analytic and numeric, no solution has been commonly recognized as final. Activities in this direction were especially intensive in the 1980s and were summarized and completed in [303]. Basically, many authors exploited a reduction to a non-linear eigenvalue problem. Several groups in 1986–87 presented numerical results on instability of all finger widths except $\lambda = 1/2$, in the limit of vanishingly-small surface tension. Another survey [489] on the selection problem was written by Saffman in 1986. One of the most cited (especially by physicists) paper on the fingering phenomenon is the review by Bensimon, *et al.* [50]. Among numerous works in this direction let us distinguish a couple of mathematical treatments which are closer to the topic of this book. Following Saffman and Taylor's proposal on the selection mechanism, Tanveer and Xie proved [549] that there is no classical steady finger solution for non-zero surface tension and $\lambda \in (0, 1/2)$. This confutes previous numerical simulations. They also established that if the classical solution for the finger problem with small non-zero surface tension exists for some $\lambda \in (0, 1)$, then it is close to a certain Saffman–Taylor solution, satisfies an algebraic decay at infinity, and belongs to some special space of analytic functions. There can be a discrete number of $\lambda \in [1/2, 1)$ for which the selection problem has a solution under this driving mechanism. An interesting alternative approach by Mineev-Weinstein [391] rejects surface tension as a selection mechanism and proposes finger formation because of a special dynamics of exterior analytic singularities in a class of solutions suggested by Dawson and Mineev-Weinstein earlier in [390].

Fig. 1.18: P.G. Saffman, from [578]

1.9.11   Modern period

Fig. 1.19: S. Richardson,
from [478]

The development of the Hele-Shaw problem can be rather easily seen before 1970. Splashes of interest followed by non-active periods. The number of authors was restricted, though there were many brilliant scientists among them. The last decades of the XXth century were characterized by a real boom of interest from all fields of natural sciences and engineering. It is just impossible to summarize all recent achievements and challenges in the topic. So we restrict ourself to only a partial overview of results that led finally to what is now known as the *Laplacian Growth* problem, applications of conformal mapping and some potential theoretic remarks.

While the title 'Modern period' is rather vague and questionable, it is commonly accepted that the modern period of the developments in the Hele-Shaw problem started from a 1972 seminal paper [470] by Stanley Richardson (1943–2008) who suddenly passed away March 12, 2008 after having suffered a heart attack. Dr. Stan Richardson was born near Macclesfield in 1943. He studied at Cambridge, completing the four years of the Mathematical Tripos in three, before going on to do his Ph.D. After 6 years in Cambridge he came to Edinburgh. He was a lecturer and then, a reader in the School of Mathematics, University of Edinburgh from 1971. He published 22 papers which represent a master work and which established Stan as a world class figure in the field.

Further developments after Richardson's contribution are the subject of the present monograph.

1.9.12   Acknowledgements for Section 1.9

In 2007 Mark Mineev and Björn Gustafsson asked Alexander Vasil'ev to present a historical overview of Hele-Shaw flows at the Banff International Research Station (Canada) meeting. Working on that lecture he discovered many interesting and unknown (to him) facts about persons who contributed to this interesting and challenging topic, first of all about H.S. Hele-Shaw. After discussions with colleagues at BIRS, he decided to take a risk and to share his lecture with a wider audience, also adding information that was given during these discussions. This was realized in [566] and partly entered Section 1.9.

Chapter 2

Rational and Other Explicit Strong Solutions

In this chapter we shall construct several explicit solutions to the Hele-Shaw problem, more precisely, to the Polubarinova–Galin equation, starting with the classical ones of Polubarinova-Kochina [438], [439], Galin [199] and Saffman, Taylor [488], [489]. Some properties of polynomial and rational solutions will be discussed, and it will be proved that the property of the conformal map to the fluid domain of being a polynomial or a rational function is preserved under the time evolution. The same is true also when logarithmic singularities are allowed. From these properties one easily deduces local existence and uniqueness of solutions within such classes.

After that we shall consider angular Hele-Shaw flows and give some new families of explicit solutions in terms of hypergeometric functions that contain, as particular cases, those constructed earlier by Ben Amar *et al.* [43], [44], [45], Arnéodo *et al.* [25], Kadanoff [292], etc.

2.1 Classical solutions of the Polubarinova–Galin equation

It is possible to construct many explicit solutions to the Hele-Shaw problem using the nonlinear Polubarinova–Galin equation (1.22). The main idea is to use a special form of the parametric univalent function $f(\zeta, t)$. The simplest solution is the expansion/shrinking of a disk centered at the injection/suction point, which we take to be the origin. This is the only case when the fluid can be completely removed (see [199], [269]). The solution has the obvious form

$$f(\zeta, t) = \sqrt{\frac{|\Omega_0| + tQ}{\pi}}\, \zeta.$$

Here Ω_0 is a disk centered at the origin and $t \in [0, \infty)$ in the case of injection $(Q > 0)$ and $t \in [0, -|\Omega_0|/Q]$ in the case of suction $Q < 0$. In the case of injection it is possible to start with $\Omega_0 = \emptyset$.

2.1.1   Polubarinova and Galin's cardioid

The first non-trivial solution for the problem with suction $(Q < 0)$ was constructed by Polubarinova-Kochina [438], [439] and Galin [199]. They chose a quadratic mapping

$$f(\zeta, t) = a_1(t)\zeta + a_2(t)\zeta^2,$$

$\zeta \in \mathbb{D}$, with real coefficients $a_1(t)$ and $a_2(t)$. This mapping being substituted into equation (1.22) gives the following system for the coefficients

$$a_1^2(t)a_2(t) = a_1^2(0)a_2(0),$$

$$a_1^2(t) + 2a_2^2(t) = a_1^2(0) + 2a_2^2(0) + \frac{Qt}{\pi}.$$

For any initial condition such that $|a_2/a_1| < 1/2$ the solution $f(\zeta, t)$ is a univalent map locally in time $t \in [0, t_0)$. The blow-up time t_0 occurs exactly when the equality $|a_2/a_1| = 1/2$ is reached, that corresponds to the vanishing boundary derivative of f and cusp formation at the boundary. This evolution is shown in Figure 2.1. As is observed, cusp formation occurs before the moving boundary reaches the sink. This phenomenon is general for polynomial solutions. It seems that Galin knew that, but did not prove it correctly. A correct proof appeared in [269]. See also Section 3.3.1 below.

Considering a general polynomial,

$$f(\zeta, t) = \sum_{k=1}^{n} a_k(t)\zeta^k, \quad a_n(0) \neq 0, \; a_1(t) > 0,$$

one substitutes it into equation (1.22) and obtains a system of $2n-1$ real equations for the coefficients $a_k(t)$, which represent $2n - 1$ unknown real quantities. The equations are more specifically first-order differential equations. The first one is

$$\frac{d}{dt} \sum_{k=1}^{n} k|a_k|^2 = \frac{Q}{\pi}$$

and the last ones combine into the complex equation

$$n\bar{a}_n \frac{da_1}{dt} + a_1 \frac{d\bar{a}_n}{dt} = 0,$$

which is equivalent to

$$\frac{d}{dt}(\bar{a}_n a_1^n) = 0.$$

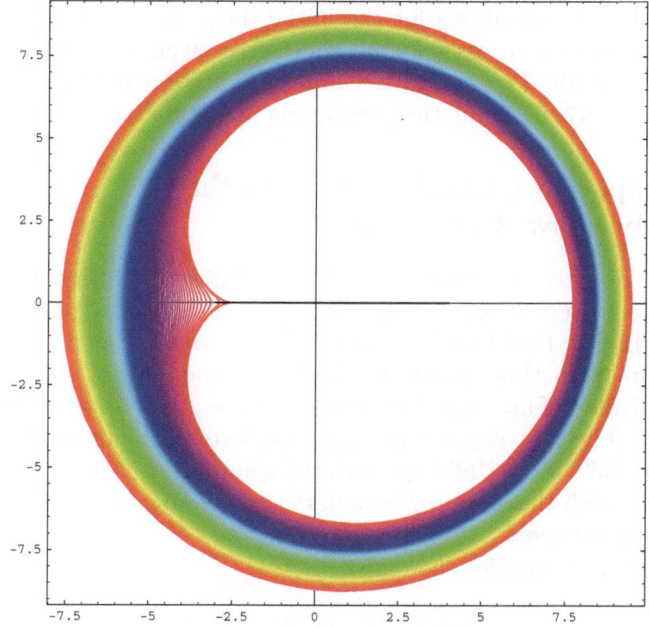

Fig. 2.1: Polubarinova and Galin's cardioid, from [566]

The mentioned equations are just two special cases of the Richardson moment conservation law (1.34), namely

$$\frac{d}{dt}M_0 = \frac{Q}{\pi}, \quad \frac{d}{dt}M_{n-1} = 0.$$

On integration they give

$$\sum_{k=1}^{n} k|a_k(t)|^2 = \sum_{k=1}^{n} k|a_k(0)|^2 + \frac{Qt}{\pi}, \tag{2.1}$$

$$\bar{a}_n(t)a_1^n(t) = \bar{a}_n(0)a_1^n(0). \tag{2.2}$$

Note that if, by a rotation $e^{i\alpha}f(e^{-i\alpha}\zeta, t)$, one normalizes the coefficient $a_n(0)$ to be real and positive, then this property will be preserved in time, by (2.2).

The differential equations obtained can always be solved for the derivatives $\frac{da_k}{dt}$, and combining this fact with the standard theory for ordinary differential equations, the local solvability of (1.22) follows in the polynomials case (see more precisely Section 2.2).

If the boundary reaches the sink at the moment t_0, then the kernel of the family $\Omega(t)$ degenerates: $\mathrm{Ker}\{\Omega(t)\} = \{0\}$ for $t \to t_0$. The Carathéodory kernel theorem implies that $\lim_{t \to t_0} f(\zeta, t) \equiv 0$, which contradicts (2.1)–(2.2) $(a_n(0)a_1^n(0) \neq 0)$.

Some sufficient conditions for the initial data $(a_1(0), \ldots, a_n(0))$ for a polynomial strong solution to exist for all time were given in [334]. Several explicit solutions similar to the Polubarinova–Galin cardioid were obtained by Vinogradov and Kufarev in 1947 [570] but their work was unfortunately forgotten.

2.1.2 Examples of rational solutions of the Polubarinova–Galin equation

After the first non-trivial Polubarinova–Galin solution in Section 2.1.1 many other explicit solutions were constructed. Among them we distinguish a solution by Saffman and Taylor, that will be discussed in Section 2.5. It deals with a flow in a narrow channel. In this section we shall give examples of solutions by means of rational univalent functions. One finds them, e.g., in a paper by Hohlov and Howison [269]. The first explicit rational solutions were obtained by Kufarev in 1948–1950, see [330, 331]. Unlike polynomial solutions, rational solutions can produce evolutions such that the free boundary reaches the sink under suction before the total fluid is removed.

Let $Q < 0$ and consider the map

$$f(\zeta, t) = a(t)\frac{\zeta(1 - b(t)\zeta)}{1 - c(t)\zeta}, \tag{2.3}$$

where

$$a(t) = -\frac{2\alpha^4 - \alpha^2 + \frac{Qt}{\pi}}{2\alpha^3}, \quad b(t) = \frac{\alpha^3 + \frac{\alpha Qt}{\pi}}{2\alpha^4 - \alpha^2 + \frac{Qt}{\pi}}, \quad c(t) = \frac{1}{\alpha},$$

and $\alpha = \alpha(t) < -1$, which is the finite pole of f, is the root of the algebraic equation

$$2\alpha^6 - \left(5 + \frac{2Qt}{\pi}\right)\alpha^4 + \left(\frac{Qt}{\pi}\right)^2 = 0,$$

satisfying the condition $\lim_{t \to -\pi/Q} \alpha(t) = -1$. The initial domain is given by the mapping

$$f(\zeta, 0) = \frac{\zeta(4 - \sqrt{\frac{5}{2}}\zeta)}{2\zeta - \sqrt{10}}.$$

The solution $f(\zeta, t)$ exists and is univalent during the time interval $[0, -\pi/Q)$. At the final moment, $t = -\pi/Q$, the moving boundary reaches the sink at the origin and the residual fluid occupies the disk $|z + 1| < 1$, see Figure 2.2.

The equations for determining the parameters $a(t)$, $b(t)$, $c(t)$ can be explained in terms of quadrature identities satisfied by $\Omega(t)$, see Section 3.3.1, and also [112, 116, 416, 560].

A slight modification of the above example is obtained by making the final disk larger, let it for example be $|z + 1| < 2$. Let us also reset time so that this

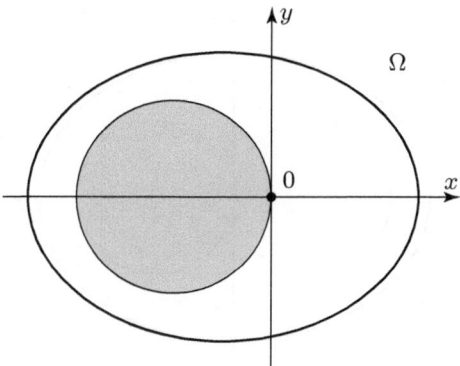

Fig. 2.2: Rational solution

disk is what we have at time $t = 0$. Then further suction (at $z = 0$) is possible, and when this takes place the conformal map onto the fluid region will again, for $t \neq 0$, be of the form (2.3), with the parameters $a(t)$, $b(t)$, $c(t)$ adapted to the new situation. However, the conformal map onto $\Omega(t)$ at time $t = 0$ is just the Möbius transformation

$$f(\zeta, 0) = \frac{3\zeta}{\zeta + 2}.$$

What we want to say with this example is that the structure of f as a rational function may change at one moment of time ($t = 0$ in the example), despite the solution actually being fully smooth. In the present case the solution $f(\zeta, t)$ will exist in some time interval $t < t_0$, where $t_0 > 0$, and the breakdown at $t = t_0$ is caused by the development of a cusp, like in the cardioid case.

Another example is the rational map

$$f(\zeta, t) = a(t) \frac{\zeta(1 - b(t)\zeta^2)}{1 - c(t)\zeta^2},$$

with the parameters a, b, c chosen so that the final domain in a blow-up time consists of two equal disks touching at the sink. Due to complicated details we give only a sketch in Figure 2.3.

2.1.3 The Huntingford example

The cusp in the Polubarinova–Galin cardioid (Section 2.1.1) really stops the solution, i.e., there is no way to continue the suction without changing the mathematical model. In local coordinates centered at the cusp, the equation of the boundary is of type $y = cx^{3/2} + \cdots$, i.e., the cusp is of order $3/2$. Under exceptional conditions there may also appear cusps (of higher order) which do not stop further suction. These can then only be of order $(4k + 1)/2$ for some integer $k \geq 1$, not

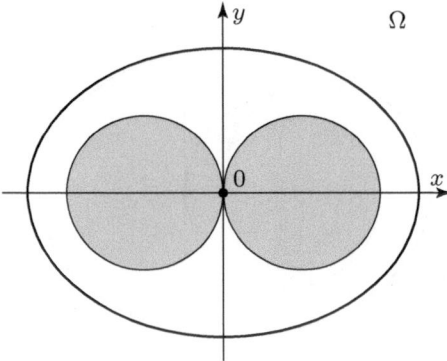

Fig. 2.3: Symmetric rational solution

of any order $(4k - 1)/2$ (see further Section 3.1.4, in particular Remark 3.1.2). Examples of such cusps were found by S. Howison [278] in the reverse geometry, i.e., with the fluid region containing the point of infinity. A similar example for bounded fluid regions has been elaborated by C. Huntingford [284]. This example is quite elementary but still illustrates several interesting features of the Hele-Shaw problem.

   We start by considering in general a polynomial solution of the Polubarinova–Galin equation (1.22) of degree three with real coefficients. Writing

$$f(\zeta, t) = a_1(t)\zeta + a_2(t)\zeta^2 + a_3(t)\zeta^3,$$

where $a_j \in \mathbb{R}$, Richardson's formula (1.37) for the moments becomes (supressing t)

$$\begin{cases} M_0 = a_1^2 + 2a_2^2 + 3a_3^2, \\ M_1 = a_1^2 a_2 + 3a_1 a_2 a_3, \\ M_2 = a_1^3 a_3. \end{cases}$$

Recall that M_1 and M_2 are constants of motion. Using this to eliminate a_2 and a_3 the solution becomes fully explicit provided one uses a_1 to parametrize time. This gives

$$f(\zeta) = a_1\zeta + \frac{a_1^{-2} M_1 \zeta^2}{1 + 3a_1^{-4} M_2} + a_1^{-3} M_2 \zeta^3. \tag{2.4}$$

It is easy to see that increasing a_1 means growing fluid domains (injection), as long as f is univalent. From (2.4) one may guess a general asymptotic behavior of the coefficients in the Hele-Shaw problem, namely that $a_n \sim a_1^{-n} M_{n-1}$ for large a_1. This has indeed been confirmed by Y.L. Lin [360] in the polynomial case, see Section 5.5.

The shape of the image domain $\Omega = f(\mathbb{D})$ is determined solely by the two ratios

$$A_2 = \frac{a_2}{a_1}, \quad A_3 = \frac{a_3}{a_1},$$

and with these fixed the parameter a_1 just measures the length scale in the physical plane. The case $A_2 < 0$ can be obtained from $A_2 > 0$ by reflection (considering $-f(-\zeta)$ instead), so we may as well assume that $A_2 \geq 0$. Then

$$f(\zeta) = a_1(\zeta + A_2\zeta^2 + A_3\zeta^3),$$
$$f'(\zeta) = a_1(1 + 2A_2\zeta + 3A_3\zeta^2),$$

and one easily checks that the range of coefficients for which f is locally univalent is the triangle in the right half of the (A_2, A_3)-plane delimited by the lines

$$L_1 : \quad A_2 = 0,$$
$$L_2 : \quad A_3 = 1/3,$$
$$L_3 : \quad 1 - 2A_2 + 3A_3 = 0.$$

In fact, within this triangle, which has corners at $(0, 1/3)$, $(1, 1/3)$ and $(0, -1/3)$, the derivative f' has on L_2 a complex conjugate pair of zeros on the unit circle, while on L_3 it has one root at -1, the other root outside the unit circle.

The range of univalence of f is only slightly smaller: close to the corner $(1, 1/3)$ the line L_3 is to be replaced by the nearby segment of the ellipse

$$E : \quad A_2^2 + 4\left(A_3 - \frac{1}{2}\right)^2 = 1,$$

with $\frac{1}{5} \leq A_3 \leq \frac{1}{3}$; see [106, 284] for details. The left end point of this A_3-interval corresponds to the point $(A_2, A_3) = (\frac{4}{5}, \frac{1}{5})$, at which the ellipse is tangent to L_3, and the right end point corresponds to $(A_2, A_3) = (\frac{2\sqrt{2}}{3}, \frac{1}{3})$, at which the ellipse intersects L_2. At the latter point, the image domain $\Omega = f(\mathbb{D})$ has two cusps, of order $3/2$, and in addition one double point. This means that f is simultaneously on the edge of losing local univalence (because of cusps), and global univalence (due to the double point). At the first mentioned point, $(A_2, A_3) = (\frac{4}{5}, \frac{1}{5})$, $\partial\Omega$ has a $5/2$-power cusp.

Any Hele-Shaw trajectory is by (2.4) described in the (A_2, A_3)-plane by a trajectory

$$\begin{cases} A_1 = \frac{a_1^{-3} M_1 \zeta^2}{1 + 3a_1^{-4} M_2}, \\ A_2 = a_1^{-4} M_2, \end{cases} \tag{2.5}$$

where a_1 is used as a parameter. As $a_1 \to \infty$ the trajectory always approaches $(0, 0)$, provided the solution does not break down during the course. The only way a solution can break down in the present picture is that the point (A_2, A_3) moves out from the region of univalence. In particular, if it moves out from the triangle of local univalence it will be serious, because even though one cannot see from the formulas that anything bad happens, what does happen is that the meaning of the Polubarinova–Galin equation (1.22) changes. For example, outside the line L_3 and with f given by (2.4), increasing a_1 will no longer mean injection, it will mean suction of the, then multi-sheeted, domain $f(\mathbb{D})$.

Now, a Hele-Shaw evolution may, in the injection case, be started at any point on L_2 with $0 \leq A_2 < \frac{2\sqrt{2}}{3}$ (the univalent regime), hence with two 3/2-cusps on the boundary. However, if A_2 is very close to $\frac{2\sqrt{2}}{3}$, then the two cusps are so close together that shortly after the solution has been created the region just outside the cusps becomes trapped by a double point being developed. This means that the solution breaks down as a univalent solution. It can still be continued as a locally univalent solution, but even as such it will later break down. This is a general phenomenon, see [239] and Section 4.2 below.

The borderline case between the above two regimes is an interesting trajectory investigated by C. Huntingford [284]. The harmonic moments for this trajectory are

$$
\begin{cases}
M_1 = \frac{32}{25}, \\
M_2 = \frac{1}{5}.
\end{cases}
$$

It starts at the point $(A_2, A_3) = (\frac{16}{15}\sqrt[4]{\frac{3}{5}}, \frac{1}{3})$, which lies to the left of $(\frac{2\sqrt{2}}{3}, \frac{1}{3})$, hence in the region of univalence. So there are initially two cusps on the boundary. As a_1 increases these resolve, but at the same time the appearance of a double point just outside is threatening. However, this borderline case means that at the same time as this double point is going to appear, the (almost) trapped region outside the initial cusps has shrunk down to be empty. Thus the solution barely escapes from a double point being created, instead there will be a kind of triple collision, which manifests itself in form of a 5/2-cusp. In the (A_2, A_3)-plane this is located exactly at the point of tangency between the ellipse and the triangle, i.e., at $(A_2, A_3) = (\frac{4}{5}, \frac{1}{5})$. Also this cusps resolves, and the solution then goes on forever $(a_1 \to \infty)$ as a smooth and univalent solution.

The appearance of cusps above is reflected in the motion of the two zeros, ω_1 and ω_2, of $f'(\zeta)$. At the starting point $(A_2, A_3) = (\frac{16}{15}\sqrt[4]{\frac{3}{5}}, \frac{1}{3})$, these form a complex conjugate pair on the unit circle. Then, as a_1 increases, ω_1, ω_2 move out from the unit circle and later collide on the real axis. After that one of the zeros, say ω_1, moves back to the unit circle and reaches it again (at $\omega_1 = -1$). This is when the Hele-Shaw trajectory has reached the point $(A_2, A_3) = (\frac{4}{5}, \frac{1}{5})$, corresponding to the 5/2-power cusp. After that, the root ω_1 moves away from the unit circle along the real axis, captures and collides again with the other root ω_2, after which they leave the real axis and finally move towards infinity in the asymptotic directions of the positive and negative imaginary axes. All of this can be seen by examining [284] carefully. The final asymptotics is in accordance with the general asymptotic theory for roots, see Section 5.5, in particular Theorem 5.5.1.

2.2 Existence of rational solution in general

Let us now discuss rational solutions in general. When speaking about a strong solution of a differential equation one generally means that all functions and boundaries appearing are smooth enough and that the equations involved hold in a pointwise sense. For the Hele-Shaw problem it is convenient to introduce the notion of a smooth family of domains [554]. We call a family of domains $\{\Omega(t)\}$ smooth if the boundaries $\partial\Omega(t)$ are smooth (C^∞) for each t, and the normal velocity V_n continuously depends on t at any point of $\partial\Omega(t)$.

Then a *strong solution* of the Hele-Shaw problem is defined to be a smooth family $\{\Omega(t) : 0 \le t < t_0\}$ such that (1.16)–(1.19) hold in a pointwise sense (the function p will be uniquely determined by $\Omega(t)$ and will be smooth up to $\partial\Omega(t)$). If the domains $\Omega(t)$ are simply connected it is equivalent to require the Polubarinova–Galin equation (1.22) to hold.

In the above definition the interval $[0, t_0)$ may be replaced by any open, closed or half-open interval.

Given a domain $\Omega(0)$ with smooth boundary it is known that in the well-posed case $Q > 0$ there exists a strong solution of (1.16)–(1.19) on some interval $[0, t_0)$. In the ill-posed case $Q < 0$ such a statement is true only if $\partial\Omega(0)$ is analytic (see, e.g., [554]).

Since we do not know any reasonably short proof of these general existence results we shall not include any such proof here, but just refer to the literature: [180, 361, 467, 569]. Instead we shall discuss some general properties of solutions in the simply connected case, and also provide an elementary proof of existence of solutions when the initial domain is the conformal image of $\mathbb{D}$ under a rational function. We shall first make some general observations.

Assume that $f(\zeta, t)$ is analytic and univalent in a neighbourhood of $\overline{\mathbb{D}}$ for each t and is normalized by

$$f(0, t) = 0, \quad f'(0, t) > 0. \tag{2.6}$$

It is useful to introduce the notations

$$\varphi(\zeta, t) = \frac{\dot{f}(\zeta, t)}{\zeta},$$
$$g(\zeta, t) = f'(\zeta, t). \tag{2.7}$$

In view of (2.6), both of φ and g are holomorphic functions in $\mathbb{D}$ with strictly positive values at the origin. The Polubarinova–Galin equation (1.22) becomes

$$\mathrm{Re}\,[\overline{\varphi} \cdot g] = \frac{Q}{2\pi}. \tag{2.8}$$

On dividing by $|g|^2$ we get

$$\mathrm{Re}\,[\frac{\varphi(\zeta, t)}{g(\zeta, t)}] = \frac{Q}{2\pi|g(\zeta, t)|^2} \quad (\zeta = e^{i\theta}). \tag{2.9}$$

Here the left member is a harmonic function in $\mathbb{D}$ and (2.9) gives its boundary values on $\partial\mathbb{D}$. This Dirichlet problem can be solved explicitly in terms of a Poisson integral, and taking also the imaginary part into account we get φ solved in terms of g as

$$\varphi = \mathcal{F}(g),$$

where $\mathcal{F}$ is the nonlinear operator defined by

$$\mathcal{F}(g)(\zeta) = \frac{Qg(\zeta)}{4\pi^2} \int_0^{2\pi} \frac{1}{|g(e^{i\theta})|^2} \frac{e^{i\theta} + \zeta}{e^{i\theta} - \zeta} d\theta \tag{2.10}$$

(We suppress t from notation whenever convenient.)

Now we observe the following properties of $\mathcal{F}(g)$:

(i) If g is holomorphic in $\mathbb{D}_R$ for some $R > 1$ then so is $\mathcal{F}(g)$.

(ii) If g is a rational function with poles of order k_j at some points ζ_j (outside $\overline{\mathbb{D}}$) then the same is true for $\mathcal{F}(g)$. One of the points may be the point of infinity.

It is important for these conclusions that g is holomorphic in a neighbourhood of $\overline{\mathbb{D}}$ and has no zeros there (whereas the univalence of f is not needed in itself).

To prove (i) and (ii) we first write (2.10) as

$$\mathcal{F}(g)(\zeta) = \frac{Qg(\zeta)}{4\pi^2 i} \int_{\partial\mathbb{D}} \frac{1}{g(z)\overline{g(\frac{1}{\bar{z}})}} \frac{z + \zeta}{z - \zeta} \frac{dz}{z}. \tag{2.11}$$

With $\zeta \in \mathbb{D}$ the integrand is holomorphic in some neighbourhood of $\partial\mathbb{D}$. It follows that the path of integration can be replaced by a contour slightly outside $\partial\mathbb{D}$. This shows that $\varphi = \mathcal{F}(g)$ is analytic in some neighbourhood of $\overline{\mathbb{D}}$ (to start with).

Next we go back to (2.8) and write it as

$$\mathrm{Re}\left(\overline{\varphi} \cdot g - \frac{Q}{2\pi}\right), = 0$$

holding on $\partial\mathbb{D}$. It will be convenient to introduce the notation, for any function h defined in a neighborhood of the unit circle,

$$h^*(\zeta) = \overline{h(1/\bar{\zeta})}. \tag{2.12}$$

Thus, if h is holomorphic, so is h^*. Now, spelling out the real part of the previous equation and using that $\bar{g} = g^*$ on $\partial\mathbb{D}$ (similarly for φ) makes (2.8) appear in the form

$$\varphi^* g + \varphi g^* - \frac{Q}{\pi} = 0. \tag{2.13}$$

This holds on $\partial\mathbb{D}$, but since the left member is holomorphic in some neighbourhood of $\partial\mathbb{D}$, (2.13) actually holds identically in any such (connected) neighbourhood.

Since now both φ and g are holomorphic in a neighbourhood of $\overline{\mathbb{D}}$ and g has no zeros there it follows from (2.13) that the singularities of φ outside $\overline{\mathbb{D}}$, i.e., the singularities of φ^* in $\mathbb{D}$, can be no worse than the singularities of g^* in $\mathbb{D}$. This proves (i) and (ii).

From the above remarks we shall deduce the following theorem.

Theorem 2.2.1. *Assume $f(\zeta,0)$ is a rational function which is holomorphic and univalent in some neighbourhood of $\overline{\mathbb{D}}$ and is normalized by (2.6). Then in some time interval around $t = 0$ there exists a rational solution $f(\zeta,t)$ of (1.22). Each $f(\zeta,t)$ is analytic and univalent in a neighbourhood of $\overline{\mathbb{D}}$ and normalized by (2.6). The pole structure of $f(\zeta,t)$ is the same as that of $f(\zeta,0)$, but all poles except the one at infinity may move. Poles cannot collide or disappear, the sole exception being that the pole at infinity may disappear for one value of t.*

Remark 2.2.1. We shall see later (see Theorem 3.1.2 and Corollary 2.4.2) that, in the well-posed case $Q > 0$, the radius of analyticity $R(t)$ of $f(\zeta,t)$, i.e., the largest number R such that $f(\zeta,t)$ is holomorphic in $\mathbb{D}_R$ is a strictly increasing function of t. If the solution exists for all $0 \le t < \infty$ we shall even have that $R(t) \to \infty$ as $t \to \infty$. In addition, each individual pole moves away from the origin, see Corollary 2.4.1. Thus, the poles of $f(\zeta,t)$ will not cause any breakdown of the solution. If the solution breaks down in finite time this will be because univalence will be lost, either due to zeros of $f'(\zeta,t)$ reaching $\partial\mathbb{D}$ or of $f(\zeta,t))$ taking the same value twice on $\partial\mathbb{D}$.

This remark applies in general to strong solutions of (2.9), not only when $f(\zeta,0)$ is rational. See also Section 2.4.

Proof. In order to avoid the proof being overloaded with summation signs and indices, let us assume for simplicity that $f(\zeta,0)$ has only two poles, one finite pole and one pole at infinity:

$$f(\zeta,0) = \sum_{k=1}^{m} \frac{b_k}{(\zeta - a)^k} + \sum_{j=0}^{n} c_j \zeta^j,$$

where $b_m \neq 0$, $m, n \ge 1$. The general case is obtained by replacing a by a_l, letting m depend on l and summing over l. Then we make the 'Ansatz'

$$f(\zeta,t) = \sum_{k=1}^{m} \frac{b_k(t)}{(\zeta - a(t))^k} + \sum_{j=0}^{n} c_j(t)\zeta^j. \tag{2.14}$$

Here it is necessary to postulate $n \ge 1$, even if $c_1 = 0$, because the Hele-Shaw evolution will in any case create a pole at infinity. This gives

$$f'(\zeta,t) = -\sum_{k=1}^{m} \frac{kb_k(t)}{(\zeta - a(t))^{k+1}} + \sum_{j=1}^{n} jc_j(t)\zeta^{j-1},$$

$$\dot{f}(\zeta, t) = \sum_{k=1}^{m} \frac{\dot{b}_k(t)}{(\zeta - a(t))^k} + \sum_{k=1}^{m} \frac{k\dot{a}(t)b_k(t)}{(\zeta - a(t))^{k+1}} + \sum_{j=0}^{n} \dot{c}_j(t)\zeta^j$$

$$= \frac{m\dot{a}(t)b_m(t)}{(\zeta - a(t))^{m+1}} + \sum_{k=1}^{m} \frac{\dot{b}_k(t) + (k-1)\dot{a}(t)b_{k-1}(t)}{(\zeta - a(t))^k} + \sum_{j=0}^{n} \dot{c}_j(t)\zeta^j.$$

By the properties (i), (ii) of $\mathcal{F}$, $\mathcal{F}(f')$ will be of the form

$$\mathcal{F}(f')(\zeta, t) = \sum_{k=1}^{m+1} \frac{B_k(t)}{(\zeta - a(t))^k} + \sum_{j=1}^{n} C_j(t)\zeta^{j-1}$$

for suitable coefficients $B_k(t)$ and $C_j(t)$. It is not hard to see, for example from the formula (2.11), that these finitely many coefficients depend smoothly on the coefficients of f' (see [233] for details).

Now the Polubarinova–Galin equation in terms of present notation is

$$\dot{f}(\zeta, t) = \zeta \mathcal{F}(f')(\zeta).$$

Inserting here the above expressions for $\dot{f}$ and $\mathcal{F}(f')$, and identifying coefficients gives a system of differential equations for $a(t)$, $b_k(t)$, $c_j(t)$, and one sees immediately from the last expression for $\dot{f}$ that this system can be solved for the time derivatives $\dot{a}(t)$, $\dot{b}_k(t)$, $\dot{c}_j(t)$ as long as $b_m(t) \neq 0$. Thus, the Polubarinova–Galin equation reduces to a finite-dimensional system of ordinary differential equations of standard form, which by Picard's theorem has a unique solution, at least for a short two-sided interval around $t = 0$. This means that the 'Ansatz' (2.14) was successful so that the rational solution (2.14) survives in the same form for a little while. This proves the theorem, except for the statement about collision, which will be discussed in Section 2.4 (see Theorem 2.4.1).   □

2.3   A non-existence result for polynomial lemniscates

Given the considerably rich family of rational solutions discussed previously, it may be rather surprising that Laplacian growth is not at all compatible with a common class of domains with analytic boundaries, namely polynomial lemniscates. As a matter of fact, lemniscate domains are in a certain sense completely opposite to conformal images of the unit disk under rational univalent functions: they are pre-images of the unit disk (or, as below, the exterior disk) under non-univalent (in most cases) rational functions.

The theorem below deals with the exterior problem, governed by (1.56), (1.57), and polynomial lemniscates.

Theorem 2.3.1 ([306]). *Suppose that a smooth family of closed connected curves* $\Gamma(t)$ *moves by Laplacian growth with a source/sink at infinity, and that these curves* *are polynomial lemniscates* $\{z \in \mathbb{C} : |P(z,t)| = 1\}$, *where*

$$P(z,t) = a(t) \prod_{j=1}^{n} [z - \lambda_j(t)], \qquad (2.15)$$

with $a(t) > 0$ *and with all the* $\lambda_j(t)$ *inside* $\Gamma(t)$. *Then,* $n = 1$ *and* $\lambda_1 = \text{const}$, *i.e.,* $\Gamma(t)$ *is a family of concentric circles.*

Proof. Let $\Omega(t) = \{z \in \mathbb{C} : |P(z,t)| > 1\} \cup \{\infty\}$ be the unbounded fluid domain and let $\mathbb{D}^* = \overline{\mathbb{C}} \setminus \overline{\mathbb{D}}$ be the exterior disk. It is easy to see that the function $\varphi(z,t) = \sqrt[n]{P(z,t)}$, with the branch for the nth root chosen so that $\varphi'(\infty,t) > 0$, maps $\Omega(t)$ conformally onto $\mathbb{D}^*$ with $\varphi(\infty,t) = \infty$. Thus $P(z,t) = \varphi(z,t)^n$, and it follows that the inverse conformal map $f(\cdot,t) = \varphi^{-1}(\cdot,t) : \mathbb{D}^* \to \Omega(t)$ satisfies

$$P(f(\zeta,t),t) = \zeta^n \quad (\zeta \in \mathbb{D}), \qquad (2.16)$$

and also that f is normalized as in (1.56).

Differentiating (2.16) with respect to t and ζ gives

$$\dot{P} + P' \dot{f} = 0,$$
$$P' \cdot f' = n\zeta^{n-1}.$$

The latter equation can also be written $P' \cdot \zeta f' = n\zeta^n$, and inserting the above identities (holding in $\mathbb{D}$, up to the boundary) into the Polubarinova equation on the form (1.57), namely

$$\text{Re}\left[\dot{f} \cdot \overline{\zeta f'} \right] = -\frac{Q}{2\pi} \quad (\zeta \in \partial\mathbb{D}),$$

gives

$$\text{Re}\left(\dot{P}\overline{P} \right) = \frac{Q}{2\pi n} |P'|^2. \qquad (2.17)$$

With P interpreted as the composition $P \circ f$, as in the above computations, (2.17) holds on $\partial\mathbb{D}$, but since the independent variable ζ appears only implicitly in $f(\zeta,t)$, (2.17) can also be interpreted as an equation holding directly on $\Gamma(t)$. It can then be rewritten as

$$\frac{d}{dt} \left(|P(z,t)|^2 \right) = \frac{Q}{n\pi} |P'(z,t)|^2 \quad (z \in \Gamma(t)). \qquad (2.18)$$

Now the theorem follows from the following.

Lemma 2.3.1. *With* $P(z,t)$ *as in* (2.15), *assume that an equation of type* (2.18) *holds on* $\Gamma(t)$, *namely*

$$\frac{d}{dt} \left(|P(z,t)|^2 \right) - c(t) |P'(z,t)|^2 = 0, \qquad (2.19)$$

where the function $c(t)$ *depends on* t *only. Then,* $n = 1$, $\lambda_1 = \lambda_1(t) = \text{const}$ *and* $\Gamma(t)$ *is a family of concentric circles centered at* λ_1.

Proof of the lemma. For each fixed t the left member of (2.19) is a polynomial (in z and $\bar{z}$, or in x and y), which vanishes on the zero-locus $\Gamma(t)$ of the polynomial $|P(z,t)|^2 - 1$. Since, by our hypotheses, the latter polynomial is irreducible it follows from the Hilbert Nullstellensatz (actually from a simple argument using the Euclid algorithm) that this latter polynomial is a factor in the first polynomial. Taking degrees into account it follows that the quotient must be a constant, i.e., we have

$$\frac{d}{dt}\left(|P(z,t)|^2\right) - c(t)\,|P'(z,t)|^2 = b(t)\left(|P(z,t)|^2 - 1\right). \tag{2.20}$$

Equation (2.20) holds for all $z \in \mathbb{C}$ and for an interval of times t, and for each t both sides are real-analytic functions in z and $\bar{z}$. Hence, we can "polarize" (2.20), i.e., replace $\bar{z}$ by an independent complex variable ξ. (This is due to a simple observation: real-analytic functions of two variables are nothing else but restrictions of holomorphic functions in z, ξ-variables to the plane $\{\xi = \bar{z}\}$. Hence, if two real-analytic functions coincide on that plane, they coincide in $\mathbb{C}^2$ as well.) Denoting by $P^{\#}(z,t) = \overline{P(\bar{z},t)}$ the polynomial whose coefficients are obtained from P by complex conjugation, we have (2.20) in a "polarized" form holding for $(z,\xi) \in \mathbb{C}^2$:

$$\frac{d}{dt}\left(P(z,t)P^{\#}(\xi,t)\right) - c(t)\left(P'(z,t) \cdot \left(P^{\#}\right)'(\xi,t)\right) \tag{2.21}$$
$$= b(t)\left(P(z,t)P^{\#}(\xi,t) - 1\right).$$

Since

$$P(z,t) = a(t)\prod_{1}^{n}\left(z - \lambda_j(t)\right),$$

$$P^{\#}(\xi,t) = a(t)\prod_{1}^{n}\left(\xi - \overline{\lambda_j(t)}\right),$$

dividing by $P(z,t)P^{\#}(\xi,t)$ we obtain:

$$2\frac{\dot{a}(t)}{a(t)} - \sum_{1}^{n}\left(\frac{\dot{\lambda}_j(t)}{z - \lambda_j(t)} + \frac{\overline{\dot{\lambda}_j}(t)}{\xi - \overline{\lambda_j}(t)}\right) - c(t)\left[\sum_{1}^{n}\frac{1}{z - \lambda_j(t)}\right] \cdot \left[\sum_{1}^{n}\frac{1}{\xi - \overline{\lambda_j}(t)}\right]$$
$$= b(t)\left(1 - \frac{1}{P(z,t)P^{\#}(\xi,t)}\right). \tag{2.22}$$

Evaluating the residue at $z = \lambda_k(t)$ (for any $k = 1,\ldots,n$) in (2.22) gives, for all ξ:

$$-\dot{\lambda}_k(t) - c(t)\sum_{j=1}^{n}\frac{1}{\xi - \overline{\lambda_j}(t)} = -\frac{b(t)}{P'\left(\lambda_k(t),t\right)P^{\#}(\xi,t)}. \tag{2.23}$$

Letting $\xi \to \infty$ in (2.23) implies that $\dot{\lambda}_k(t) = 0$. In other words, the "nodes" $\lambda_k(t)$ $(k = 1, \ldots, n)$ of the lemniscates $\Gamma(t)$ do not move with time. So,

$$P(z,t) = a(t) \prod_{j=1}^{n} (z - \lambda_j) = a(t) \, Q(z). \tag{2.24}$$

Substituting (2.24) into (2.21), we obtain

$$\frac{d}{dt} \left(a(t)^2 \right) Q(z) Q^{\#}(\xi) - c(t) a(t)^2 Q' \left(Q^{\#} \right)' $$
$$= b(t) \left(a(t)^2 Q(z) Q^{\#}(\xi) - 1 \right). \tag{2.25}$$

Comparing the leading terms (i.e., the coefficients of $z^n \xi^n$) in (2.25) yields

$$\frac{d}{dt} \left(a(t)^2 \right) = b(t) \, a(t)^2. \tag{2.26}$$

Therefore,

$$c(t) \, a(t)^2 Q' \left(Q^{\#} \right)' = b(t), \tag{2.27}$$

and thus $\deg Q' = 0$, i.e., $n = \deg P = 1$. The proofs of the lemma and the theorem are now complete. □

2.4 Logarithmic solutions and dynamics of poles

We shall here give an alternative approach for proving the preservation of rationality of the conformal map under the Polubarinova–Galin equation. This will actually give more, namely it will allow for inclusion of logarithmic terms, and it also will give more direct information on how the poles move. It is based on rewriting the Polubarinova–Galin equation in terms of the derivative of the mapping function, for which we keep (see (2.7)) the notation

$$g(\zeta, t) = f'(\zeta, t).$$

Of course, f can be recaptured from g by

$$f(z,t) = \int_0^z g(\zeta,t) d\zeta.$$

Allowing f to have logarithmic terms (besides its rational terms) means exactly that g is a rational function. Or perhaps better to say, $g(\zeta)d\zeta$ is a rational differential, in other words an Abelian differential on the Riemann sphere. The terminology *Abelian domain* for the image domain $\Omega = f(\mathbb{D})$ has been used [560] for this case. The treatment below is taken from [239].

When $g = f'$ is rational we shall write it in the form

$$g(\zeta, t) = b(t) \frac{\prod_{k=1}^{m}(\zeta - \omega_k(t))}{\prod_{j=1}^{n}(\zeta - \zeta_j(t))}, \tag{2.28}$$

where $m \geq n$, $|\omega_k| > 1$, $|\zeta_j| > 1$, and repetitions are allowed among the ω_k, ζ_j to account for multiple zeros and poles. The argument of $b(t)$ is chosen so that $g(0, t) > 0$. The assumption $m \geq n$ means that $g(\zeta)d\zeta$, as a differential, has at least a double pole at infinity, which the Hele-Shaw evolution in any case will force it to have because the source/sink at the origin creates a pole of f at infinity.

Recall that the Löwner–Kufarev version (1.24), (2.33)

$$\dot{f}(\zeta, t) = \zeta f'(\zeta, t) P(\zeta, t) \quad (\zeta \in \mathbb{D}), \tag{2.29}$$

of the Polubarinova–Galin equation involves the Poisson-integral

$$P(\zeta, t) = P_g(\zeta, t) = \frac{Q}{4\pi^2 i} \int_{\partial \mathbb{D}} \frac{1}{|g(z, t)|^2} \frac{z + \zeta}{z - \zeta} \frac{dz}{z} \quad (\zeta \in \mathbb{D}). \tag{2.30}$$

In terms of P_g, the operator $\mathcal{F}$ used in Section 2.2 is $\mathcal{F}(g) = g P_g$. We recall also the notation (2.12) for holomorphic reflection.

Theorem 2.4.1. *The form* (2.28) *of g is stable in time under* (2.29), (2.30), *with the sole exception that when $m = n$ the pole of f at infinity may disappear at one moment of time (then the value of m drops below n at that particular instant). As a consequence, if the initial domain $\Omega(0)$ is Abelian ($g(\cdot, 0)$ is rational), then there exists a unique solution of* (2.29), (2.30), *within the same class ($g(\cdot, t)$ rational), for some time interval around $t = 0$.*

Proof. We shall not give the proof in full generality because it is more informative to just explain the ideas in a generic case (see [239] for a complete proof). We take

$$g(\zeta, t) = b \frac{(\zeta - \omega_1)(\zeta - \omega_2)}{(\zeta - \zeta_1)(\zeta - \zeta_2)},$$

where b, ω_1, ω_2, ζ_1, ζ_2 all depend on t. Thus $m = n = 2$, referring to the general expression (2.28). We assume that the points ω_1, ω_2, ζ_1, ζ_2 are all distinct.

We start by taking the derivative (with respect to ζ) of (2.29). This gives $\dot{g} = (\zeta g P_g)'$, and after dividing by g,

$$\frac{\partial}{\partial t} \log g(\zeta, t) = \zeta P_g(\zeta, t) \frac{\partial}{\partial \zeta} \log g(\zeta, t) + \frac{\partial}{\partial \zeta}(\zeta P_g(\zeta, t)). \tag{2.31}$$

Now everything is expressed in g, or $\log g$, and f itself has disappeared. One may notice that $\log |g|$, and even its restriction to $\partial \mathbb{D}$, contains all information of f. In fact, given $u = \log|g|$ on $\partial\mathbb{D}$ we can extend it harmonically to $\mathbb{D}$, then form its harmonic conjugate v, normalized by $v(0) = 0$, and finally define f by

$$f(\zeta) = \int_0^\zeta \exp(u(z) + \mathrm{i}\,v(z))dz.$$

The function $(\log|g|)|_{\partial\mathbb{D}}$ is a "free" real-valued function, i.e., it is subject to no constraints besides regularity (real-analyticity is what is needed to start a Hele-Shaw evolution in a full neighborhood of $t = 0$).

In the above case we have

$$\log g(\zeta, t) = \log b + \log(\zeta - \omega_1) + \log(\zeta - \omega_2) - \log(\zeta - \zeta_1) - \log(\zeta - \zeta_2),$$

$$\frac{\partial}{\partial t} \log g(\zeta, t) = \frac{\dot{b}}{b} - \frac{\dot{\omega}_1}{\zeta - \omega_1} - \frac{\dot{\omega}_2}{\zeta - \omega_2} + \frac{\dot{\zeta}_1}{\zeta - \zeta_1} + \frac{\dot{\zeta}_2}{\zeta - \zeta_2},$$

$$\frac{\partial}{\partial \zeta} \log g(\zeta, t) = \frac{1}{\zeta - \omega_1} + \frac{1}{\zeta - \omega_2} - \frac{1}{\zeta - \zeta_1} - \frac{1}{\zeta - \zeta_2}.$$

Before inserting into (2.31) we need to compute P_g. This can be done by a simple residue calculus in (2.30), using that $|g(\zeta, t)|^2 = g(\zeta, t)g^*(\zeta, t)$ ($\zeta \in \partial\mathbb{D}$) extends to a rational function in ζ. However, the calculation becomes more transparent if everything is done at an algebraic level, by which it essentially reduces to an expansion in partial fractions.

The rational function $Q/2\pi g(\zeta, t)g^*(\zeta, t)$ has poles at the zeros of g and g^*, i.e., at ω_1, ω_2, ω_1^* and ω_2^*, and it tends to a finite limit as $\zeta \to \infty$. In addition it is symmetric under the involution $h \mapsto h^*$. Expanding it in partial fractions therefore gives something of the form

$$\frac{Q}{2\pi g(\zeta, t)g^*(\zeta, t)} = C + \frac{A_1}{\zeta - \omega_1} + \frac{\overline{A}_1\zeta}{1 - \overline{\omega}_1\zeta} + \frac{A_2}{\zeta - \omega_2} + \frac{\overline{A}_2\zeta}{1 - \overline{\omega}_2\zeta}, \qquad (2.32)$$

for coefficients C, A_1, A_2, which depend on b, ω_1, ω_2, ζ_1, ζ_2. More precisely, the coefficients are easily seen to be rational functions in these parameters and their complex conjugates.

Now, $P_g(\zeta, t)$ is by definition (2.30) that holomorphic function in $\mathbb{D}$ whose real part has boundary value $Q/2\pi g(\zeta, t)g^*(\zeta, t)$ and whose imaginary part vanishes at the origin. The function (2.32) itself certainly has the right boundary behaviour on $\partial\mathbb{D}$, but it is not holomorphic in $\mathbb{D}$. On the other hand, the two types of polar parts occurring in (2.32) have the same real parts on the boundary:

$$\mathrm{Re}\,\frac{A_k}{\zeta - \omega_k} = \mathrm{Re}\,\frac{\overline{A}_k\zeta}{1 - \overline{\omega}_k\zeta} \quad \text{on } \partial\mathbb{D}, \quad k = 1, 2.$$

Therefore, without changing the real part on the boundary we can make the function (2.32) holomorphic in $\mathbb{D}$ by a simple exchange of polar parts. In addition, one can add a purely imaginary constant to account for the normalization of P at the origin. The result is that

$$P_g(\zeta, t) = A_0 + \frac{2A_1}{\zeta - \omega_1} + \frac{2A_2}{\zeta - \omega_2} \tag{2.33}$$

for a suitable constant A_0. Since $P_g(0, t)$ is real, the imaginary part of A_0 is given by

$$\text{Im}\, A_0 = \text{Im}\, \left(\frac{2A_1}{\omega_1} + \frac{2A_2}{\omega_2} \right). \tag{2.34}$$

Now we are ready to insert everything into (2.31). The result is

$$\frac{\dot{b}}{b} - \frac{\dot{\omega}_1}{\zeta - \omega_1} - \frac{\dot{\omega}_2}{\zeta - \omega_2} + \frac{\dot{\zeta}_1}{\zeta - \zeta_1} + \frac{\dot{\zeta}_2}{\zeta - \zeta_2}$$

$$= \left(A_0 \zeta + \frac{2A_1 \zeta}{\zeta - \omega_1} + \frac{2A_2 \zeta}{\zeta - \omega_2} \right) \cdot \left(\frac{1}{\zeta - \omega_1} + \frac{1}{\zeta - \omega_2} - \frac{1}{\zeta - \zeta_1} - \frac{1}{\zeta - \zeta_2} \right)$$

$$+ \frac{\partial}{\partial \zeta} \left(A_0 \zeta + \frac{2A_1 \zeta}{\zeta - \omega_1} + \frac{2A_2 \zeta}{\zeta - \omega_2} \right).$$

This is an identity between two rational functions. The crucial observation is that the right-hand side has only simple poles, despite the apparent occurrence of double poles at ω_1 and ω_2. However, a closer look shows that these cancel out.

Thus the right-hand side has exactly the same pole structure as the left-hand side, and also the same behavior at infinity. This means that at each instant, the derivatives $\dot{b}$, $\dot{\omega}_1$, $\dot{\omega}_2$, $\dot{\zeta}_1$, $\dot{\zeta}_2$ can be chosen to match the equation. The result is a system of five ordinary differential equations in which $\dot{b}$, $\dot{\omega}_1$, $\dot{\omega}_2$, $\dot{\zeta}_1$, $\dot{\zeta}_2$ are set equal to certain rational expressions in b, ω_1, ω_2, ζ_1, ζ_2 and their complex conjugates. This system of rational differential equations certainly has a unique local solution, which essentially proves the theorem. □

Remark 2.4.1. The above proof still works if $\zeta_1 = \zeta_2$, and then gives $\dot{\zeta}_1 = \dot{\zeta}_2$. On the other hand, the identifications needed for (2.31) do not work out if $\omega_1 = \omega_2$; no finite values of $\dot{\omega}_1$, $\dot{\omega}_2$ will satisfy the rational system obtained in this case. However, this does not mean that the solution g itself breaks down; the coefficients of g behave better than the zeros. In fact, zeros of g can in general collide, but immediately after they will always disintegrate again. Thus, zeros are almost always simple, while poles can have higher multiplicity, and the orders of the poles are then preserved in time. Collisions between poles do not occur.

From the final step of the proof, or actually directly from (2.31), one can read off that $\dot{\omega}_1$, $\dot{\omega}_2$, $\dot{\zeta}_1$, $\dot{\zeta}_2$ simply are the residues of the poles of the right-hand side of (2.31). Since the last term in (2.31), being a pure derivative, has no residues we get the following general formulas.

Corollary 2.4.1. *The dynamics of the zeros and poles of g are, in case these are simple, given by*

$$\dot{\omega}_k(t) = -\operatorname{Res}_{\zeta=\omega_k}\left[\zeta P_g(\zeta,t)\frac{\partial}{\partial\zeta}\log g(\zeta,t)\right],$$

$$\dot{\zeta}_j(t) = \operatorname{Res}_{\zeta=\zeta_j}\left[\zeta P_g(\zeta,t)\frac{\partial}{\partial\zeta}\log g(\zeta,t)\right].$$

The latter equation can also be written

$$\frac{d}{dt}\log\zeta_j(t) = -P_g(\zeta_j,t).$$

Remark 2.4.2. When P_g is evaluated outside the unit disk, like in the corollary, we always understand the analytic continuation of P_g from the unit disk, for example the rational function (2.33) in the case discussed in the proof.

For the reflected Poisson integral $P^*(\zeta,t) = \overline{P(1/\bar{\zeta},t)}$, which is holomorphic in the exterior unit disk, we have

$$P(\zeta,t) + P^*(\zeta,t) = \frac{Q}{\pi g(\zeta,t)g^*(\zeta,t)}, \tag{2.35}$$

since the real part of P equals $Q/2\pi|g|^2$ on $\partial\mathbb{D}$. From this it follows easily that $P^*(\zeta,t)$ is nothing else than the Poisson integral for the exterior disk $\mathbb{D}^* = \overline{\mathbb{C}}\setminus\overline{\mathbb{D}}$:

$$P^*(\zeta,t) = \frac{Q}{4\pi^2 i}\int_{\partial\mathbb{D}^*}\frac{1}{|g(z,t)|^2}\frac{z+\zeta}{z-\zeta}\frac{dz}{z} \quad (\zeta\in\mathbb{D}^*). \tag{2.36}$$

The orientation of the unit circle in (2.36) is the opposite of that in (2.30). Note that $P^* > 0$ in $\mathbb{D}^*$ with $P^*(\infty) = P(0)$, and also that the right-hand member in (2.35) vanishes at the poles ζ_j of g. Hence it follows that

$$P^*(\zeta_j,t) = -P(\zeta_j,t).$$

This shows (see also Theorem 3.1.2 below, and [249, 247, 547]) that the poles always move away from the origin. More precisely we have

Corollary 2.4.2. *For each $1 \le j \le n$ we have*

$$\frac{d}{dt}\log\zeta_j(t) = P_g^*(\zeta_j,t),$$

hence in the injection case $Q > 0$,

$$\frac{d}{dt}\log|\zeta_j(t)| = \operatorname{Re}P_g^*(\zeta_j,t) > 0$$

along with the estimates

$$\frac{|\zeta_j(t)|-1}{|\zeta_j(t)|+1} \le \frac{d\log|\zeta_j(t)|}{d\log g(0,t)} \le \frac{|\zeta_j(t)|+1}{|\zeta_j(t)|-1}. \tag{2.37}$$

Proof. It remains only to prove (2.37), and since

$$\frac{d\log|\zeta_j(t)|}{d\log g(0,t)} = \frac{d\log|\zeta_j(t)|/dt}{d\log g(0,t)/dt} = \frac{\operatorname{Re}P^*(\zeta_j,t)}{P(0,t)} = \frac{\operatorname{Re}P(\zeta_j^*,t)}{P(0,t)}$$

this follows easily from standard Harnack inequalities applied to the positive harmonic function $\operatorname{Re}P(\zeta,t)$ in $\mathbb{D}$, applied at the point $\zeta = \zeta_j^*$. $\qquad\square$

Remark 2.4.3. Note that $g(0,t) = a_1(t) > 0$, where

$$f(\zeta,t) = a_1(t)\zeta + a_2(t)\zeta^2 + \cdots.$$

By (2.29),

$$\frac{d}{dt}\log a_1(t) = \frac{\dot a_1}{a_1} = P(0,t) > 0,$$

assuming $Q > 0$, so $a_1(t)$ then is a strictly increasing function of t (see more precisely Section 4.1). One may view

$$\tau = \log g(0,t) = \log a_1(t) \tag{2.38}$$

as a new and "natural" time parameter which makes the power series of f start

$$f(\zeta) = e^\tau\zeta + \cdots,$$

as in the general theory of Löwner chains. See [446] for example.

2.5   Saffman–Taylor fingers in channel geometry

The most famous solutions to the original Hele-Shaw problem are the travelling-wave fingers of Saffman and Taylor (1958), see [488, 489]. When a low viscosity fluid (for example, water) is injected into a more viscous one, such as glycerin, an instability occurs. In fact, Hele-Shaw (1898) [265] proposed the model of the air injection into a narrow channel. An important reason for studying this problem is that it is closely related to many technologically relevant ones, such as a flow in porous media. One of the features of the channel model is that we should change the Dirichlet problem (1.16), (3.64) to a mixed boundary problem[1] for the potential function p.

Let us consider an infinite channel with parallel sides

$$\operatorname{Re}z \in (-\infty,\infty), \quad \operatorname{Im}z \in (-\pi,\pi),$$

in which an inviscid fluid is injected from the left (or the viscous fluid is extracted from the right) at a constant rate $-Q > 0$, see Figure 2.4. The function $p(z,t)$ is

---

[1]This type of boundary conditions, known also as Robin's boundary conditions, named so after the French mathematical physicist *Gustave Robin* (1855–1897) by Bergman and Schiffer [51], appeared in connection with the third type of boundary conditions (after Dirichlet's and Neumann's). Robin completed a doctoral thesis in 1886 under Emile Picard and it is most probable that this attribution does not correspond to Robin's own works (see [231]) though his name in this context is widely used nowadays.

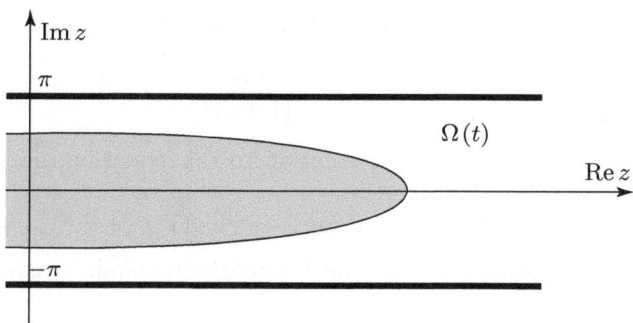

Fig. 2.4: The Saffman–Taylor finger

harmonic in the region $\Omega(t)$ occupied by the viscous fluid and vanishes on the free boundary $\Gamma(t)$. It satisfies the condition of non-penetration at the walls $\operatorname{Im} z = \pm\pi$. Therefore, we have the following mixed boundary value problem;

$$
\begin{aligned}
\Delta p &= 0 \quad \text{in } \Omega(t), \\
p|_{\Gamma(t)} &= 0, \\
\left.\frac{\partial p}{\partial n}\right|_{\operatorname{Im} z = \pm\pi} &= 0, \\
\left.\frac{\partial p}{\partial n}\right|_{\Gamma(t)} &= -V_n,
\end{aligned}
$$

with the normalization at infinity $p \sim \frac{Q}{2\pi}\operatorname{Re} z$, as $\operatorname{Re} z \to +\infty$. We choose an auxiliary parametric domain $D = \mathbb{D}\setminus(-1,0]$ and construct the conformal univalent mapping $z = f(\zeta, t)$ from D onto $\Omega(t)$ assuming that the slit along the negative axis is mapped onto the walls. For the flow outside the bubble we require $\arg f(\zeta, t) \sim -\arg\zeta$. The pressure in terms of this auxiliary variable ζ is written as just

$$
(p \circ f)(\zeta) = -\frac{Q}{2\pi}\log|\zeta|.
$$

Applying the standard technique, as was done in Section 1.4.2, we come to the Polubarinova–Galin equation for the free boundary $\Gamma(t)$:

$$
\operatorname{Re}\left[\dot{f}(\zeta, t)\overline{\zeta f'(\zeta, t)}\right] = \frac{Q}{2\pi}, \quad \zeta = e^{i\theta}, \quad \theta \in (-\pi, \pi).
$$

We are looking for travelling-wave solutions $f(\zeta, t) = At + h(\zeta)$ with $A > 0$. The slit in D is mapped onto the walls of the channel, therefore taking into account possible singularities at the points $\zeta = 0, 1$ we have $h(\zeta) \sim -\log\zeta$ as $\zeta \to 0$ and $h(\zeta) \sim \log(1 + \zeta)$ as $\zeta \to -1$. Substituting $f(\zeta, t)$ into the Polubarinova–Galin equation we have

$$
\operatorname{Re}(A\zeta h'(\zeta)) = Q/2\pi, \quad \zeta = e^{i\theta}.
$$

Differentiating with respect to θ leads to

$$
\operatorname{Im}\left(\zeta(\zeta h'(\zeta))'\right) = 0.
$$

The singularities of h suggest the form of the function

$$\zeta(\zeta h'(\zeta))' = \frac{c_1\zeta}{(1+\zeta)^2},$$

where c_1 is some real constant. The solution to this equation, neglecting a horizontal shift, is

$$h(\zeta) = (c_2 - c_1)\log\zeta + c_1\log(1+\zeta).$$

To determine the constants we use the Polubarinova–Galin equation again and obtain

$$A\operatorname{Re}\left(c_2 - \frac{c_1}{2}\right) = \frac{Q}{2\pi}.$$

One can choose the ratio $\lambda \in (0,1)$ of the width of the finger at $\operatorname{Re}\zeta \to -\infty$ as a parameter and derive $c_1 = 2(1 - \lambda)$, $c_2 = 1 - 2\lambda$, $A = -\frac{Q}{2\pi\lambda}$. Finally,

$$f(\zeta, t) = \frac{Q}{2\pi\lambda}t - \log\zeta + 2(1 - \lambda)\log(1+\zeta).$$

This function gives a travelling-wave solution, moving with the speed $\frac{-Q}{2\pi\lambda}$, for any value $\lambda \in (0, 1)$. The way in which Saffman and Taylor derived the solution shows that it is the only possible form for a steady, travelling wave. In terms of analytic functions this corresponds to the uniqueness of the solution to the mixed boundary value problem with given singularities.

Curiously enough, Saffman–Taylor's work seems to have been underestimated when it appeared. For example, a MR review says that "... the authors' analysis does not seem to be completely rigorous, mathematically. Many details are lacking. Besides, the authors do not seem to be aware of the fact that there exists a vast amount of literature concerning viscous fluid flow into porous (homogeneous and non-homogeneous) media in Russian and Romanian". Nowadays, the Saffman–Taylor fingers are widely known in many fields of mechanics, chemistry and industry.

Saffman and Taylor found experimentally that an unstable planar interface evolves through finger competition to a steady translating finger with $\lambda = 1/2$. Recently, Tanveer and Xie [548, 549] proved that even a small surface tension effect implies non-existence of a strong solution when the relative finger width λ is smaller than $1/2$. They also solved [548] the selection problem for $\lambda > 1/2$.

2.6 Corner flows

Having handled these first steps many authors have been constructing non-trivial solutions. We should say that mostly these explicit solutions are either polynomials and rational functions, or else, logarithmic solutions linked to Saffman–Taylor fingers. Another type of explicit solutions was proposed by, e.g., Howison, King [282], Cummings [118], who reduced the problem to solving the Poisson equation eliminating time by applying the Baiocchi transformation. The solutions were given making use of the Riemann P-function and hypergeometric functions.

Corner flows of an inviscid incompressible fluid were studied intensively, e.g., in [300, 301, 384, 422, 559] (see also the references therein). In particular, we mention here papers [25, 43, 44, 45, 292, 473, 474, 552]. A solution constructed by Kadanoff [292] is directly linked with ours.

In this section we shall construct explicit solutions in an infinite corner of arbitrary angle such that the viscous fluid is glued to one of the walls, the interface extends to infinity along it and has fluid-wall angle $\pi/2$ at a moving contact point at the other wall. These solutions will present a logarithmic deformation of the trivial (circular) solution. In the case of a right angle we get Kadanoff's solution [292]. Then we present an analogous solution in the corner with a source at its vertex. Finally, we construct self-similar solutions in a wedge analogous to Saffman–Taylor fingers.

2.6.1 Mathematical model

In this subsection we deal with a general case of corner flows. We suppose that the viscous fluid occupies a simply connected domain $\Omega(t)$ in the phase z-plane. The boundary $\Gamma(t)$ consists of two walls $\Gamma_1(t)$ and $\Gamma_2(t)$ of the corner and a free interface $\Gamma_3(t)$ between them at a moment t. The inviscid fluid (or air) fills the complement to $\Omega(t)$. The simplifying assumption of constant pressure at the interface between the fluids means that we omit the effect of surface tension. The velocity must be bounded close to the contact wall-fluid point that yields the contact angle between the walls of the corner and the moving interface to be $\pi/2$ (see Figure 2.5). A limiting case corresponds to one finite contact point and the other tending to infinity. By a shift we can place the point of the intersection of the wall extensions at the origin. To simplify matters, we set the corner of angle

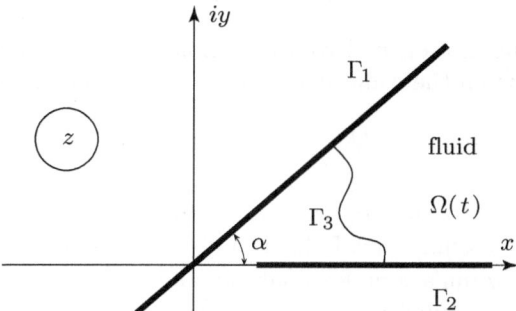

Fig. 2.5: $\Omega(t)$ is the phase domain within an infinite corner and the homogeneous sink/source at ∞

α between the walls so that the positive real axis x contains one of the walls and fix this angle as $\alpha \in (0, \pi]$.

Let us mention here that the model can be studied in the presence of surface tension and the macroscopic contact angle between the walls and the free interface can then be different from $\pi/2$. Let us denote it by β. The contact angle β at a moving contact line obeys interesting properties that was studied by Ablett [5] (see also [107], [163]) in a particular case of water in contact with a paraffin surface. It turns out that the steady angle β depends on the velocity of the contact line. The angle β increases with the velocity increased for the liquid advancing over the surface up to a certain value β_0 and, then, remains the same for a greater velocity. Reciprocally, β decreases with the velocity increased for the liquid receding over the surface up to a certain value β_1 (different from β_0) and, then, remains the same for a greater velocity.

In our zero surface tension model we have Robin's boundary value problem for fluid pressure $p(z, t) \equiv p(x, y, t)$

$$\Delta p = 0 \quad \text{in the flow region } \Omega(t), \tag{2.39}$$

and the fluid velocity $\mathbf{V}$ averaged across the gap is $\mathbf{V} = -\nabla p$. The free boundary conditions

$$p\big|_{\Gamma_3} = 0, \quad \frac{\partial p}{\partial t}\Big|_{\Gamma_3} = |\nabla p|^2 \tag{2.40}$$

are imposed on the free boundary $\Gamma_3 \equiv \Gamma_3(t)$. This implies that the normal velocity v_n of the free boundary Γ_3 outwards from $\Omega(t)$ is given by

$$\frac{\partial p}{\partial n}\Big|_{\Gamma_3} = -V_n. \tag{2.41}$$

On the walls $\Gamma_1 \equiv \Gamma_1(t)$ and $\Gamma_2 \equiv \Gamma_2(t)$ the boundary conditions are given as

$$\frac{\partial p}{\partial n}\Big|_{\Gamma_1 \cup \Gamma_2} = 0. \tag{2.42}$$

We suppose that the motion is driven by a homogeneous source/sink at infinity. Since the angle between the walls at infinity is also α, the pressure behaves about infinity as

$$p \sim \frac{Q}{\alpha} \log |z|, \quad \text{as } |z| \to \infty,$$

where Q corresponds to the constant strength of the source $(Q > 0)$ or sink $(Q < 0)$. Finally, we assume that $\Gamma_3(0)$ is a given analytic curve.

We introduce Robin's complex analytic potential $W(z, t) = p(z, t) + i\psi(z, t)$, where $-\psi$ is the stream function. Let us consider an auxiliary parametric complex ζ-plane, $\zeta = \xi + i\eta$. We set $D = \{\zeta : |\zeta| > 1, 0 < \arg \zeta < \alpha\}$, $D_3 = \{z : z = e^{i\theta}, \theta \in (0, \alpha)\}$, $D_1 = \{z : z = re^{i\alpha}, r > 1\}$, $D_2 = \{z : z = r, r > 1\}$, $\partial D = D_1 \cup D_2 \cup D_3$, and construct a conformal univalent time-dependent map

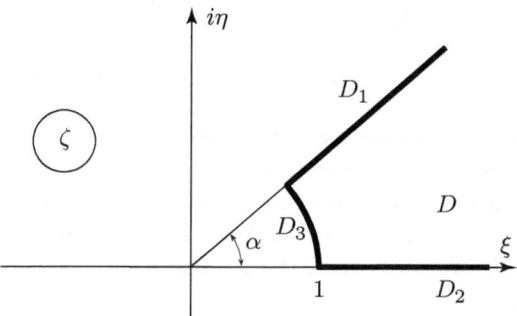

Fig. 2.6: The parametric domain D

$z = f(\zeta, t)$, $f : D \to \Omega(t)$, so that being continued onto ∂D, $f(\infty, t) \equiv \infty$, and the circular arc D_3 of ∂D is mapped onto Γ_3 (see Figure 2.6). This map has an expansion

$$f(\zeta, t) = \zeta \sum_{n=0}^{\infty} a_n(t) \zeta^{-\frac{\pi n}{\alpha}}$$

near infinity and $a_0(t) > 0$. The function f parameterizes the boundary of the domain $\Omega(t)$ by $\Gamma_j = \{z : z = f(\zeta, t), \zeta \in D_j\}$, $j = 1, 2, 3$.

The normal unit vector in the outward direction is given by

$$\mathbf{n} = -\zeta \frac{f'}{|f'|} \text{ on } \Gamma_3, \quad \mathbf{n} = -i \text{ on } \Gamma_2, \text{ and } \mathbf{n} = ie^{i\alpha} \text{ on } \Gamma_1.$$

Therefore, the normal velocity is obtained as

$$V_n = \mathbf{V} \cdot \mathbf{n} = -\frac{\partial p}{\partial n} = \begin{cases} -\mathrm{Re}\left(\dfrac{\partial W}{\partial z} \dfrac{\zeta f'}{|f'|}\right) & \text{for } \zeta \in D_3, \\ 0 & \text{for } \zeta \in D_1, \\ 0 & \text{for } \zeta \in D_2. \end{cases} \tag{2.43}$$

The superposition $W \circ f$ is the solution to the mixed boundary problem (2.39), (2.40), (2.42) in D, therefore, it is Robin's function given by $(W \circ f)(\zeta) = \frac{Q}{\alpha} \log \zeta$. On the other hand,

$$V_n = \begin{cases} \mathrm{Re}\left(\dot{f}\overline{\zeta f'}/|f'|\right) & \text{for } \zeta \in D_3, \\ \mathrm{Im}\left(\dot{f} e^{-i\alpha}\right) & \text{for } \zeta \in D_1, \\ -\mathrm{Im}\left(\dot{f}\right) & \text{for } \zeta \in D_2. \end{cases} \tag{2.44}$$

The first lines of (2.43), (2.44) give that

$$\mathrm{Re}\left(\dot{f}\,\overline{\zeta f'}\right) = -\frac{Q}{\alpha}, \quad \text{for } \zeta \in D_3. \tag{2.45}$$

The remaining lines of (2.43), (2.44) imply

$$\text{Im}\,(\dot{f}e^{-i\alpha}) = 0 \quad \text{for } \zeta \in D_1, \quad \text{Im}\,(\dot{f}) = 0 \quad \text{for } \zeta \in D_2. \tag{2.46}$$

One of the typical properties of the problem (2.39)–(2.42) is that starting with an analytic boundary component $\Gamma_3(0)$, the one-parameter evolutionary chain of solutions develops possible cusps at a finite blow-up time t_0. Another typical scenario is fingering. The strong solution exists locally in time in the case of an analytic boundary Γ_3. We only refer the reader to some relevant works [179, 277, 280, 452, 470, 488].

2.6.2   Logarithmic perturbations of the trivial solution

We consider the case of a sink at infinity ($Q < 0$). The simplest explicit solution in this case is

$$f(\zeta, t) = \sqrt{-\frac{2Qt}{\alpha}}\,\zeta,$$

that produces a circular dynamics of the free boundary. Our aim is to perturb this trivial solution by a function independent of t, say we are looking for a solution in the form

$$f(\zeta, t) = \sqrt{-\frac{2Qt}{\alpha}}\,\zeta + \zeta g(\zeta),$$

where $g(\zeta)$ is regular in D with the expansion

$$g(\zeta) = \sum_{n=0}^{\infty} \frac{a_n}{\zeta^{\frac{\pi n}{\alpha}}}$$

near infinity. The branch is chosen so that g, being continued symmetrically into the reflection of D, is real at real points. Equation (2.45) implies that the function g satisfies

$$\text{Re}\,(g(\zeta) + \zeta g'(\zeta)) = 0, \quad \zeta \in D_3.$$

Taking into account the expansion of g we are looking for a solution satisfying the equation

$$g(\zeta) + \zeta g'(\zeta) = \frac{\zeta^{\frac{\pi}{\alpha}} - 1}{\zeta^{\frac{\pi}{\alpha}} + 1}, \quad \zeta \in D, \tag{2.47}$$

although other forms may be possible. The general solution to (2.47) can be given in terms of the Gauss hypergeometric function $\mathbf{F} \equiv {}_2\mathbf{F}_1$ as

$$\zeta g(\zeta) = \zeta - 2\zeta \mathbf{F}\left(\frac{\alpha}{\pi}, 1, 1 + \frac{\alpha}{\pi}; -\zeta^{\frac{\pi}{\alpha}}\right) + C.$$

We note that f' vanishes only when $\zeta^{\frac{\pi}{\alpha}} = (2/(1 + \sqrt{-2Qt/\alpha})) - 1$, therefore, the function f is locally univalent, the cusp problem appears only at the initial time $t = 0$ and the solution exists during infinite time. The resulting function is homeomorphic on the boundary ∂D, hence it is univalent in D. This presents an example (apart from the trivial one) of the long-time existence of the solution in the problem with suction (ill-posed problem). To complete our solution we need to determine the constant C. First of all we choose the branch of the function $_2\mathbf{F}_1$, so that the points of the ray $\zeta > 1$ have real images. This implies that $\operatorname{Im} C = 0$.

We continue verifying the asymptotic properties of the function $f(e^{i\theta}, t)$ as $\theta \to \alpha - 0$. The slope is

$$\lim_{\theta \to \alpha - 0} \arg[ie^{i\theta} f'(e^{i\theta}, t)] = \alpha + \frac{\pi}{2} + \lim_{\theta \to \alpha - 0} \arg\left(\sqrt{-\frac{2Qt}{\alpha} + \frac{e^{i\frac{\pi\theta}{\alpha}} - 1}{e^{i\frac{\pi\theta}{\alpha}} + 1}}\right) = \alpha + \pi.$$

To calculate the shift we choose C such that

$$\lim_{\theta \to \alpha - 0} \operatorname{Im}\left[e^{-i\alpha} f'(e^{i\theta}, t)\right] = 0.$$

Using the properties of hypergeometric functions we have

$$\lim_{\gamma \to 0+0} \operatorname{Im} \mathbf{F}\left(\frac{\alpha}{\pi}, 1, 1 + \frac{\alpha}{\pi}; e^{i\gamma}\right) = \frac{\alpha}{2}.$$

Therefore, $C = \alpha$. We present numerical simulation in Figure 2.7.

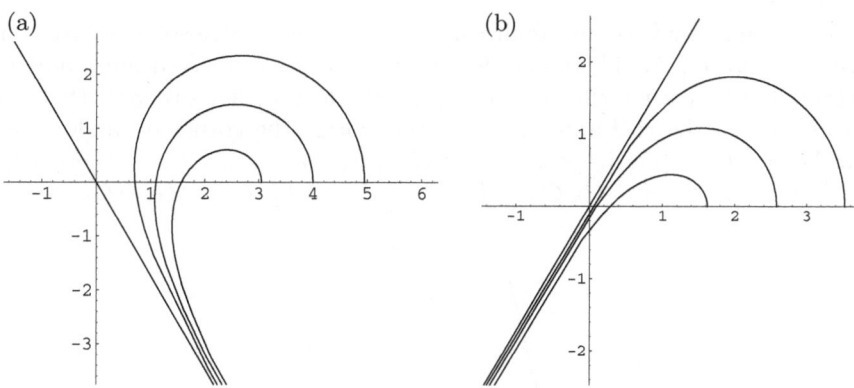

Fig. 2.7: Evolution in the corner of angle: (a) $\alpha = 2\pi/3$; (b) $\alpha = \pi/3$

The special case with angle $\alpha = \pi/2$ was considered by Kadanoff [292]. The hypergeometric function reduces to an arctangent and we obtain

$$f(\zeta, t) = (\sqrt{-4Qt/\pi} + 1)\zeta + i \log \frac{1 + i\zeta}{1 - i\zeta} + \frac{\pi}{2}, \quad Q < 0.$$

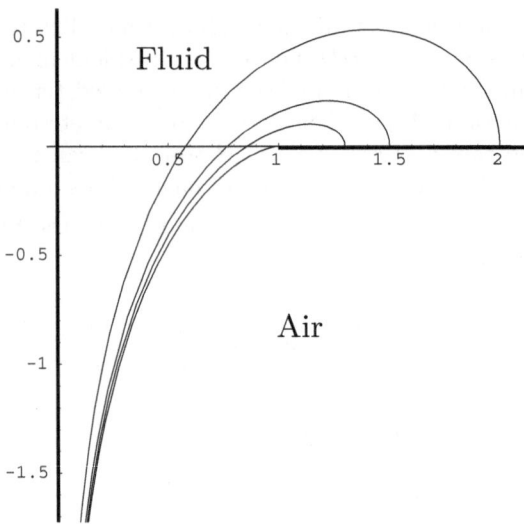

Fig. 2.8: Kadanoff's solution

This function maps the domain $\{|\zeta| > 1,\ 0 < \arg \zeta < \pi/2\}$ onto the infinite domain bounded by the imaginary axis (Γ_1), the ray $\Gamma_2 = \{r : r \geq \sqrt{-4Qt/\pi+1}\}$ of the real axis and an analytic curve Γ_3 which is the image of the circular arc, see Figure 2.8.

By the analogy with an infinite sink we are able to give solutions for a finite source (see Figure 2.9). The phase domain is a simply connected finite domain at the vertex of the corner which is a source. We locate the corner so that one of the walls lies on the real axis and the other forms the corner of angle α at the origin. We set $G = \{\zeta : |\zeta| < 1,\ 0 < \arg \zeta < \alpha\}$, $G_3 = \{z : z = e^{i\theta},\ \theta \in (0, \alpha)\}$, $G_1 = \{z : z = re^{i\alpha},\ r < 1\}$, $G_2 = \{z : z = r,\ r < 1\}$, $\partial G = G_1 \cup G_2 \cup G_3$, and

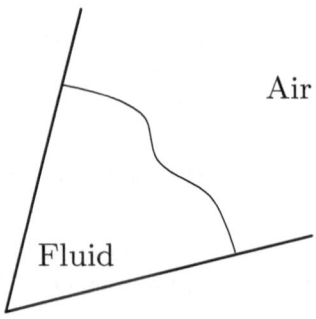

Fig. 2.9: Finite source

construct a conformal univalent time-dependent map $z = f(\zeta, t)$, $f : G \to \Omega(t)$. This map has an expansion

$$f(\zeta, t) = \zeta \sum_{n=0}^{\infty} a_n(t)\zeta^{\frac{\pi n}{\alpha}}$$

near the origin and $a_0(t) > 0$. The equations for this function at the boundary of G are

$$\mathrm{Re}\,(\dot f\, \overline{\zeta f'}) = \frac{Q}{\alpha}, \quad \text{for } \zeta \in G_3,$$

where $Q > 0$, and

$$\mathrm{Im}\,(\dot f e^{-i\alpha}) = 0 \quad \text{for } \zeta \in G_1, \quad \mathrm{Im}\,(\dot f) = 0 \quad \text{for } \zeta \in G_2.$$

We give a solution analogous to the infinite case by

$$f(\zeta, t) = \sqrt{\frac{2Qt}{\alpha}}\zeta - \zeta + 2\zeta \mathbf{F}\left(\frac{\alpha}{\pi}, 1, 1 + \frac{\alpha}{\pi}; -\zeta^{\frac{\pi}{\alpha}}\right).$$

The numerical simulation is presented in Figure 2.10

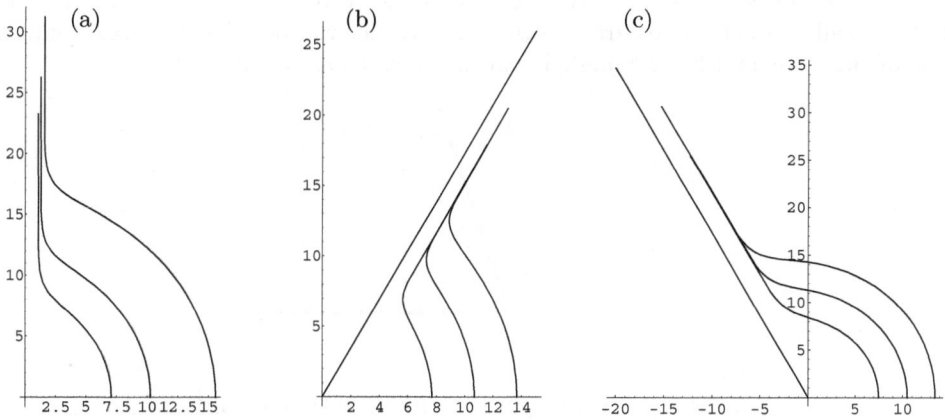

Fig. 2.10: Long-pin dynamics of the advancing fluid in the corner of angle:
(a) $\alpha = \pi/2$; (b) $\alpha = \pi/3$; (c) $\alpha = 2\pi/3$

Remark 2.6.1. By the proposed method we can perturb several known self-similar solutions even for more general flows. The idea is as follows. Let $f_0(\zeta, t) = H(t)F(\zeta)$ be a known solution to the problem, the basic equation of which is the Polubarinova–Galin equation $\mathrm{Re}\,(\dot f\, \overline{\zeta f'}) = \mathrm{const}$ (positive or negative) and where ζ belongs to a circular component of the parametric domain. We are looking for a new solution of the form $f(\zeta, t) = f_0(\zeta, t) + g(\zeta)$, where $g(\zeta)$ is an analytic function with an appropriate expansion. Then, on the circular component this function satisfies the equation

$$\mathrm{Re}\, \frac{\zeta g'(\zeta)}{F(\zeta)} = 0.$$

So one must solve the equation $\zeta g'(\zeta) = F(\zeta)P(\zeta)$, where $P(\zeta)$ is a function with vanishing real part at the points of the circular component.

2.6.3   Self-similar bubbles

In this subsection we discuss deformation of two-dimensional bubbles in a corner flow in which there is a replacement of two immiscible fluids one of which is viscous and the other is effectively inviscid. We shall give self-similar (homothetic) drop-shaped solutions in a corner that include Ben Amar's solution [43] as well as those constructed in [25, 552] as particular cases. Y. Tu [555] also analysed viscous fingering in corners applying the hodograph method for the complex velocity potential. In the symmetric case this leads to Ben Amar's solution [43] given in terms of hypergeometric functions, whereas in the non-symmetric case no explicit solution was given.

The bubbles are assumed to originate at the vertex as in Figure 2.11 and the bubble-wall contact angles are $\beta \in (0, \alpha/2)$. We let the positive real axis contain one of the walls and fix the angle between the walls as $\alpha \in (0, 2\pi)$.

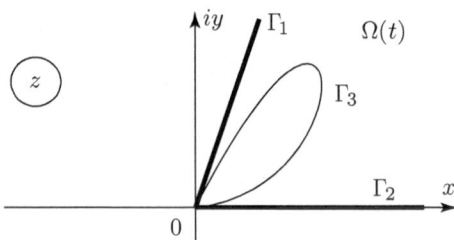

Fig. 2.11: $\Omega(t)$ is the phase domain within an infinite corner and the homogeneous sink/source at ∞

Mathematically, this model is described by Robin's boundary value problem (2.39)–(2.42), where the potential function $p(z, t)$ behaves near infinity as

$$p \sim \frac{Q}{\alpha} \log |z|, \quad \text{as } |z| \to \infty,$$

and where Q is the constant strength of the source $(Q > 0)$ or sink $(Q < 0)$.

Let us consider an auxiliary parametric complex ζ-plane, $\zeta = \xi + i\eta$. We set $D = \{\zeta : |\zeta| > 1, 0 < \arg \zeta < \pi\}$, $D_3 = \{z : z = e^{i\theta}, \theta \in (0, \pi)\}$, $D_1 = \{z :$

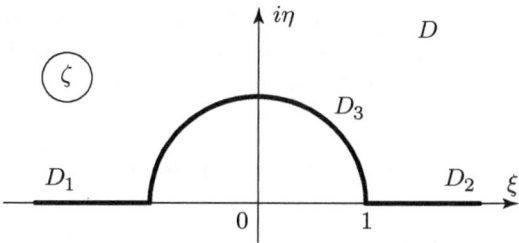

Fig. 2.12: The parametric domain D

$z = -r, r > 1$}, $D_2 = \{z : z = r, r > 1\}$, $\partial D = D_1 \cup D_2 \cup D_3$. Construct a conformal univalent time-dependent map $z = f(\zeta, t)$, $f : D \rightarrow \Omega(t)$, such that being continued onto ∂D, $f(\infty, t) \equiv \infty$, and the circular arc D_3 of ∂D is mapped onto Γ_3 (see Figure 2.12). This map has an expansion $f(\zeta, t) = \zeta^{\alpha/\pi} \sum_{k=0}^{\infty} a_k(t) \zeta^{-k}$ near infinity, and $a_0(t) > 0$. The function f parameterizes the boundary of the domain $\Omega(t)$ by $\Gamma_j = \{z : z = f(\zeta, t), \zeta \in D_j\}$, $j = 1, 2, 3$.

The free boundary condition is expressed in terms of the function f as in the preceding subsection by

$$\mathrm{Re}\,(\dot{f}\,\overline{\zeta f'}) = -\frac{Q}{\pi}, \quad \text{for } \zeta \in D_3, \tag{2.48}$$

and the wall conditions imply that

$$\begin{aligned} \mathrm{Im}\,(\dot{f}e^{i\alpha}) &= 0 \quad \text{for } \zeta \in D_1; \\ \mathrm{Im}\,(\dot{f}) &= 0 \quad \text{for } \zeta \in D_2. \end{aligned} \tag{2.49}$$

We are going to construct an analogue of the Saffman–Taylor fingers for the corner flows (*self-dilating drops* whose interface contains the vertex). Analytic solutions were discovered first in the case $\alpha = \pi/2$ in [552] and then for general values of angles in [43, 44, 45]. We give a generalization that, in fact, presents possible self-similar solutions, and in particular, we obtain exact solutions for non-symmetric drops.

To simplify matters we scale the angles α, β by $\alpha \rightarrow \alpha\pi$, $\beta \rightarrow \beta\pi/2$. Let us analyse the auxiliary mapping $f(\zeta, t)$. In the case of self-dilating solutions the phase domain $\Omega(t)$ is a dilation of an initial domain $\Omega(0)$. Then the solution $f(\zeta, t)$ to the equations (2.48)–(2.49) is represented as $f(\zeta, t) = G(t)F(\zeta)$. Since Q does not depend on t, the equation (2.48) implies that $G(t) = C\sqrt{t}$, where C is a constant. Reducing the mapping f to a regular function we represent it as

$$f(\zeta, t) = \sqrt{t}\zeta^{\alpha}g(\zeta),$$

where $g(\zeta)$ is an analytic function which is regular at infinity.

The boundary Γ_3 starts and ends at the origin under the same bubble-wall contact angles $\beta \in (0, \alpha)$, and forms a self-similar drop-shaped bubble. Therefore, the function $g(\zeta)$ can be represented as

$$g(\zeta) = \left(1 - \frac{1}{\zeta^2}\right)^\beta h(\zeta),$$

where $h(\zeta)$ is a regular function in the closure of D. We differentiate equation (2.48) with respect to θ, taking into account $\zeta = e^{i\theta}$, $\theta \in (0, \pi)$ and obtain

$$\mathrm{Im}\left[(2\alpha + 1)\frac{\zeta g'(\zeta)}{g(\zeta)} + \frac{\zeta^2 g''(\zeta)}{g(\zeta)}\right] = 0, \quad \zeta = e^{i\theta}.$$

In terms of the function h we have $\mathrm{Im}\, G(\zeta) = 0$, where

$$G(\zeta) \equiv \frac{2\beta(2\alpha + 1)}{\zeta^2 - 1} + \frac{4\beta(\beta - 1)}{(\zeta^2 - 1)^2} - \frac{6\beta}{\zeta^2 - 1} + \left((2\alpha + 1) + \frac{4\beta}{\zeta^2 - 1}\right)\frac{\zeta h'(\zeta)}{h(\zeta)} + \frac{\zeta^2 h''(\zeta)}{h(\zeta)}.$$

Equations (2.49) imply that the equation $\mathrm{Im}\, G(\zeta) = 0$ is satisfied on the whole boundary $D_1 \cup D_2 \cup D_3$. The function $h(\zeta)$ is regular at ± 1, therefore

$$G(\zeta) \sim \frac{1}{(\zeta^2 - 1)^2} \quad \text{as } \zeta \to \pm 1.$$

Taking into account the regularity of $h(\zeta)$ near infinity we propose that the function G has the form

$$G(\zeta) = \frac{4\beta(\beta - 1)\zeta^2}{(\zeta^2 - 1)^2},$$

although other forms may be possible. Our intention is to obtain a complex differential equation for which we can construct explicit solutions. So we have it in the form

$$\frac{4\beta(\alpha - \beta)}{\zeta^2 - 1} + \left((2\alpha + 1) + \frac{4\beta}{\zeta^2 - 1}\right)\frac{\zeta h'(\zeta)}{h(\zeta)} + \frac{\zeta^2 h''(\zeta)}{h(\zeta)} = 0.$$

Changing variables $w = 1/\zeta^2$, $Y(w) \equiv h(1/\sqrt{w})$ we come to the hypergeometric equation

$$(1 - w)wY'' + (1 - \alpha - (1 + 2\beta - \alpha)w)\, Y' - \beta(\beta - \alpha)Y = 0. \qquad (2.50)$$

The general solution of (2.50) can be given in terms of the Gauss hypergeometric function $\mathbf{F} = {}_2\mathbf{F}_1$. We thus have two linearly independent solutions

$$h_1(\zeta) = \mathbf{F}\left(\beta - \alpha, \beta, 1 - \alpha; \frac{1}{\zeta^2}\right), \quad h_2(\zeta) = \frac{1}{\zeta^{2\alpha}}\mathbf{F}\left(\beta, \beta + \alpha, 1 + \alpha; \frac{1}{\zeta^2}\right).$$

Finally, we find $f(\zeta, t)$ in the form

$$f(\zeta, t) = \sqrt{t}\zeta^\alpha \left(1 - \frac{1}{\zeta^2}\right)^\beta (C_1 h_1(\zeta) + C_2 h_2(\zeta)), \qquad (2.51)$$

for real constants C_1, C_2 and we choose the branch so that $f(r) > 0$ and $h(r) > 0$ for $r > 1$. Since the primitive

$$\int \mathrm{Im}\left(\left.|f|^2 G(e^{i\theta})\right|_{h=C_1 h_1 + C_2 h_2}\right) d\theta = \mathrm{Re}\, \dot{f}(e^{i\theta}, t)\overline{e^{i\theta} f'(e^{i\theta}, t)}$$

is constant, we can choose C_1, C_2 such that it is exactly $-Q/\pi > 0$ and $f(\zeta, t)$ satisfies the equation (2.48) in the arc $\{e^{i\theta}, \theta \in (0, \pi)\}$. By construction we have that the function f maps the rays $(-\infty, -1]$ and $[1, \infty)$ onto the walls Γ_1 and Γ_2 respectively. In order to check the univalence of f we note that given $Q < 0$ and f of the form (2.51), we choose the constants C_1, C_2 as mentioned above. The function f is starlike with respect to the origin because $Q < 0$ and, hence, univalent. If the constant C_2 vanishes, then the equality $f(-\bar{\zeta}, t) = e^{i\alpha\pi}\overline{f(\zeta, t)}$ is easily verified. This means that the solution is symmetric with respect to the bisectrix of the phase angle, namely the ray $z = re^{i\alpha/2}$, $r > 0$.

In Figures 2.13, 2.14 we present asymmetric drops in angles $\pi/3$ and $2\pi/3$ (a, c), as well as the symmetric case (b).

In the case $\alpha = 1/2$ the hypergeometric functions reduce to a simpler form:

$$h_1(\zeta) = \frac{1}{2}\left(\left(1 + \frac{1}{\zeta}\right)^{1-2\beta} + \left(1 - \frac{1}{\zeta}\right)^{1-2\beta}\right),$$

$$h_2(\zeta) = \frac{1}{2(1 - 2\beta)}\left(\left(1 + \frac{1}{\zeta}\right)^{1-2\beta} - \left(1 - \frac{1}{\zeta}\right)^{1-2\beta}\right),$$

and we have

$$f(\zeta, t) = \sqrt{(t/\zeta)}\left(A(\zeta + 1)^{1-\beta}(\zeta - 1)^\beta + B(\zeta - 1)^{1-\beta}(\zeta + 1)^\beta\right), \qquad (2.52)$$

where $\beta \in (0, 1/2)$, $Q = -4AB(1 - 2\beta)\sin(\frac{\pi}{2}(1 - 2\beta))$, $A, B > 0$. We remark here that the map $f(\zeta, t)$ becomes non-univalent for other choices of A, B, β.

The function $f(\zeta, t)$ obviously satisfies the equations (2.48), (2.49). It maps D onto $\Omega(t)$ that is a complement of a bubble for any time t. The boundary Γ_3 starts and ends at the origin under the same bubble-wall angle $\pi\beta/2$, and forms a self-similar drop-shaped bubble. If $A = B$, then the bubble is symmetric with respect to the bisectrix of the corner (Figures 2.13(b), 2.14(b) and 2.15) and the solution is known [25], [552]. If $A \neq B$, then we have non-symmetric dynamics (Figures 2.13(a, c), 2.14(a, c), 2.16, 2.17). It is interesting that although the bubble-wall angles are the same, we have a two-parameter $(A/B, \beta)$ continuum of possible developments of fingers.

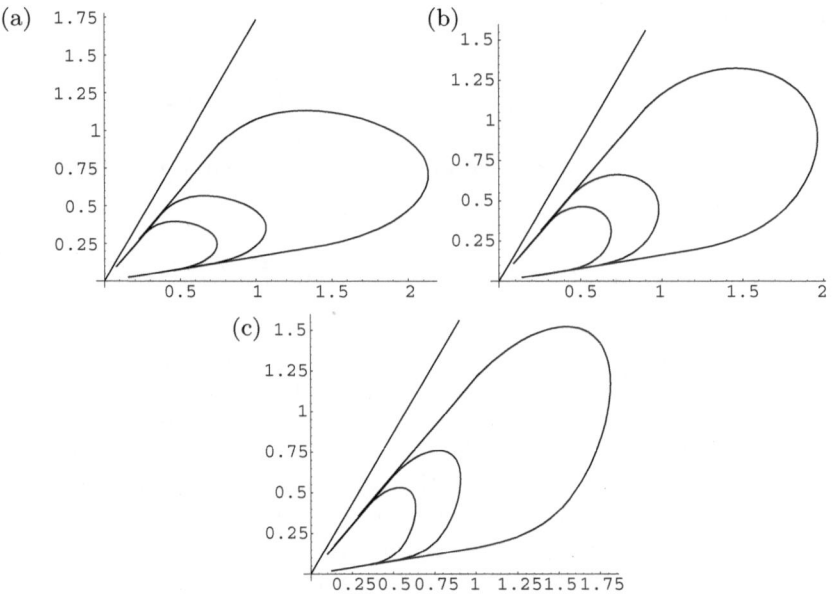

Fig. 2.13: Finger dynamics in the wedge angle $\pi/3$ and the bubble-wall
angles $\pi/20$: (a) $C_1 = 1$, $C_2 = 0.9$; (b) $C_1 = 1$, $C_2 = 0$; (c)
$C_1 = 1$, $C_2 = -1$

For angles greater than π the procedure is the same. A corner of angle π
implies other linearly independent solutions of the equation (2.50):

$$h_1(\zeta) = \frac{1}{\zeta^2}\mathbf{F}\left(\beta, \beta+1, 2; \frac{1}{\zeta^2}\right),$$

$$h_2(\zeta) = \frac{-2\log\zeta}{\zeta^2}\mathbf{F}\left(\beta, \beta+1, 2; \frac{1}{\zeta^2}\right)$$

$$+ \sum_{k=1}^{\infty} \frac{\prod\limits_{j=0}^{k-2}(\beta+j)^2(\beta-1)(\beta+k-1)}{\zeta^{2k+2}(k!)^2(k+1)}\left(2\left(\sum_{j=1}^{k-1}\frac{1}{\beta+j} - \sum_{j=2}^{k}\frac{1}{j}\right)\right.$$

$$\left.+ \frac{1}{\beta} + \frac{1}{\beta+k} - 1 - \frac{1}{k+1}\right) - \frac{1}{\beta(\beta+1)},$$

that can be treated similarly.

Most of the results presented in this section are found in [368], [369].

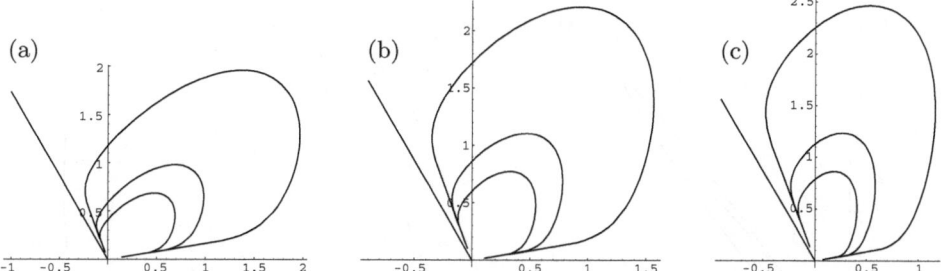

Fig. 2.14: Finger dynamics in the wedge angle $2\pi/3$ and the bubble-wall angle $\pi/20$: (a) $C_1 = 1$, $C_2 = 0.9$; (b) $C_1 = 1$, $C_2 = 0$; (c) $C_1 = 1$, $C_2 = -1$

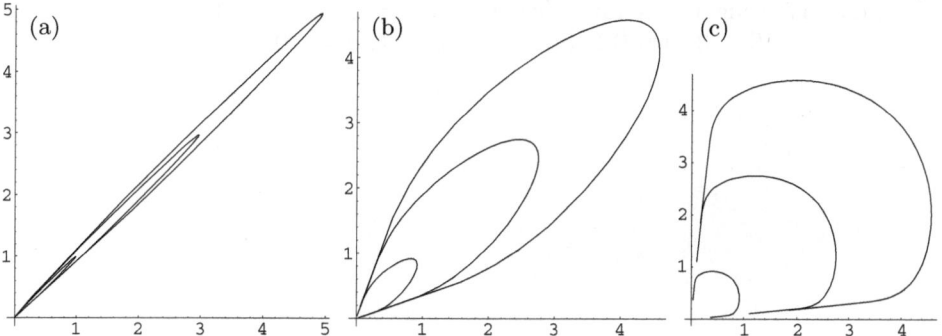

Fig. 2.15: Finger dynamics: (a) $A = 1$, $B = 1$, $\beta = 0.16$; (b) $A = 1$, $B = 1$, $\beta = 0.1$; (c) $A = 1$, $B = 1$, $\beta = 0.05$

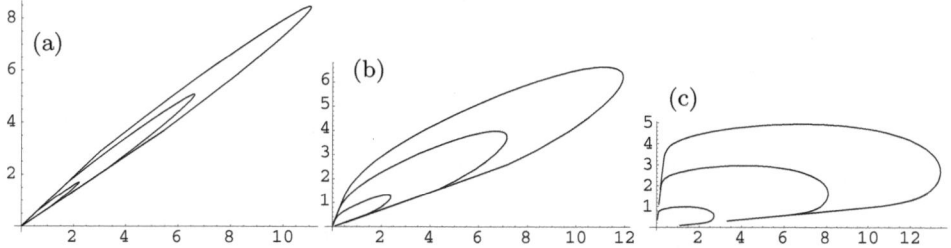

Fig. 2.16: Finger dynamics: (a) $A = 1$, $B = 3$, $\beta = 0.16$; (b) $A = 1$, $B = 3$, $\beta = 0.1$; (c) $A = 1$, $B = 3$, $\beta = 0.05$

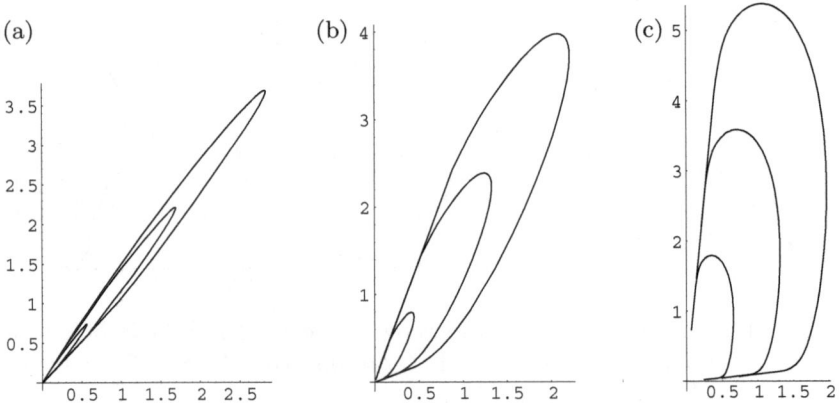

Fig. 2.17: Finger dynamics: (a) $A = 1$, $B = 1/3$, $\beta = 0.16$; (b) $A = 1$, $B = 3$, $\beta = 0.1$; (c) $A = 1$, $B = 1$, $\beta = 0.05$

Chapter 3

Weak Solutions and Related Topics

3.1 Variational inequality weak solutions

In the previous chapter we discussed strong solutions, which for their definition require smooth boundaries and smooth dependence on t. This section is devoted to *weak solutions*, more precisely variational inequality weak solutions, which are closely related to potential theory (they can be viewed as instances of partial balayage) and also to quadrature domains and related topics.

3.1.1 Definition of a weak solution

For the well-posed version ($Q > 0$) of the Hele-Shaw problem (without surface tension) there is a good notion of a weak solution. It is based on the *Baiocchi transform*, replacing the pressure p in (1.16) by

$$u(z,t) = \int_0^t p(z,\tau)\, d\tau. \tag{3.1}$$

This type of transformation, with time t replaced by the vertical coordinate y, was used by C. Baiocchi in [29] to obtain a variational inequality formulation of the so-called dam problem. For the Hele-Shaw problem, weak or variational inequality formulations (in somewhat different disguises) were obtained around 1980 by Elliott, Janovský [173], Sakai [490], [493], Gustafsson [234]. See also [174].

To arrive at the concept of a weak solution, let $\Omega(t)$ be a strong solution of (1.16)–(1.19) with $Q > 0$, i.e., $\Omega(t)$ is a smooth family of domains such that (1.16)–(1.19) hold. For simplicity we shall take $Q = 1$, so that $p(\cdot,t)$ has the singularity

$$p(z,t) = -\frac{1}{2\pi} \log |z| + \text{harmonic}.$$

Using the Reynolds transport theorem (1.1) and Green's formula we have, with V_n denoting the normal velocity of the boundary,

$$\frac{d}{dt} \int_{\Omega(t)} \varphi d\sigma = \int_{\partial\Omega(t)} V_n \varphi ds = - \int_{\partial\Omega(t)} \frac{\partial p}{\partial n} \varphi ds$$

$$= - \int_{\partial\Omega(t)} p \frac{\partial \varphi}{\partial n} ds - \int_{\Omega(t)} \varphi \Delta p d\sigma + \int_{\Omega(t)} p \Delta \varphi d\sigma \geq \varphi(0),$$

for any test function $\varphi \in C^\infty(\mathbb{C})$ which is subharmonic in $\Omega(t)$. Here we used that $-\Delta p = \delta_0$ in $\Omega(t)$ and that $p \geq 0$.

We have already remarked that $\{\Omega(t)\}$ is a monotone increasing family. Integrating the above inequality from s to t, where $s < t$, gives

$$\int_{\Omega(t)} \varphi d\sigma - \int_{\Omega(s)} \varphi d\sigma \geq (t - s)\varphi(0),$$

for φ as above. In particular,

$$\int_{\Omega(t)} \varphi d\sigma - \int_{\Omega(0)} \varphi d\sigma \geq t\varphi(0), \tag{3.2}$$

which already is the weak formulation given by Sakai [490]. By approximation, any integrable subharmonic function φ in $\Omega(t)$ is allowed in (3.2). Sakai shows that given $\Omega(0)$ and $t > 0$ there is a unique, up to null-sets, domain $\Omega(t)$ satisfying (3.2) for these φ. If φ is harmonic in $\Omega(t)$, then both $\pm\varphi$ are subharmonic so we get (3.2) with equality. Therefore, (3.2) contains (1.35) as a special case (with $Q = 1$).

To go further, we keep $t > 0$ fixed and define

$$u(z, t) = \frac{1}{2\pi} \int_{\Omega(t)} \log|\zeta - z| d\sigma_\zeta - \frac{1}{2\pi} \int_{\Omega(0)} \log|\zeta - z| d\sigma_\zeta - \frac{t}{2\pi} \log|z| \tag{3.3}$$

for any $z \in \mathbb{C}$. Notice that this is the difference between the left and the right member in (3.2) with φ chosen to be

$$\varphi(\zeta) = \frac{1}{2\pi} \log|\zeta - z|.$$

Since this φ is integrable and subharmonic in $\Omega(t)$, (3.2) gives that

$$u \geq 0 \quad \text{everywhere.} \tag{3.4}$$

For z outside $\Omega(t)$ also $\varphi(\zeta) = -\frac{1}{2\pi} \log|\zeta - z|$ is subharmonic in $\Omega(t)$, so we obtain $u \leq 0$ outside $\Omega(t)$, hence

$$u(z, t) = 0 \quad \text{for } z \notin \Omega(t). \tag{3.5}$$

Finally, by definition (3.3) u satisfies

$$\chi_{\Omega(t)} = \chi_{\Omega(0)} + t\delta_0 + \Delta u \tag{3.6}$$

in the sense of distributions.

Equations (3.4)–(3.6) comprise the requirements we shall have on a weak solution. The function u is a kind of potential (indeed, it is the logarithmic potential of the measure $\chi_{\Omega(0)} - \chi_{\Omega(t)} + t\delta$) and it is uniquely determined by $\Omega(t)$, even at every point if the natural representative, given by (3.3), is chosen. Away from the origin Δu is a bounded function by (3.6), therefore u is continuously differentiable in the spacial variables. Since u attains its minimum ($u = 0$) on $\mathbb{C} \setminus \Omega(t)$, in particular on $\partial\Omega(t)$, it follows that also $\nabla u = 0$ there.

As an open set just satisfying (3.4)–(3.6), $\Omega(t)$ is not always uniquely determined. Indeed, equation (3.6) allows for arbitrary changes as to null-sets of $\Omega(t)$, whereas (3.5) is more sensitive. A point $z \in \Omega(t)$ may be removed from $\Omega(t)$ only if $u(z,t) = 0$, while points may be added as long as the area does not increase and $\Omega(t)$ remains an open set. As a matter of normalization, in order to make $\Omega(t)$ uniquely determined as a set, we shall take it to be *saturated* with respect to area measure. This means that we add to $\Omega(t)$ any disk $\mathbb{D}_r(z)$ such that $\mathbb{D}_r(z) \setminus \Omega(t)$ has area measure zero. This gives a domain which contains $\Omega(t)$, has the same area as $\Omega(t)$, and which cannot be further enlarged keeping these properties.

In summary we state the following definition.

Definition 3.1.1. A weak solution of the Hele-Shaw problem (1.16)–(1.19) with $Q = 1$ is a family $\{\Omega(t) : 0 \le t < t_0\}$ of bounded saturated open sets containing the origin such that there exists, for each t, a function $u = u(z,t)$ so that (3.4)–(3.6) hold.

When u exists satisfying (3.4)–(3.6) it is given by (3.3), because being $= 0$ in a neighbourhood of infinity it is the logarithmic potential of $(-\Delta u)$.

It is clear from the above derivation that a strong solution (if it exists) is always a weak solution. In the case a weak solution is derived from a strong solution we may differentiate (3.6) with respect to t to obtain that $\delta_0 + \Delta \frac{\partial u}{\partial t} = 0$ in $\Omega(t)$. Using that $u = |\nabla u| = 0$ on $\partial\Omega(t)$, showing that u grows only quadratically near $\partial\Omega(t)$, it also follows that $\frac{\partial u}{\partial t} = 0$ on $\partial\Omega(t)$. Thus, $\frac{\partial u}{\partial t} = p$ (in (1.16)), so the function u is indeed the Baiocchi transform of p. Note that $u(z,0) = 0$.

3.1.2 Existence and uniqueness of weak solutions

For weak solutions we have the following remarkably good existence theorem.

Theorem 3.1.1. *Given any bounded open set Ω_0 there exists a unique weak solution $\{\Omega(t) : 0 \le t < \infty\}$ with $\Omega(0) = \Omega_0$.*

Proof. Let $t > 0$ be fixed. In order to construct $\Omega(t)$ we shall relax (3.4)–(3.6) further, to

$$u \geq 0, \tag{3.7}$$

$$\chi_{\Omega(0)} + t\delta_0 + \Delta u \leq 1, \tag{3.8}$$

$$\int u(1 - \chi_{\Omega(0)} - t\delta_0 - \Delta u)d\sigma = 0. \tag{3.9}$$

Here $\Omega(t)$ has disappeared from the formulation, but u remains and $\Omega(t)$ can finally be recovered again. The system (3.7)–(3.9) is sometimes called a *linear complementarity problem* because it states that two linear inequalities are to hold and that at each point there shall be equality in at least for one of them.

There are several ways of constructing the solution of (3.7)–(3.9). The most direct way of obtaining u is simply to say that u shall be the smallest among all functions satisfying (3.7), (3.8) alone. This function will satisfy (3.9) as well.

To see that such a smallest function exists we choose a function ψ satisfying

$$\Delta\psi = \chi_{\Omega(0)} + t\delta_0 - 1 \quad \text{in } \mathbb{C},$$

for example

$$\psi(z) = -\frac{1}{2\pi} \int\limits_{\Omega(0)} \log|\zeta - z|d\sigma_\zeta - \frac{t}{2\pi}\log|z| - \frac{1}{4}|z|^2, \tag{3.10}$$

and set

$$v = u + \psi. \tag{3.11}$$

Then v is to be the smallest among all functions satisfying

$$v \geq \psi,$$
$$-\Delta v \geq 0.$$

We can think of ψ as an obstacle function, and the problem becomes that of finding the smallest superharmonic function v passing above the obstacle. It is well known from general potential theory (see, e.g., [24], [463]) that such a v exists. It is the lower semicontinuous regularization of the pointwise infimum of all superharmonic functions $\geq \psi$. Superharmonic functions are usually normalized to be lower semicontinuous.

Thus v, and hence u, are now constructed as solutions of suitable *obstacle problems*. Now we have to show that u satisfies (3.9). For this we continue to work with v. Suppose we have the strict inequality $v(z) > \psi(z)$ at some point z. Then since v is lower semicontinuous and ψ is continuous (outside the origin) there is an $\varepsilon > 0$ and a disk $\mathbb{D}_r(z) = \{w : |w - z| < r\}$, $(r > 0)$, such that $v \geq \psi + \varepsilon$ in all $\mathbb{D}_r(z)$. Therefore, if v is not already harmonic in $\mathbb{D}_r(z)$, then it can be made smaller by making a Poisson modification of it (i.e., by replacing v in $\mathbb{D}_r(z)$ by the harmonic function with the same boundary values on $\mathbb{D}_r(z)$). If the radius $r > 0$ is sufficiently small, then the modified v will be $\geq \psi$ in $\mathbb{D}_r(z)$.

From this we realize that $\Delta v = 0$ in the open set $\{v > \psi\}$, or that

$$\Delta u = \Delta \psi = \chi_{\Omega(0)} + t\delta_0 - 1$$

in $\{u > 0\}$. This proves (3.9).

Thus we have produced a solution u of (3.7)–(3.9). One easily sees that $u = 0$ outside some large disk $\mathbb{D}_R = \mathbb{D}_R(0)$. In fact, by comparing with an expanding disk solution one sees that if $\Omega(0) \subset \mathbb{D}_{R_0}$, then any $R > 0$ with $\pi R_0^2 + t < \pi R^2$ will do. Thus R depends on t.

To prove that u is unique and to obtain some further properties of it, let us indicate two other ways of constructing u, or v. They are based on minimizing Dirichlet integrals (measuring energies) of u and v, keeping one of the inequalities (3.7) or (3.8) as a side condition, namely as follows.

(i) Minimize $\int_{\mathbb{D}_R} |\nabla u|^2 d\sigma$ among all u vanishing on $\partial \mathbb{D}_R$ and satisfying

$$\Delta u + \chi_{\Omega(0)} + t\delta_0 \leq 1;$$

(ii) Minimize $\int_{\mathbb{D}_R} |\nabla v|^2 d\sigma$ among all functions v which agree with ψ on $\partial \mathbb{D}_R$ and satisfy $v \geq \psi$ everywhere.

In order to have finite integrals above one should, for the first integral, smooth out δ_0 a little (e.g., replace δ_0 by $\tilde{\delta} = \frac{1}{|\mathbb{D}_\varepsilon|} \delta_{\mathbb{D}_\varepsilon}$ for some small $\varepsilon > 0$). The proper settings then are that one works in the Sobolev space $H^{-1}(\mathbb{D}_R)$ for measures or mass distributions, and correspondingly in $H_0^1(\mathbb{D}_R)$ for potentials, like u. Both problems (i) and (ii) then have unique solutions which can be characterized by their variational formulations (*variational inequalities*).

The variational formulation for (ii) is

$$\int_{\mathbb{D}_R} \nabla v \cdot \nabla (w - v) d\sigma \geq 0,$$

to hold for all $w \geq \psi$ having the same boundary values as ψ on $\partial \mathbb{D}_R$. Using Stokes' theorem this becomes

$$\int_{\mathbb{D}_R} \Delta v \, (w - v) d\sigma \leq 0,$$

for all w as above. Since any $w \geq v$ is allowed here, we get $\Delta v \leq 0$. Choosing next $w = \psi$ gives

$$\int_{\mathbb{D}_R} \Delta v \, (v - \psi) d\sigma \geq 0,$$

hence actually,

$$\int_{\mathbb{D}_R} \Delta v \, (v - \psi) d\sigma = 0.$$

Thus, in the two inequalities $\Delta v \leq 0$ and $v \geq \psi$ there are nowhere strict inequalities in both. The problem (i) is treated similarly. In both cases we find that u and v satisfy linear complementary problems, which expressed in terms of u are (3.7)–(3.9). Conversely, one can go backward in the above reasoning, so the complementary problem (3.7)–(3.9) is equivalent to the two variational inequalities and the two minimum problems. In particular, it follows that the solution u of (3.7)–(3.9) is unique and that it has finite Dirichlet integral $\int |\nabla u|^2 d\sigma$ (after having replaced δ_0 by a mollified version $\tilde{\delta}$).

Next, invoking general regularity theory for the obstacle problem [193], [310], it follows that u actually is in the higher-order Sobolev space $H^{2,p}(\mathbb{D}_R)$ for any $p < \infty$. Now define $\Omega(t)$ to be the largest open set in which

$$\Delta u + \chi_{\Omega(0)} + t\delta_0 = 1.$$

In other words, $\Omega(t)$ is the complement of the closed support of the distribution $1 - \Delta u - \chi_{\Omega(0)} - t\delta_0$. By (3.9), $u = 0$ outside $\Omega(t)$. It is known (see for example Lemma A.4 in Ch. II of [310]) that this implies that $\Delta u = 0$ almost everywhere outside $\Omega(t)$ (when $u \in H^{2,p}(\mathbb{D}_R)$). Therefore,

$$\Delta u + \chi_{\Omega(0)} + t\delta_0 = \chi_{\Omega(0)} + 2\pi t\delta_0 < 1$$

almost everywhere outside $\Omega(t)$, hence actually $\Delta u + \chi_{\Omega(0)} + t\delta_0 = 0$ there (because $\chi_{\Omega(0)} + t\delta_0 \geq 1$, where it is non-zero). Finally, we conclude that

$$\Delta u + \chi_{\Omega(0)} + t\delta_0 = \chi_{\Omega(t)},$$

which means that we have established all properties of a weak solution. Note also that $\Omega(t)$ was defined to be saturated and that it contains the "noncoincidence set" $\{u > 0\} = \{v > \psi\}$. $\square$

3.1.3   General properties of weak solutions

It is clear from the way weak solution were introduced that a strong solutions always is a weak solution. This guarantees the uniqueness of a strong solution by the uniqueness of the weak one. However, a weak solution need not be a strong one, e.g., because a weak solution may change topology, a possibility which is not allowed even in the concept of a strong solution.

Recall that, in the weak solution, time appears only as a parameter. No derivative with respect to t occurs and one may jump to compute $\Omega(t)$ for only $t > 0$ directly, without computing it for any smaller values of t.

The following proposition shows that a weak solution has the monotonicity and semigroup properties one expects. We here allow weak solutions which are just open sets (not necessarily connected).

Proposition 3.1.1. *Let $\{\Omega(t)\}$ and $\{\tilde{\Omega}(t)\}$ be weak solutions with the initial open sets $\{\Omega(0)\}$ and $\{\tilde{\Omega}(0)\}$ respectively, and let $u = u_t$ and $\tilde{u} = \tilde{u}_t$ be the corresponding potentials. Then,*

- (i) *if $\Omega(0)$ is connected, then so is $\Omega(t)$ for any $t > 0$;*
- (ii) *if $0 < s < t$, then $u_s \leq u_t$ and $\Omega(s) \subset \Omega(t)$;*
- (iii) *if $\Omega(0) \subset \tilde{\Omega}(0)$, then $\Omega(t) \subset \tilde{\Omega}(t)$ for all $t > 0$;*
- (iv) *if $\tilde{\Omega}(0) = \Omega(s)$ for some $s > 0$, then $\tilde{\Omega}(t) = \Omega(t+s)$ for all $t > 0$.*
- (v) *The inequality (3.2) holds for all integrable subharmonic functions φ in $\Omega(t)$. In particular, $|\Omega(t)| = |\Omega(0)| + t$.*

Proof. (i) Let D be a connected component of $\Omega(t)$. Then $u = 0$ on ∂D. If D does not meet $\Omega(0)$, then $\Delta u = 1$ in D, which in view of the maximum principle contradicts $u \geq 0$. Thus every component of $\Omega(t)$ intersects $\Omega(0)$.

(ii) u_s is the smallest function satisfying $u_s \geq 0$, $\chi_{\Omega(0)} + s\delta_0 + \Delta u_s \leq 1$, and similarly for u_t. Since $\chi_{\Omega(0)} + s\delta_0 \leq \chi_{\Omega(0)} + t\delta_0$, it follows immediately that $u_s \leq u_t$. Outside $\Omega(t)$ we have $u_t = 0$, hence also $u_s = 0$, and so

$$
\begin{aligned}
\chi_{\Omega(s)} &= \chi_{\Omega(0)} + s\delta_0 + \Delta u_s \\
&= \chi_{\Omega(0)} + s\delta_0 + \Delta u_t \\
&\leq \chi_{\Omega(0)} + t\delta_0 + \Delta u_t \\
&= \chi_{\Omega(t)} = 0
\end{aligned}
$$

there. Thus $\Omega(s) \subset \Omega(t)$.

(iii) This is proved the same way as (ii).

(iv) Fix $s > 0$, $t > 0$, and set $w = u_{s+t} - u_s$. Then

$$
\Delta w = \chi_{\Omega(s+t)} - \chi_{\Omega(s)} - t\delta_0.
$$

By (ii), $w \geq 0$ and $w = 0$ outside $\Omega(s+t)$. Thus $\{\Omega(s+t), 0 \leq t < \infty\}$ is the weak solution with the initial domain $\Omega(s)$, which is exactly what is stated in (iv).

(v) By (ii), $\Omega(0) \subset \Omega(t)$, so (3.2) makes sense. The definition of a weak solution means exactly that (3.2) holds for all φ of the form $\varphi(\zeta) = \log|\zeta - z|$ for $z \in \mathbb{C}$, and $\varphi(\zeta) = -\log|\zeta - z|$ for $z \notin \Omega(t)$ and it is known [490] that the positive linear combinations of these are dense in the set of integrable subharmonic functions in $\Omega(t)$. Now (v) follows. $\square$

3.1.4 Regularity of the boundary

Let $\{\Omega(t) : 0 \leq t < \infty\}$ be a weak solution. Then $\Omega(s) \subset \Omega(t)$ for all $0 \leq s < t$, but it is not always true that $\overline{\Omega(0)} \subset \Omega(t)$. If we, for example, choose the initial domain $\Omega(0)$ such that $\partial\Omega(0)$ has positive area measure, then it will take $\partial\Omega(t)$ a finite time to move through $\partial\Omega(0)$. Even if $\Omega(0)$ has a piecewise smooth boundary, but has a corner with the interior angle smaller than $\pi/2$, then it turns out [311, 312, 497, 503, 202] that $\partial\Omega(t)$ stays at the corner for some positive amount of time.

On the other hand, if $\Omega(0)$ is C^1-smooth (not only piecewise), then $\overline{\Omega(0)} \subset \Omega(t)$. The following regularity theorem is mainly due to Sakai [491], [492], and shows that the situation outside $\overline{\Omega(0)}$ is rather pleasant. The last statement is taken from [247].

Theorem 3.1.2. *Assume* $\overline{\Omega(0)} \subset \Omega(t)$. *Then* $\partial\Omega(t)$ *consists of finitely many analytic curves, which may have finitely many singularities in the form of inward cusps or double points, but no other singularities. In case the* $\Omega(t)$ *are simply connected, the Riemann map* $f(\zeta, t)$ *parameterizing the phase domain* $\Omega(t)$ *extends analytically to a disk* $\mathbb{D}_{R(t)}$ *where the radius of analyticity* $R(t) > 1$ *is nondecreasing as a function of* t.

Remark 3.1.1. The most difficult part of the proof is actually to show that $\Omega(t)$ is finitely connected. We shall not take this difficulty here, but just prove the theorem in the case $\Omega(t)$ is finitely connected, say simply connected (the arguments of the proof will work for any isolated component of $\partial\Omega(t)$).

A different approach, which will be discussed in Section 3.3.4, to the regularity of $\partial\Omega(t)$ uses the exponential transform. This approach gives directly a real-analytic defining function of the boundary. See specifically Corollary 3.3.1.

Proof. So we assume that $\Omega(t)$ is simply connected and let $f : \mathbb{D} \to \Omega(t)$ be the Riemann map, $z = f(\zeta, t)$. Using the potential u in (3.4)–(3.6) we can define a one-sided Schwarz function, defined in $\overline{\Omega(t)} \setminus \Omega(0)$, by

$$S(z, t) = \bar{z} - 4\frac{\partial u}{\partial z}.$$

We see immediately from (3.6) that $S(z, t)$ is analytic in $\Omega(t) \setminus \overline{\Omega(0)}$. Since u is continuously differentiable away from the origin, $u \geq 0$ attains its minimum on $\mathbb{C} \setminus \Omega(t)$, and $|\nabla u| = 0$ there, $S(z, t)$ is continuous up to $\partial\Omega(t)$ with $S(z, t) = \bar{z}$ on $\partial\Omega(t)$.

The conjugate of $S(z)$ can be interpreted as the anticonformal reflection in $\partial\Omega(t)$ and we use it to extend f in the following way. We extend the function f by

$$f(1/\bar{\zeta}, t) = \overline{S(f(\zeta, t))},$$

for those $\zeta \in \mathbb{D}$ for which $f(\zeta, t) \in \Omega(t) \setminus \overline{\Omega(0)}$. This defines f analytically in $\mathbb{D}$ and in an annulus $1 < |\zeta| < R(t)$. Here we take $R(t) > 1$ largest possible, which means that $R(t) = 1/r(t)$ where $0 < r(t) < 1$ is the smallest radius such that $f^{-1}(\Omega(0), t) \subset \mathbb{D}_{r(t)}$. Across $\partial \mathbb{D}$ we have a certain form of continuity because of the continuity of $S(z, t)$. Indeed, as $|\zeta| \to 1$ with $\zeta \in \mathbb{D}$ we have

$$|f(\zeta, t) - f(1/\bar{\zeta}, t)| = |f(\zeta, t) - \overline{S(f(\zeta, t), t)}| \to 0, \qquad (3.12)$$

by the continuity and boundary properties of $S(z, t)$ obtained above. Now the function $f(\zeta, t)$ is defined in $\mathbb{D}$ as well as in the annulus $1 < |\zeta| < R(t)$, hence, almost everywhere in the disk $|\zeta| < R(t)$.

Next we shall use (3.12) to prove that the distributional derivative $\partial f(\zeta, t)/\partial \bar{\zeta}$ vanishes in $|\zeta| < R(t)$. Given a test function φ with compact support in $|\zeta| < R(t)$ we have

$$\left\langle \frac{\partial f}{\partial \bar{\zeta}}, \varphi \right\rangle = -\int_{\mathbb{C}} f(\zeta, t) \frac{\partial \varphi}{\partial \bar{\zeta}} d\sigma_\zeta$$

$$= -\frac{1}{2i} \int_{\mathbb{D}} f(\zeta, t) \frac{\partial \varphi}{\partial \bar{\zeta}} d\bar{\zeta} d\zeta - \frac{1}{2i} \int_{|\zeta|>1} f(\zeta, t) \frac{\partial \varphi}{\partial \bar{\zeta}} d\bar{\zeta} d\zeta$$

$$= -\frac{1}{2i} \lim_{\varepsilon \downarrow 0} \left(\int_{|\zeta|=1-\varepsilon} f(\zeta, t)\varphi(\zeta) \, d\zeta - \int_{|\zeta|=1+\varepsilon} f(\zeta, t)\varphi(\zeta) \, d\zeta \right)$$

$$= -\frac{1}{2i} \lim_{\varepsilon \downarrow 0} \int_{|\zeta|=1-\varepsilon} (f(\zeta, t) - f(1/\bar{\zeta}, t))\varphi(\zeta) \, d\zeta = 0.$$

In the above curve integrals we take the counterclockwise direction on the circles. Thus, the function $f(\zeta, t)$ is analytic in the disk $|\zeta| < 1/r(t)$.

For any pair of numbers s, t such that $0 < s < t \le T$, we have that the function $h(\zeta, s, t) = f^{-1}(f(\zeta, s), t)$ maps the unit disk into itself and $h(0, s, t) = 0$. A simple application of the Schwarz Lemma to the function h shows that

$$f^{-1}(\Omega(0), t) \subset \mathbb{D}_{r(s)}.$$

Therefore, $r(t) \le r(s)$, hence $R(s) \le R(t)$.

We have $f(\partial \mathbb{D}, t) = \partial \Omega(t)$ as sets, f is univalent in $\mathbb{D}$ but need not be univalent on $\bar{\mathbb{D}}$. Therefore, $\partial \Omega(t)$ is analytic with possible singularities as stated. □

Remark 3.1.2. In the proof we did not take into account the requirement that $u \ge 0$. This requirement actually excludes the most generic types of cusp singularities to appear. For example, it turns out that u always becomes strictly negative in some region behind a 3/2-cusp (i.e., a cusp with local equation of type $y = x^{3/2} + \cdots$).

It can be shown, see [278, 492, 510], that only cusps of order $(4n + 1)/2$ ($n = 1, 2, \ldots$) can show up in the injection problem. Equivalently, these are the cusps which allow further suction in the ill-posed direction of the Hele-Shaw problem. The more detailed analysis is quite subtle, see [492].

3.1.5   Connections to nonlinear PDE and the Stefan problem

As mentioned, the weak solutions treated in this chapter are variational inequality weak solutions. There are also other types of weak solutions to Hele-Shaw problems. For example, Crandall and Lions [108] introduced the notion of *viscosity solution*, which now is a standard tool in the theory of nonlinear elliptic and parabolic equations, see [109], [81]. In particular, Caffarelli and Vázquez [81] have proved the existence and uniqueness of the viscosity solution for the porous medium equation, and later, Kim [308] has adapted this for the Hele-Shaw problem.

In addition, there are solution concepts related to enthalpy solutions of the Stefan problem, see below, and to related phase field models, developed by G. Caginalp and others [84, 82, 83].

In this section we shall in particular indicate the connection of Hele-Shaw problem to the Stefan problem. This will also explain why the variational inequality weak formulation of the Hele-Shaw problem just becomes a series elliptic problems, with no time derivatives involved.

Let us start with a remark concerning the linear complementarity problem (3.7)–(3.9), namely that it is possible to reformulate it as just one single nonlinear partial differential equation. In fact, a general such equation, in two variables and of order two, is something of the form

$$F(x, y, u, Du, D^2u) = 0, \tag{3.13}$$

where Du, D^2u denote the set of all partial derivatives of order one and two, respectively. Now taking F to be

$$F(x, y, u, Du, D^2u) = \min\{u, 1 - \chi_{\Omega(0)} - t\delta_0 - \Delta u\} \tag{3.14}$$

gives exactly the system (3.7)–(3.9), proving the claim. Moreover, this F has the right monotonicity properties to fit into the framework of viscosity solutions, as discussed, e.g., in [109].

Next we turn to the Stefan problem, which describes heat conduction with a phase change (e.g., ice/water), see [174] for example. We then denote the temperature by p, because it will correspond to the pressure in the Hele-Shaw problem. Let H denote the *enthalpy*, or heat content, per unit volume. If there are no phase changes H is just a linear function of temperature p. In case phase changes do occur H will, in the sharp interface case, be a multi-valued function of p. The relationship between these two quantities can be expressed in terms of a graph, or alternatively in terms of a set-valued function. Using the second option, we have in the typical case

$$H \in \beta(p),$$

where β is the set-valued function

$$\beta(p) = \begin{cases} c_1 p & \text{for } p < 0, \\ [0, L] & \text{for } p = 0, \\ L + c_2 p & \text{for } p > 0. \end{cases}$$

Here $c_1, c_2 > 0$ are specific heats and $L > 0$ is the latent heat (the amount of heat needed to melt a unit volume, of for example ice, at fixed temperature $p = 0$). (Formally we should have written $\{c_1 p\}$ etc., but we avoid that when $\beta(p)$ consists of only one value.)

The equation for heat balance is, in the simplest case,

$$\frac{\partial H}{\partial t} = \Delta p + \delta_0.$$

Here we have chosen a source term δ_0 in order to make the equation comparable to what we have for Hele-Shaw flow with injection. Integration of this equation with respect to time, from $t = 0$ to some arbitrary $t > 0$, gives

$$H - H_0 = \Delta u + t\delta_0, \tag{3.15}$$

where H_0 is the enthalpy at time $t = 0$ and

$$u(t) = \int_0^t p(\tau) d\tau, \tag{3.16}$$

the dependence on space variables being suppressed from notation.

Now we claim that the Hele-Shaw problem appears, after some minor modifications, as the degenerate limit $c_1 = c_2 = 0$ of the above problem. Choosing $L = 1$ for simplicity, β becomes

$$\beta(p) = \begin{cases} 0 & \text{for } p < 0, \\ [0, 1] & \text{for } p = 0, \\ 1 & \text{for } p > 0. \end{cases}$$

The Hele-Shaw problem is really a one-phase moving boundary problem, whereas the Stefan problem in general has two phases. To account for the fact that p is never strictly negative in the Hele-Shaw problem, we may modify the above β to be

$$\beta(p) = \begin{cases} \emptyset & \text{for } p < 0, \\ [0, 1] & \text{for } p = 0, \\ 1 & \text{for } p > 0. \end{cases}$$

As a further change, it is natural from a purely mathematical point of view to extend β so that it describes a maximal monotone graph. There are several ways of doing this, but from our perspective the choice

$$\beta(p) = \begin{cases} \emptyset & \text{for } p < 0, \\ [-\infty, 1] & \text{for } p = 0, \\ 1 & \text{for } p > 0 \end{cases} \tag{3.17}$$

is the best one. If initially $p \geq 0$ and $H \geq 0$, the above two changes of β will make
no difference for the solution.

So let us now stick to the choice (3.17) and try to connect to the Hele-Shaw
problem. In the dynamical equation (3.15) the initial enthalpy will then be

$$H_0 = \chi_{\Omega(0)}.$$

Now (3.15) combined with (3.17) becomes

$$\Delta u + t\delta_0 + \chi_{\Omega(0)} \in \beta(p).$$

Since $p = \frac{\partial u}{\partial t}$ by (3.16), this is in principle a parabolic problem for u. However, it is
easy to check directly from (3.16) that under some light smoothness assumptions
$\beta(u) \subset \beta(p)$. And in case p is monotone increasing, which it is in the well-posed
version of the Hele-Shaw problem, we also have the opposite inclusion. Hence we
have in fact

$$\beta(u) = \beta(p).$$

This is the crucial observation, which is what makes the, in principle parabolic,
Hele-Shaw problem to decouple (in the weak formulation) into just series of elliptic
problems, namely (for each individual t),

$$\Delta u + t\delta_0 + \chi_{\Omega(0)} \in \beta(u). \tag{3.18}$$

One easily checks that (3.18) is just another way of writing the complementarity
system (3.7)–(3.9), or the equation (3.13), with F given by (3.14).

3.2   Balayage techniques

3.2.1   Weak solutions as partial balayage

Returning to the general treatment of weak solutions it may be apparent that the
expression $\chi_{\Omega(0)} + t\delta_0$ always appears as one quantity. The weak solution itself
is the family $\{\Omega(t)\}$, or better $\{\chi_{\Omega(t)}\}$. Moreover, time t only plays the role of a
parameter, and for any fixed $t > 0$ the whole construction really amounted to the
construction of a map

$$\chi_{\Omega(0)} + t\delta_0 \mapsto \chi_{\Omega(t)}.$$

This map finally came out to be just the addition of the term Δu, where u solves
the complementary problem (3.7)–(3.9).

For a further systematic treatment of weak solutions it is really advantageous to take this operator theoretic point of view. In addition, everything looks more natural if we replace the number one in the right member of (3.8) by a more general function, say ρ, which will have the interpretation of a density. In some applications it is appropriate to allow ρ to take negative values, but then the negative part ρ_- should not be too bad; one may ask it for example to have compact support and its potential U^{ρ_-} to be a continuous function. We shall use letters μ or similar for what used to be $\chi_{\Omega(0)} + t\delta_0$ (also a density, or a measure denoted as a density).

Assume that μ is a measure (possibly signed) with compact support and that ρ is large enough at infinity, in particular that

$$\int d\mu < \int \rho d\sigma.$$

Then we define

$$\mathrm{Bal}\,(\mu, \rho) = \mu + \Delta u,$$

where u is the smallest function satisfying

$$u \geq 0 \quad \text{in } \mathbb{C}, \tag{3.19}$$

$$\mu + \Delta u \leq \rho \quad \text{in } \mathbb{C}. \tag{3.20}$$

Such a function exists as before, being the unique solution of an obstacle problem. We also define a corresponding saturated set Ω by

$$\Omega = \{\text{the largest open set in which } \mu + \Delta u = \rho\}. \tag{3.21}$$

This is a bounded open set, and the complementary condition

$$\{u > 0\} \subset \Omega \tag{3.22}$$

holds.

The interpretation of Bal is that it performs a kind of balayage[2] – *partial balayage*. Indeed, let $\nu = \mathrm{Bal}\,(\mu, \rho)$, and let, as a matter of general notation,

$$U^\mu(z) = -\frac{1}{2\pi} \int \log|z - \zeta|\, d\mu(\zeta) \tag{3.23}$$

be the logarithmic potential of μ. In case $d\mu = \chi_D d\sigma$ for some set $D \subset \mathbb{C}$ we write simply U^D for the potential. Then since $\nu = \mu + \Delta u$ and $u = 0$ outside Ω (in particular, in a neighbourhood of infinity) it follows that

$$u = U^\mu - U^\nu.$$

---

[2]The idea of balayage (sweeping in English translation) goes back [203] to Johann Carl Friedrich Gauss, 1777–1855, see [145] for historical accounts in general. The method of successive sweeping operations was invented by Poincaré [433], [434] for solving the Dirichlet problem.

The vanishing of u outside Ω, therefore, means that μ and ν are graviequivalent in a certain sense, which explains the word "balayage" (sweeping of measures without changing exterior potentials). For details on partial balayage we refer to [243, 238, 526, 201, 483]. See also [148] and [365].

In view of the minimization problem (i) in the proof of Theorem 3.1.1, it is clear that the operator $\mu \mapsto \nu = \mathrm{Bal}\,(\mu, \rho)$ has the effect that it replaces any measure μ by a measure ν satisfying $\nu \leq \rho$ (everything denoted as densities) using as little work

$$\int |\nabla u|^2 d\sigma = \text{energy of } \nu - \mu$$

as possible. The result of the whole thing is a measure ν and an open set Ω, such that

$$U^\nu = U^\mu \quad \text{outside } \Omega,$$

$$\nu = \rho \quad \text{in } \Omega$$

(by (3.22) and (3.21) respectively). Thus ν has the desired potential U^μ outside Ω and the desired density ρ in Ω.

In terms of $\mathrm{Bal}\,(\mu, \rho)$, a weak solution of the Hele-Shaw problem now is just a family of (saturated) open sets $\{\Omega(t) : 0 \leq t < \infty\}$ satisfying

$$\mathrm{Bal}\,(\chi_{\Omega(0)} + t\delta_0, 1) = \chi_{\Omega(t)}. \tag{3.24}$$

By (iv) of Proposition 3.1.1 we also have, for arbitrary $s < t$,

$$\mathrm{Bal}\,(\chi_{\Omega(s)} + (t - s)\delta_0, 1) = \chi_{\Omega(t)}.$$

This is then an instance of a general property of $\mathrm{Bal}\,$, namely that

$$\mathrm{Bal}\,(\mu_1 + \mu_2, \rho) = \mathrm{Bal}\,(\mathrm{Bal}\,(\mu_1, \rho) + \mu_2, \rho).$$

Take $\rho = 1$, $\mu_1 = \chi_{\Omega(0)} + t\delta_0$, $\mu_2 = 2\pi(t - s)\delta_0$ to get the previous statement. A more general statement is also true:

$$\mathrm{Bal}\,(\mu_1 + \mu_2, \rho_1) = \mathrm{Bal}\,(\mathrm{Bal}\,(\mu_1, \rho_2) + \mu_2, \rho_1), \tag{3.25}$$

whenever $\rho_1 \leq \rho_2 + \mu_2$.

Similarly, parts (ii) and (iii) in Proposition 3.1.1 are special cases of the implication

$$\mu_1 \leq \mu_2 \Rightarrow \mathrm{Bal}\,(\mu_1, \rho) \leq \mathrm{Bal}\,(\mu_2, \rho). \tag{3.26}$$

Given μ, taking here $\mu_1 = \min(\rho, \mu)$, $\mu_2 = \mu$, gives the lower bound in the estimate

$$\min(\rho, \mu) \leq \mathrm{Bal}\,(\mu, \rho) \leq \rho, \tag{3.27}$$

because $\mathrm{Bal}\,(\min(\rho, \mu), \rho) = \min(\rho, \mu)$. The upper bound is just by definition.

The inequality (3.27) can be viewed as a regularity statement for the functions u and v in (i), (ii) (in the proof of Theorem 3.1.1). With, for example, $\rho = 1$,

$\mu = \chi_{\Omega(0)} + t\delta_0$, we get, for u, that $0 \le \Delta u \le 1$ away from the origin, which gives the previously used regularity $u \in H^{2,p}$ for all $p < \infty$.

The general structure of $\mathrm{Bal}\,(\mu, \rho)$ is

$$\mathrm{Bal}\,(\mu, \rho) = \rho\chi_\Omega + \mu\chi_{\mathbb{C}\setminus\Omega}. \tag{3.28}$$

Indeed, (3.28) is true in Ω by definition (3.21) of Ω, and outside Ω we have $u = 0$, hence $\Delta u = 0$ there, at least under some regularity assumptions (e.g., $\mu \in L^\infty$); the general case can be handled by an approximation argument [526, 201]. Thus $\mathrm{Bal}\,(\mu, \rho) = \mu$ outside Ω.

In the special case of Hele-Shaw dynamics ($\rho = 1$, $\mu = \chi_{\Omega(0)} + t\delta_0$) we have $\mu \ge \rho$ everywhere where μ does not vanish. This guarantees

$$\mathrm{Bal}\,(\mu, \rho) = \rho\chi_\Omega,$$

as is immediate from (3.28) together with the definition (3.21) of Ω. For any kind of injection Hele-Shaw problem, if the sources are located within the initial domain $\Omega(0)$, and if the accumulated sources up to time $t > 0$ are represented by the measure $\mu = \mu(t) \ge 0$, then the weak solution $\Omega(t)$ is given by

$$\mathrm{Bal}\,(\chi_{\Omega(0)} + \mu, 1) = \chi_{\Omega(t)}.$$

If there are sources outside $\Omega(0)$ and these are sufficiently weak (meaning that $\mu(t) < 1$ outside $\Omega(0)$ and for some time $t > 0$), then there will also be the second term in the right member of (3.28),

$$\mathrm{Bal}\,(\chi_{\Omega(0)} + \mu, 1) = \chi_{\Omega(t)} + \mu\chi_{\mathbb{C}\setminus\Omega(t)},$$

corresponding to some kind of "mushy region".

As a useful application of (3.25) we have the following. Given $t > 0$ choose $r > 0$ so small that $\pi r^2 < t$ and let

$$\tilde{\delta} = \frac{1}{|\mathbb{D}_r|}\chi_{\mathbb{D}_r} = \mathrm{Bal}\left(\delta_0, \frac{1}{|\mathbb{D}_r|}\right)$$

be the Dirac measure swept out to a uniform density on $\mathbb{D}_r$. Then also

$$t\tilde{\delta} = \mathrm{Bal}\left(t\delta_0, \frac{t}{|\mathbb{D}_r|}\right),$$

for any $t > 0$. Since $\frac{t}{|\mathbb{D}_r|} > 1$, (3.25) with $\rho_1 = 1$, $\rho_2 = \frac{t}{|\mathbb{D}_r|}$ shows that

$$\mathrm{Bal}\,(t\delta_0 + \chi_{\Omega(0)}, 1) = \mathrm{Bal}\left(\mathrm{Bal}\left(t\delta_0, \frac{t}{|\mathbb{D}_r|}\right) + \chi_{\Omega(0)}, 1\right)$$

$$= \mathrm{Bal}\,(t\tilde{\delta} + \chi_{\Omega(0)}, 1),$$

i.e., the Hele-Shaw evolutions with δ_0 and $\tilde{\delta}$ are exactly the same.

3.2.2   Existence and non-branching of backward weak solutions

In this section we discuss the existence and uniqueness of weak and strong solutions backward in time. We are keeping the assumption $Q = 1$, so speaking about backward solutions is just another way of saying that we are considering the suction case.

As for strong solutions, F.R. Tian [553, 554], see also [361], proved the local backward existence, uniqueness for an analytic smooth initial boundary, and the fact that if the initial boundary is not analytic (but still smooth), then the backward strong solution will not exist. Also for the existence of a backward weak solution $\{\Omega(t)\}$, on some interval $-\varepsilon < t < 0$ and satisfying $\overline{\Omega(t)} \subset \Omega(0)$, we shall need the analyticity of the initial boundary. In the balayage picture, this analyticity allows for some "inverse balayage", by which the mass of the initial domain becomes represented in an equipotential way by a more concentrated body of higher density. On performing modifications on this body one can then achieve many kinds of domain variations, of Laplacian growth type, of the original domain.

The crucial step of inverse balayage is contained in the following lemma.

Lemma 3.2.1. *Let $\Omega \subset \mathbb{C}$ be a bounded domain with smooth analytic boundary. Then, for any $\beta > 0$ sufficiently small there exists a subdomain D with $\overline{D} \subset \Omega$ and a measure μ supported on $\overline{D}$ satisfying $d\mu \geq (1 + \beta)d\sigma$ so that*

$$\mathrm{Bal}\,(\mu, 1) = \chi_\Omega. \tag{3.29}$$

Proof. Recall that (3.29) means that there exists a function $u \geq 0$ satisfying

$$\begin{cases} \Delta u = \chi_\Omega - \mu & \text{in } \mathbb{C}, \\ u = 0 & \text{outside } \Omega. \end{cases}$$

Near $\partial\Omega$ this u will have to be C^1, hence it will have to solve the Cauchy problem

$$\begin{cases} \Delta u = 1, & \text{in } \Omega, \text{ near } \partial\Omega, \\ u = 0, & \text{on } \partial\Omega, \\ \nabla u = 0, & \text{on } \partial\Omega. \end{cases}$$

Since $\partial\Omega$ is analytic, this problem has a (unique) local solution, by the Cauchy–Kovalevskaya theorem. This is true in any number of dimensions, but in our two-dimensional case the solution can actually be gotten directly from the Schwarz function $S(z)$ of $\partial\Omega$, namely as

$$u(z) = \frac{1}{2}\mathrm{Re} \int_{z_0}^{z} (S(\zeta) - \bar{\zeta})d\zeta,$$

where z_0 is any point on $\partial\Omega$.

The function u will grow quadratically with the distance from $\partial\Omega$,

$$u(z) \sim \frac{1}{2}\mathrm{dist}(z, \partial\Omega)^2,$$

and we define D by taking ∂D to be a level set $u = \alpha$ for u, with $\alpha > 0$ so small so that $0 < u < \alpha$ in $\Omega \setminus \overline{D}$ and so that the normal derivative in the direction out from D satisfies

$$\frac{\partial u}{\partial n} \leq c_1 < 0.$$

By this we have defined D, and u is defined in $\overline{\Omega} \setminus D$. We extend this u by zero outside Ω.

Inside D we continue u in a different way, we let it solve the Dirichlet problem

$$\Delta u = -\beta, \quad \text{in } D,$$
$$u = \alpha, \quad \text{on } \partial D,$$

with $\beta > 0$ sufficiently small, so that the outward normal derivative of $u|_D$ satisfies

$$c_1 < c_2 \leq \frac{\partial u}{\partial n} < 0.$$

We now have

$$\Delta u = \chi_\Omega - \mu, \tag{3.30}$$

where μ is a positive measure on $\overline{D}$, which has density $1 + \beta$ in D and which also has a contribution on ∂D coming from the jump of the normal derivative of u. Since $u \geq 0$ by construction, u has the required properties stated in the beginning of the proof, hence the lemma is proved. $\qquad\square$

Using the lemma it is now easy to prove the following theorem on weak backward solutions.

Theorem 3.2.1. *Assume that $\Omega(0)$ has a smooth analytic boundary and contains the origin. Then there exists, for some $\varepsilon > 0$, a weak solution $\{\Omega(t) : -\varepsilon < t < 0\}$, with $\overline{\Omega(t)} \subset \Omega(0)$, such that*

$$\mathrm{Bal}\left(\chi_{\Omega(s)} + (t - s)\delta_0, 1\right) = \chi_{\Omega(t)}$$

whenever $s < t$. In particular,

$$\mathrm{Bal}\left(\chi_{\Omega(t)} - t\delta_0, 1\right) = \chi_{\Omega(0)}.$$

Proof. Choose D and μ as in Lemma 3.2.1 (applied to $\Omega(0)$). We may assume that $0 \in D$. Next recall that Hele-Shaw evolutions are not affected by smoothing out δ_0 to

$$\tilde{\delta} = \frac{1}{|\mathbb{D}_r|}\chi_{\mathbb{D}_r}, \tag{3.31}$$

as long as the support of $\tilde{\delta}$ remains inside the fluid domains. We shall take $r > 0$ so small that $\overline{\mathbb{D}_r} \subset D$. Then $\mu + t\tilde{\delta} \geq 1$ in D for all $-\pi r^2 \beta \leq t < 0$. Thus choosing $\varepsilon = \pi r^2 \beta$ we have, for $-\varepsilon \leq t < 0$,

$$\mathrm{Bal}\,(\mu + t\tilde{\delta}, 1) = \chi_{\Omega(t)}, \tag{3.32}$$

for some domains $\Omega(t)$. This will be the weak solution.

One may notice, using (3.25), that

$$\begin{aligned}
\mathrm{Bal}\,(\mu + t\tilde{\delta}, 1) &= \mathrm{Bal}\,(\mu - \varepsilon\tilde{\delta} + (t + \varepsilon)\tilde{\delta}, 1) \\
&= \mathrm{Bal}\,(\mathrm{Bal}\,(\mu - \varepsilon\tilde{\delta}, 1) + (t + \varepsilon)\tilde{\delta}, 1) \\
&= \mathrm{Bal}\,(\chi_{\Omega(-\varepsilon)} + (t + \varepsilon)\tilde{\delta}, 1),
\end{aligned}$$

so the family $\chi_{\Omega(t)}$ really is an ordinary weak solution started with $\Omega(-\varepsilon)$. By construction of μ, taking $t = 0$ in (3.32) gives the given initial domain $\Omega(0)$, and for $t > 0$ one gets the usual forward solution. Thus (3.32) defines a weak solution on all $-\varepsilon \leq t < \infty$. $\qquad\square$

As a corollary of the above proof, or of Lemma 3.2.1, we get (local) existence of weak solutions for Hele-Shaw problems with completely general source and sinks (and even dipoles etc) provided the initial domain has smooth analytic boundary. In other words, we can perform general domain variations of Laplacian growth type. Indeed, a classical formulation of the Hele-Shaw problem driven by an arbitrary distribution ν with a compact support in the fluid region $\Omega(t)$ goes as follows: given a bounded region $\Omega(t) \subset \mathbb{C}$ with analytic boundary, one solves the Dirichlet problem

$$-\Delta p = \nu \quad \text{in } \Omega(t), \tag{3.33}$$

$$p = 0 \quad \text{on } \partial\Omega(t). \tag{3.34}$$

The solution $p = p_{\Omega(t)}$ depends on $\Omega(t)$, and the dynamic condition is that one asks $\partial\Omega(t)$ to move with the speed

$$V_n = -\frac{\partial p}{\partial n} \tag{3.35}$$

in the outward normal direction.

By the mean-value property for harmonic functions this dynamic law is not affected if ν is replaced by its convolution, $\tilde{\nu}$, with a radially symmetric and positive mollifier with small support. We shall need only that $\tilde{\nu}$ is bounded as a function, and for this purpose one can simply use the $\tilde{\delta}$ in (3.31) as a mollifier:

$$\tilde{\nu} = \nu * \tilde{\delta}$$

The appropriate notion of weak solution then amounts to requiring that

$$\mathrm{Bal}\,(\chi_{\Omega(s)} + (t - s)\tilde{\nu}, 1) = \chi_{\Omega(t)}$$

whenever $s < t$. Thus we arrive at the following corollary.

Corollary 3.2.1. *Given $\Omega(0)$ with smooth analytic boundary, and given any distribution ν with $\operatorname{supp}\nu \subset \Omega(0)$, there exists, for some $\varepsilon > 0$, a weak solution $\{\Omega(t) : -\varepsilon < t < \varepsilon\}$ in the above sense of the Hele-Shaw problem driven by ν. In terms of the measure μ in Lemma 3.2.1 (with $\Omega = \Omega(0)$) the weak solution is given by*

$$\operatorname{Bal}(\mu + t\tilde{\nu}, 1) = \chi_{\Omega(t)} \quad for \ -\varepsilon < t < \varepsilon.$$

There is no difficulty in allowing time-dependent sources/sinks. In the balayage formulation the term $(t - s)\tilde{\nu}$ is then to be replaced by the integral from s to t of the then time-dependent $\tilde{\nu}$. One example is the squeezing version of Laplacian growth, to be discussed in Section 3.2.3. In this case $\nu(t) = \chi_{\Omega(t)}$, which turns out to be equivalent to taking $\nu(t) = e^t \chi_{\Omega(0)}$. (There is no need to replace ν by a smoothed out version in this case.)

Remark 3.2.1. The weak solutions obtained in Theorem 3.2.1 and Corollary 3.2.1 will in fact be strong solutions. This follows from the fact that partial balayage is just an instance of the obstacle problem and that the solutions are constructed as perturbations of a solution with smooth analytic boundary (the one given by (3.29) for $\Omega(0)$). And it is known [78], [509], that the solution $\{\Omega(t)\}$ in such a case also will vary smoothly in t. Therefore, the solution obtained will in fact be a strong solution for the Hele-Shaw problem.

A weak solution can branch at any time in the backward direction. This occurs when a simply connected domain $G(t_0) = \Omega(t_0)$ for some $0 < t_0 < \infty$ appears as a result of a strong simply connected dynamics $\Omega(t)$ and at the same time as a result of a weak dynamics $G(t)$, where $G(t)$ for $t < t_0$ is multiply connected with some holes to be filled in as $t \to t_0^-$, see Figure 3.1. Our next result says that branching can take place only when such changes of topology occur, see [248]. The result is quite general, but for simplicity we formulate it just for the case of a point source.

Theorem 3.2.2. *Let $G(0)$ and $H(0)$ be two initial domains in $\mathbb{C}$, and $G(t)$ and $H(t)$ be the corresponding weak solutions, $0 \le t < \infty$. We assume that*

- $\overline{G(0)} \subset G(t)$ *and* $\overline{H(0)} \subset H(t)$ *for all* $t > 0$;
- $\mathbb{C} \setminus \overline{G(t)}$ *and* $\mathbb{C} \setminus \overline{H(t)}$ *are connected for each* $t > 0$;
- *there exists* $0 < t_0 < \infty$ *such that* $G(t_0) = H(t_0)$;
- *there exists* $\varepsilon > 0$ *such that* $\overline{H(0)} \subset G(t_0 - \varepsilon)$.

Then, $G(t) = H(t)$ for all $t \in (t_0 - \varepsilon, t_0]$.

Remark 3.2.2. If the initial domain $G(0)$ is bounded by a smooth analytic curve, then the strong solution exists locally in time and coincides with the weak one $G(t)$. Since in the strong case the normal velocity at the boundary does not vanish, the first assumption of the theorem is satisfied whenever the boundaries of the initial domains are smooth analytic.

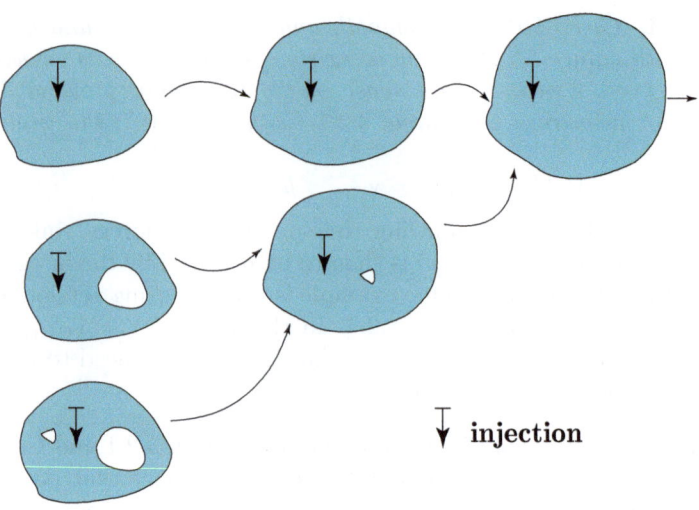

Fig. 3.1: Branching weak solutions

Proof. Let $t_0 - \varepsilon < t < t_0$) and let $u(z,t)$ and $v(z,t)$ be the functions that correspond to the domains $G(t)$ and $H(t)$ respectively. Then,

$$\Delta u = \chi_{G(t)} - \chi_{G(0)} + t\delta_0,$$
$$\Delta v = \chi_{H(t)} - \chi_{H(0)} + t\delta_0,$$

$u(z,t) \geq 0$, $v(z,t) \geq 0$ in $\mathbb{C}$ and $u(z,t) = 0$ in $\mathbb{C} \setminus G(t)$, $v(z,t) = 0$ in $z \in \mathbb{C} \setminus H(t)$.

Next consider the function

$$\gamma(z,t) = v(z,t_0) - u(z,t_0) + u(z,t).$$

One easily calculates

$$\Delta \gamma(z,t) = \chi_{G(t)} - \chi_{H(0)} + t\delta_0.$$

Under the assumption $\overline{H(0)} \subset G(t_0 - \varepsilon)$ the function $\gamma(z,t)$ is harmonic in $\mathbb{C} \setminus \overline{G(t)}$ for any $t \in (t_0 - \varepsilon, t_0)$, and $\gamma(z,t) = 0$ in $\mathbb{C} \setminus \overline{G(t_0)}$. Therefore, $\gamma(z,t) = 0$ in $\mathbb{C} \setminus \overline{G(t)}$ by harmonic continuation (using that $\mathbb{C} \setminus \overline{G(t)}$ is connected).

We have $v(z,t) \geq 0$ in $\mathbb{C}$ and $v(z,t) = 0$ in $\mathbb{C} \setminus \overline{H(t)}$. Let us set

$$w(z,t) = v(z,t_0) - v(z,t) - u(z,t_0) + u(z,t) = \gamma(z,t) - v(z,t).$$

This function is non-positive in $\mathbb{C} \setminus \overline{G(t)}$, and

$$\Delta w = \chi_{G(t)} - \chi_{H(t)}. \qquad (3.36)$$

Therefore, $\Delta w \geq 0$ in $G(t)$. Hence, $\underline{w \leq 0}$ in $\mathbb{C}$. Moreover, the function w is subharmonic in the connected set $\mathbb{C} \setminus \overline{H(t)}$. Therefore, $w < 0$ in $\mathbb{C} \setminus \overline{H(t)}$, or else, $w \equiv 0$ in $\mathbb{C} \setminus \overline{H(t)}$ by the maximum principle. Since $w(z) = 0$ for large values of z, only the second option is valid. In particular, $\Delta w = 0$ in $\mathbb{C} \setminus \overline{H(t)}$, which by the equation (3.36) implies that

$$G(t) \subset \overline{H(t)}. \tag{3.37}$$

By Proposition 3.1.1 (v) we have $|H(t)| = |G(t)|$. Since $G(t)$ and $H(t)$ are the saturated sets which satisfy (3.4)–(3.6) for $G(0)$ and $H(0)$ respectively, and $|\partial H(t)| = |\partial G(t)| = 0$, it follows from (3.37) that $G(t) = H(t)$ for all $t \in (t_0 - \varepsilon, t_0)$. This ends the proof. $\square$

One may ask what happens when a strong or weak solution in the ill-posed direction (suction) breaks down? What will the typical boundary look like, and is it possible to continue the solution in any reasonable way beyond the break-down time? Compare [270, 278, 352].

One common point of view is that a strong solution in the ill-posed direction will typically break down through the development of an (inward) cusp on the boundary, since this is what happens with most explicit solutions. On the other hand one may start with any domain whatsoever and construct a weak solution in the well-posed direction with this domain as an initial domain. The result will, after a sufficiently long time (which can be estimated), be an almost circular domain with analytic boundary. Running time backwards it follows that starting with an almost circular domain, the ill-posed Hele-Shaw evolution from this may break down in virtually any way (cf. [139]). For example, the boundary of the terminal region may have Hausdorff dimension greater than one. Conjectures concerning the typical Hausdorff dimension have been proposed in the context of DLA and some estimates have been given, see for example [89, 481].

3.2.3 Squeezing version of Hele-Shaw flow and potential theoretic skeletons

A particularly nice form of Laplacian growth is the squeezing version. This means, in the Hele-Shaw model, that starting with an initial blob of a viscous fluid between two plates one simply squeezes the plates together, or (in the ill-posed direction) separates them slowly. Thus there are no particular sources, but the mathematical description of the problem is the same as that with a uniform source distribution spread out on the actual fluid region. In practise, Hele-Shaw problems of this type show up in lubrication theory [174], and has further been studied in [339, 519, 113, 415], for example.

In its strong form the squeezing problem can be stated as follows: given $\Omega \subset \mathbb{C}$, any bounded region with analytic boundary, one solves the Dirichlet problem

$$-\Delta p = 1 \quad \text{in } \Omega, \tag{3.38}$$

$$p = 0 \quad \text{on } \partial\Omega. \tag{3.39}$$

The solution $p = p_\Omega$ depends on Ω, and the dynamic condition for an evolution $t \mapsto \Omega(t)$ of domains is that one asks $\partial\Omega(t)$ to move with the speed

$$V_n = -\frac{\partial p}{\partial n} \tag{3.40}$$

in the outward normal direction.

Extending p by zero outside $\Omega(t)$ and using that (3.40) means that the distributional contribution to Δp on $\partial\Omega(t)$ is to be equated with $\frac{d}{dt}\chi_{\Omega(t)}$, the above gives that

$$\Delta p_{\Omega(t)} = \frac{d}{dt}\chi_{\Omega(t)} - \chi_{\Omega(t)} = e^t \frac{d}{dt}(e^{-t}\chi_{\Omega(t)}).$$

Integrating with respect to t we find that

$$\Delta u = e^{-t}\chi_{\Omega(t)} - e^{-s}\chi_{\Omega(s)},$$

where

$$u = \int_s^t e^{-\tau}p_{\Omega(\tau)}d\tau$$

serves as the Baiocchi transform in the present case. For $s < t$, $\Omega(s) \subset \Omega(t)$, hence $u = 0$ outside $\Omega(t)$. Moreover, $u \geq 0$ everywhere. From this it easily follows that, in the balayage terminology,

$$e^{-t}\chi_{\Omega(t)} = \text{Bal}\,(e^{-s}\chi_{\Omega(s)}, e^{-t})$$

when $s < t$. Equivalently

$$\chi_{\Omega(t)} = \text{Bal}\,(e^{t-s}\chi_{\Omega(s)}, 1). \tag{3.41}$$

If $\Omega(0)$ is given, then $\Omega(t)$ is obtained, for any $t > 0$, as

$$\chi_{\Omega(t)} = \text{Bal}\,(e^t\chi_{\Omega(0)}, 1).$$

If the chain $\Omega(t)$ also contains negative values of t, then these $\Omega(t)$ have to satisfy

$$\chi_{\Omega(0)} = \text{Bal}\,(e^{-t}\chi_{\Omega(t)}, 1) \quad (t < 0).$$

However, $\Omega(t)$ will not be uniquely determined by this equation alone, additional requirements have to be set up (cf. Theorem 3.2.2).

Now assume that we are in the very favorable case that, given $\Omega = \Omega(0)$, a global strong solution $\{\Omega(t) : -\infty < t < \infty\}$ of the squeezing Hele-Shaw problem

(3.38)–(3.40) exists. This is a relatively rare situation, but there are a number of interesting examples, like the family of confocal ellipses described in Example 3.2.1 below. The domains $\Omega(t)$ will then be simply connected because strong solutions are not allowed to change topology, and for large t the domains will in any case be simply connected (e.g., by the inner normal theorem, Theorem 4.7.2). For such a solution (3.41) will hold for any $-\infty < s < t < \infty$.

Consider the mass distributions (measures)

$$\mu(t) = e^{-t}\chi_{\Omega(t)}.$$

By (3.41) they are all "graviequivalent". Indeed, (3.41) is equivalent to

$$U^{\mu(s)} \leq U^{\mu(t)} \quad \text{in } \mathbb{C}, \tag{3.42}$$

$$U^{\mu(s)} = U^{\mu(t)} \quad \text{outside } \Omega(t) \tag{3.43}$$

$(s < t)$. We also have

$$\mu(t) \geq 0,$$

$$\operatorname{supp}\mu(t) = \overline{\Omega(t)},$$

$$|\Omega(t)| = e^t|\Omega(0)| \to 0 \quad \text{as } t \to -\infty.$$

It easily follows from these properties that, as $t \to -\infty$, the measures $\mu(t)$ converge weakly to some limit measure μ:

$$\mu(t) \rightharpoonup \mu, \quad t \to -\infty.$$

This limit measure satisfies

$$\mu \geq 0,$$

$$|\operatorname{supp}\mu| = 0.$$

We also get from (3.41) that

$$\chi_{\Omega(t)} = \operatorname{Bal}(e^t\mu, 1)$$

for all $-\infty < t < \infty$, in particular

$$\chi_{\Omega(0)} = \operatorname{Bal}(\mu, 1).$$

In addition to the above, since all domains $\Omega(t)$ are simply connected it follows that the closed limiting set $\operatorname{supp}\mu$ has the corresponding topological property of not disconnecting $\mathbb{C}$. The generic picture is that $\operatorname{supp}\mu$ consists of finitely many points and arcs.

It is natural to think of the above μ as a *potential theoretic skeleton* for $\Omega = \Omega(0)$. Sometimes terms like *mother body* or "maternal body" are used for such a skeleton, this kind of terminology originating from the Bulgarian geophysicist D. Zidarov, see [595]. In general, one may formulate natural requirements of such a skeleton μ for a domain Ω, or rather for the corresponding mass distribution $\chi_\Omega d\sigma$, as follows (recall (3.23)).

(i) $U^\Omega = U^\mu$ outside Ω,

(ii) $U^\Omega \le U^\mu$ in $\mathbb{C}$,

(iii) $\mu \ge 0$,

(iv) $|\operatorname{supp}\mu| = 0$,

(v) $\operatorname{supp}\mu$ does not disconnect any part of Ω from $\mathbb{C} \setminus \overline{\Omega}$.

The last requirement is formulated so that it takes into account the possibility that Ω is not simply connected. A more direct formulation of the requirement is to say that for any point $z \in \Omega \setminus \operatorname{supp}\mu$ there shall exist an arc in $\mathbb{C} \setminus \operatorname{supp}\mu$ connecting z to some point in $\mathbb{C} \setminus \overline{\Omega}$.

In the above list, the requirements are (i), (iv), (v) are the most crucial ones. The two inequalities (ii) and (iii) have more subtle roles, for example (ii) is needed in order for Ω to be uniquely determined by μ (by partial balayage), and if (iii) is abandoned then μ should at least be a real-valued measure, otherwise a more restrictive version of (iv) is needed in order to keep μ reasonably well-determined by Ω.

The simplest example of a potential theoretic skeleton satisfying all requirements above is of course the appropriate point mass at the center of any disk Ω. For classical quadrature domains, as described in the next section, most requirements of a mother body are satisfied by the functional as in the right member in (3.45) (below), provided $n_k = 1$ and $c_{k0} > 0$ for all k. Still, condition (ii) need not hold. The next example is the ellipse.

Example 3.2.1. Let Ω be the ellipse defined by

$$\frac{x^2}{a^2} + \frac{y^2}{b^2} < 1,$$

where $a > b > 0$. Then the measure μ on the focal segment $[-c, c]$ $(c = \sqrt{a^2 - b^2})$ defined by

$$d\mu = \frac{2ab}{c^2}\sqrt{c^2 - x^2}dx \quad (-c < x < c)$$

is the unique potential theoretic skeleton satisfying the above requirements [525].

The corresponding squeezing Hele-Shaw evolution consists of the family of ellipses which are confocal with the initial one $\Omega(0) = \Omega$. This means that $\Omega(t)$ is the ellipse

$$\frac{x^2}{a(t)^2} + \frac{y^2}{b(t)^2} < 1,$$

determined by

$$\begin{cases} a(t)^2 - b(t)^2 = c^2, \\ a(t)b(t) = e^t ab \quad (-\infty < t < \infty). \end{cases}$$

For further discussions of potential theoretic skeletons, see for example [595], [237], [238], [506].

3.3 Quadrature domains and the exponential transform

3.3.1 Quadrature domains

Closely related to partial balayage, and hence to Hele-Shaw flow, is the notion of a quadrature domain. Ideas related to quadrature domain theory have already been used implicitly in the previous sections. Here we shall spell out some basic definitions and thus make the connections more explicit. A few general references are [132], [7], [500], [517], [246], and an overview of applications of quadrature domains to problems in fluid dynamics can be found in [116].

If $\mu \geq 0$ is a measure with compact support, then a bounded domain $\Omega \subset\subset \mathbb{C}$ containing $\operatorname{supp} \mu$ is called a quadrature domain for subharmonic functions for μ if the inequality

$$\int_\Omega \varphi \, d\sigma \geq \int_\Omega \varphi \, d\mu \qquad (3.44)$$

holds for all integrable subharmonic functions φ in Ω. See [490]. Thus equation (3.2), says that $\Omega(t)$ is a quadrature domain for subharmonic functions for the measure $\mu(t) = \chi_{\Omega(0)} + t\delta_0$, and this is an equivalent way of expressing that the family of domains $\Omega(t)$ is a weak Hele-Shaw solution. In general, a domain Ω (if assumed saturated, see before Definition 3.1.1) is a quadrature domain for subharmonic functions if and only if $\operatorname{Bal}(\mu, 1) = \chi_\Omega$. If $\operatorname{Bal}(\mu, 1)$ is not of this form, i.e., if, referring to the general structure formula (3.28), there is also a remainder term $\mu\chi_{\mathbb{C}\setminus\Omega}$, then there exists no quadrature domain for subharmonic functions for μ.

In case φ is harmonic the inequality (3.44) becomes an equality, because both φ and $-\varphi$ are then subharmonic. Replacing the inequality sign (3.44) by equality we may also consider analytic (hence complex-valued) test functions. A particularly rich theory then arises for measures of the form $\mu = \sum_{k=1}^n c_k \delta_{z_k}$, i.e., for measures with support in a finite number of points. Allowing, more generally, μ to be an arbitrary distribution with support in a finite number of points one arrives at the following classical concept of a quadrature domain: a bounded domain Ω is called a (*classical*) *quadrature domain* if there exist finitely many points $z_1, \ldots, z_m \in \Omega$ and coefficients $c_{kj} \in \mathbb{C}$ ($0 \leq j \leq n_{k-1}$, $1 \leq k \leq m$, say), such that the *quadrature identity*

$$\frac{1}{\pi} \int_\Omega h \, d\sigma = \sum_{k=1}^m \sum_{j=0}^{n_k-1} c_{kj} h^{(j)}(z_k) \qquad (3.45)$$

holds for every integrable analytic function h in Ω. The integer $n = \sum_{k=1}^m n_k$ is then called the *order* of the quadrature identity (assuming $c_{k,n_k-1} \neq 0$).

Notions of quadrature domains and identities as above were introduced in the 1970s by Davis [132] and Aharonov and Shapiro [7]. For general developments after that, see, e.g., [517], [246]. Classical quadrature domains have algebraic boundaries (see below), and are for this reason also called *algebraic domains*, see [560] for example, even though they constitute just a rather small subclass of all domains having algebraic boundary. Lemniscate domains, as discussed in Section 2.3 for example, have algebraic boundaries but are not algebraic domains in the above sense (i.e., are not quadrature domains).

The relationship between classical quadrature domains and Hele-Shaw flow is immediate from the generalized moment property (1.35): if $\Omega(0)$ is a quadrature domain as in (3.45), then all domains $\Omega(t)$ in a Hele-Shaw evolution with injection or suction at the origin are quadrature domains as well. To be precise, if (3.45) holds for $\Omega(0)$ and the injection rate is Q, then the $\Omega(t)$ satisfy the quadrature identity

$$\frac{1}{\pi} \int_{\Omega(t)} h\, d\sigma = \sum_{k=1}^{m} \sum_{j=0}^{n_k-1} c_{kj} h^{(j)}(z_k) + \frac{Qt}{\pi} h(0) \tag{3.46}$$

for any t. Some basic characterizations of quadrature domains are given in the following theorem, where we use notations from Section 1.7.

Theorem 3.3.1. *Let $\Omega \subset \mathbb{C}$ be a bounded domain. Then the following are equivalent.*

(i) *Ω is a (classical) quadrature domain.*

(ii) *The exterior part $S_-(z)$ of the Cauchy transform (1.49), (1.50) of Ω is a rational function.*

(iii) *There exists a meromorphic function $S(z)$ in Ω, continuous up to $\partial\Omega$, such that*

$$S(z) = \bar{z} \quad \text{on } \partial\Omega. \tag{3.47}$$

If Ω is simply connected a further equivalent property is:

(iv) *Any Riemann mapping function $f : \mathbb{D} \to \Omega$ is a rational function.*

Assertion (iii) can be interpreted as saying that the pair of functions $(z, S(z))$ on Ω fit together to form a meromorphic function on the *Schottky double* of Ω. This is the compact Riemann surface obtained from Ω by welding along $\partial\Omega$ another copy of Ω, a 'backside', which is provided with the opposite conformal structure. Then $\overline{S(z)}$ is meromorphic when considered as being defined on the backside, and it matches continuously on $\partial\Omega$ with the holomorphic z on the frontside. Thus the pair $(z, S(z))$ as a whole defines a meromorphic function on the Schottky double. The same is true for the opposite pair $(S(z), z)$, and since two meromorphic functions on a compact Riemann surface always are algebraically related [183] it follows that there exists a (nontrivial) polynomial $Q(z, w)$ such that

$$Q(z, S(z)) = 0 \quad (z \in \Omega).$$

In particular,

$$Q(z, \bar{z}) = 0 \quad \text{for } z \in \partial\Omega, \tag{3.48}$$

i.e., $\partial\Omega$ is an algebraic curve. This result was first obtained by D. Aharonov and H.S. Shapiro [7] using an elegant direct argument. The degree of the polynomial $Q(z, w)$ in each of the variables z and w is exactly equal to the order of the quadrature identity. See [232], and also (3.57) below, for some details on the structure of $Q(z, w)$.

Clearly the function $S(z)$ will be the Schwarz function of $\partial\Omega$. The curve (3.48) may have certain singular points, namely such points which in the simply connected case are images $f(\zeta_0)$ of points $\zeta_0 \in \partial\mathbb{D}$ such that f is not univalent in a full neighbourhood of ζ_0 ($f'(\zeta_0) = 0$, or $f(\zeta_1) = f(\zeta_0)$ for another $\zeta_1 \in \partial\mathbb{D}$). Away from these singular points $S(z)$ is analytic in a full neighbourhood of $\partial\Omega$, hence is a true Schwarz function. At singular points $S(z)$ is only a "one-sided Schwarz function". The number of singular points, plus the genus of the Schottky double (i.e., the connectivity of Ω minus one), is bounded from above by $(n-1)^2$, where n is the order of the quadrature identity. This estimate can be sharpened by using techniques from complex dynamics, cf. [307], [353]. An index theorem for singular points is given in [501].

A beautiful generalization of (iv) in the theorem to multiply connected domains has been given by D. Yakubovich [592], saying that Ω is a quadrature domain if and only if $\partial\Omega$ is traced out (on $\partial\mathbb{D}$) by the eigenvalues of some rational normal matrix function $F : \mathbb{D} \to \mathbb{C}^{N \times N}$ which is holomorphic in $\mathbb{D}$. Another generalization says that, after having selected an arbitrary model surface D of the desired conformal type, a univalent function f defined on D maps D onto a quadrature domain if and only if f extends to a meromorphic function on the Schottky double of D. In fact, this is exactly what (3.47) amounts to when pulled back to D by f. The meromorphic extension of f on the backside of D will be represented by $\overline{S \circ f}$. Choosing $D = \mathbb{D}$ in the simply connected case this exactly gives that f is rational, since the double of $\mathbb{D}$ is the Riemann sphere and the meromorphic functions there are the rational functions.

Proof. (i) implies (ii): Just choose $h(\zeta) = \frac{1}{\zeta - z}$ for $z \in \mathbb{C} \setminus \Omega$ in the quadrature identity.

(ii) implies (iii): Assuming some regularity of $\partial\Omega$ we simply define $S(z)$ by the formula (1.47) Since $S_-(z)$ is rational (by assumption) and $S_+(z)$ (see (1.47)) is always holomorphic in all of Ω, $S(z)$ then is meromorphic in Ω and the statement follows.

(iii) implies (i): Using the residue theorem we have, when $S(z)$ is meromorphic in Ω and h is analytic,

$$\frac{1}{\pi} \int_\Omega h \, d\sigma = \frac{1}{2\pi i} \int_\Omega h(z) \, d\bar{z} dz = \frac{1}{2\pi i} \int_{\partial\Omega} h(z) \bar{z} dz$$

$$= \frac{1}{2\pi i} \int_{\partial\Omega} h(z) S(z) dz = \sum_{z \in \Omega} \operatorname{Res} h(z) S(z),$$

which is a quadrature identity of the form (3.45).

(iii) implies (iv): In the presence of $S(z)$ any conformal map $f : \mathbb{D} \to \Omega$ can be extended to the Riemann sphere by

$$\overline{f(1/\bar{\zeta})} = S(f(\zeta)) \tag{3.49}$$

for $\zeta \in \mathbb{D}$ (cf. the proof of Theorem 3.1.2). This makes f meromorphic in $\overline{\mathbb{C}}$, hence rational.

(iv) implies (iii): If f is rational we can define $S(z)$ for $z = f(\zeta) \in \Omega$ by (3.49) and it is easy to see that it becomes meromorphic in Ω with $S(z) = \bar{z}$ on $\partial\Omega$. □

It is clear from the above proof that when Ω is a quadrature domain the relationship between the data in (3.45) and the poles of $S_-(z)$, $S(z)$ and $f(\zeta)$ is as follows (recall that $S_+(z)$ is holomorphic in Ω):

$$S_-(z) = \sum_{k=1}^{m} \sum_{j=0}^{n_k-1} \frac{j! c_{kj}}{(z - z_k)^{j+1}},$$

$$S(z) = \sum_{k=1}^{m} \sum_{j=0}^{n_k-1} \frac{j! c_{kj}}{(z - z_k)^{j+1}} + S_+(z), \tag{3.50}$$

$$f(\zeta) = \sum_{k=1}^{m} \sum_{j=0}^{n_k-1} \frac{b_{kj}}{(\zeta - 1/\bar{\zeta}_k)^{j+1}} + \text{regular}. \tag{3.51}$$

Here $\zeta_k \in \mathbb{D}$ are the points which are mapped onto the quadrature nodes: $f(\zeta_k) = z_k$. If $\zeta_k = 0$ for some k, this should be interpreted as $f(\zeta)$ having a pole at infinity of order n_k, and the corresponding part (the sum over j for that particular value of k) of the expression for $f(\zeta)$ takes the form

$$\sum_{j=0}^{n_k-1} b_{kj} \zeta^{j+1}.$$

The regular term in (3.51) must in fact be constant, since f is rational and all singularities have been accounted for in (3.51). A particular case is that $f(\zeta)$ is a polynomial, in which case $m = 1$, $\zeta_1 = 0$. Then the only quadrature node is $z_1 = 0$, and the order of the quadrature identity equals the degree of the polynomial, namely $n = n_1$.

The expressions for the coefficients b_{kj} in terms of c_{kj} and z_k are somewhat complicated because of the nonlinear nature of (3.49) as an equation for f, but in principle they can be derived by combining (3.49) and (3.50).

3.3.2   Hele-Shaw flow in the light of quadrature domains

Quadrature domain theory is in many cases helpful for understanding properties of Hele-Shaw evolutions, and also for construction of explicit solutions. For example, Theorem 3.3.1 gives another proof of the fact that the Polubarinova–Galin equation preserves rational mapping functions (Theorem 2.2.1). Indeed, if $f(\zeta, 0)$ is a rational function, then $\Omega(0)$ is a quadrature domain, say satisfies (3.45). Hence all the $\Omega(t)$ are quadrature domains as in (3.46), and therefore $f(\zeta, t)$ is rational for every t. In addition, from the above relationships between the poles of $f(\zeta, t)$ and the quadrature data $\{z_k\}$, $\{c_{kj}\}$, which by (3.46) remain fixed during the Hele-Shaw evolution (except for the coefficient at $z = 0$), it becomes clear that the poles of $f(\zeta, t)$ cannot collide or disappear. The only possible exception here is the pole at infinity which, being linked to the source/suction point $z = 0$, may disappear for one value of t.

Let us next revisit the first example in Section 2.1.2, and try to explain why it is possible, in the suction case, to have a rational solution $f(\zeta, t)$ of the Polubarinova–Galin equation such that the free boundary reaches the sink, whereas this is not possible in the pure polynomial case.

In Figure 2.2 the residual fluid domain after suction of fluid at the origin is the disk $\Omega = \{|z + 1| < 1\}$, for which the quadrature identity

$$\frac{1}{\pi} \int_\Omega h \, d\sigma = h(-1)$$

holds. This occurs at time $t = -\pi/Q$ in the example. From this disk the whole Hele-Shaw family can be recovered by injecting fluid at $z = 0$. This gives a family of quadrature domains $\Omega(t)$ with quadrature nodes at $z = -1$ (for the original disk) and $z = 0$ (due to injection there). Letting the time parameter t be the same as in Section 2.1.2 the quadrature identities are

$$\frac{1}{\pi} \int_{\Omega(t)} h \, d\sigma = h(-1) + \left(1 + \frac{Qt}{\pi}\right) h(0), \tag{3.52}$$

$0 \le t \le -\pi/Q$. The corresponding mapping functions $f(\zeta, t)$ will be the rational functions (2.3) with one pole at infinity, corresponding to the quadrature node $z = 0$, and one finite pole $\zeta = 1/c(t)$ having the property that the reflected point $\zeta = c(t)$ is mapped onto the other quadrature node $z = -1$. All this is perfectly fine, and the same argument can be used to show that rational solutions of the Polubarinova–Galin equation may end up in virtually any simply connected quadrature domain Ω with $0 \in \partial\Omega$.

However, these rational solutions can never be of pure polynomial type, because for polynomial solutions the fluid domains will satisfy quadrature identities of the kind

$$\frac{1}{\pi} \int_{\Omega(t)} h \, d\sigma = \left(c_0 + \frac{Qt}{\pi}\right) h(0) + c_1 h'(0) + \cdots + c_{n-1} h^{(n-1)}(0),$$

and then $z = 0$ can never be on $\partial\Omega(t)$. The corresponding mapping functions will be of the form $f(\zeta, t) = a_1(t)\zeta + \cdots + a_n(t)\zeta^n$ with $a_1(t) > 0$. An important remark

here is that the quadrature identity remains valid for the limiting domain in the
Hele-Shaw evolution, even if the Polubarinova–Galin solution breaks down there.
Notice also that the above coefficient of $h(0)$ equals the normalized area of $\Omega(t)$
(choose $h = 1$), hence vanishes only if all fluid has been sucked, which occurs only
in the shrinking disk case.

3.3.3   More general quadrature domains

Rational conformal maps thus map onto classical quadrature domains, or algebraic domains, satisfying identities of the form (3.45). Similarly, conformal maps
involving also logarithmic singularities, as studied in Section 2.4, map onto *Abelian
domains* [560]. These admit quadrature identities for which the right-hand member, besides having terms of the type appearing in (3.45), also have terms involving
line integrals of the type

$$\sum_j c_j \int_{\gamma_j} h(z)dz,$$

where c_j are complex constants and the γ_j are line segments in Ω. Even more
general quadrature domains, including line integrals $\int_\gamma h(z)\rho(z)dz$ with ρ a fairly
arbitrary density, have turned out to be useful for modeling fingering phenomena
in Hele-Shaw flow, see [1]. The corresponding solutions (for particular choices of
ρ) have been named *multi-cut solutions*.

   In the context of Hele-Shaw flow in exterior domains, as discussed in Section 1.8, unbounded quadrature domains are appropriate. These are defined in the
same way as bounded quadrature domains, namely that a relation (3.45) shall hold
for all integrable analytic functions h in the domain. See [500], [516], [514], [297],
[502] for the theory of unbounded quadrature domains in general. Considering
only the case that the complement of Ω is bounded (hence compact), it is natural
to consider Ω as a domain in the Riemann sphere with $\infty \in \Omega$. The integrability
of an analytic function h in Ω then requires that $h(z) = O(|z|^{-3})$ as $z \to \infty$. Thus
the test class of analytic functions is somewhat restricted, which opens up for the
existence of *null quadrature domains*, i.e., domains Ω for which

$$\int_\Omega h d\sigma = 0$$

holds for every integrable analytic h. Such domains have been completely classified
in [499]. Those with $\infty \in \Omega$ are simply the complements of ellipses (and disks),
disregarding some degenerate cases for which $\mathbb{C} \setminus \Omega$ has no interior points.

   As for the Hele-Shaw evolution of an unbounded quadrature domain with
a source/sink at infinity, equation (3.46) still holds, with $h(0)$ replaced by $h(\infty)$.
However, $h(\infty) = 0$ due to the integrability requirement, so this term actually disappears, which means that the quadrature identity remains completely unchanged
under the evolution. For example, a null quadrature domain remains a null quadrature domain. If a domain Ω, with $\infty \in \Omega$, can be completely emptied by suction

at infinity, then it has to be a null quadrature domain, hence $\mathbb{C} \setminus \Omega$ has to be an ellipse (if it has non-empty interior). See [279], and, for multi-dimensional cases, [140], [194].

An interesting observation is that if, in (3.45), one replaces the planar area measure $d\sigma$ by the spherical area measure, $d\sigma_z/(1 + |z|^2)^2$, the class of bounded quadrature domains remains exactly the same. In fact, in both cases Ω is a quadrature domain if and only if statement (iii) in Theorem 3.3.1 holds. Assuming that $S(z)$ is meromorphic in Ω, the calculus for the step "(iii) implies (i)" in the proof of Theorem 3.3.1 runs in the spherical case as follows:

$$\frac{1}{\pi} \int_\Omega h(z) \frac{d\sigma_z}{(1 + |z|^2)^2} = \frac{1}{2\pi i} \int_\Omega h(z) \frac{d\bar{z}dz}{(1 + z\bar{z})^2} = \frac{1}{2\pi i} \int_\Omega d\left(\frac{h(z)\bar{z}dz}{1 + z\bar{z}}\right)$$

$$= \frac{1}{2\pi i} \int_{\partial\Omega} \frac{h(z)\bar{z}dz}{1 + z\bar{z}} = \sum \text{Res}_\Omega \frac{h(z)S(z)dz}{1 + zS(z)}.$$

This becomes a quadrature identity when the residues are evaluated.

When using the spherical metric the point of infinity is no more special than any other point, hence unbounded quadrature domains can be treated on the same footing as bounded quadrature domains. For Hele-Shaw flow, one runs into considering Hele-Shaw flow on Riemannian manifolds, or weighted Hele-Shaw flow. See further remarks in Section 3.5.2. Similar remarks apply to the hyperbolic metric (Poincaré metric), with area element $d\sigma_z/(1 - |z|^2)^2$, in case $\Omega \subset \mathbb{D}$.

3.3.4 The exponential transform

Recall the definition (1.49) of the Cauchy transform, which written in the form

$$C_\Omega(z = \frac{1}{2\pi i} \int_\Omega \frac{d\zeta}{\zeta - z} \wedge d\bar{\zeta}$$

suggests a natural polarization to a "double Cauchy transform", defined by

$$C_\Omega(z, w) = \frac{1}{2\pi i} \int_\Omega \frac{d\zeta}{\zeta - z} \wedge \frac{d\bar{\zeta}}{\bar{\zeta} - \bar{w}} = -\frac{1}{\pi} \int_\Omega \frac{d\sigma_\zeta}{(\zeta - z)(\bar{\zeta} - \bar{w})}.$$

The exponential of it,

$$E_\Omega(z, w) = \exp C_\Omega(z, w),$$

is known as the *exponential transform* of Ω (any bounded domain). It originally arose in operator theory, see [88], but later turned out to be relevant in the theory of quadrature domains, [456], [457], [458], and it also provides a tool for proving regularity of free boundaries, see [240]. We shall briefly explain some of these connections.

Clearly, $E(z, w) = E_\Omega(z, w)$ is holomorphic in z, antiholomorphic in w, when the variable in question is outside Ω, and it is not hard to check that the *interior exponential transform*, defined by

$$H(z,w) = -\frac{\partial^2 E(z,w)}{\partial \bar{z} \partial w} \quad (z,w \in \Omega),$$

is similarly holomorphic/antiholomorphic in Ω. Furthermore, one has

$$H(z,w) = \frac{E(z,w)}{|z-w|^2} \quad (z,w \in \Omega),$$

$|E(z,w)| \leq 2$ everywhere, and $E(z,w) \to 1$ as $z \to \infty$ or $w \to \infty$. All these statements are elementary, and full details are given in [240] (for example). From the connections to operator theory in a Hilbert space [375], or by direct arguments [242], it follows that $H(z,w)$ is positive definite in the sense that

$$\sum_{k,j} H(z_k, z_j)\lambda_k \bar{\lambda}_j \geq 0 \tag{3.53}$$

for every finite sequences $z_k \in \Omega$, $\lambda_k \in \mathbb{C}$, with equality holding only if all the λ_k are zero.

From the above it follows that the function $1 - E(z,w)$ can be represented as a double Cauchy integral,

$$1 - E(z,w) = \frac{1}{\pi^2} \int_\Omega \int_\Omega H(u,v) \frac{d\sigma_u}{u-z} \frac{d\sigma_v}{\bar{v}-\bar{w}} \quad (z,w \in \mathbb{C}), \tag{3.54}$$

and also that $1 - E(z,w)$ is positive semidefinite. It need not be strictly definite, in fact, it is indefinite exactly if Ω is a quadrature domain (see again [242]).

Assuming some minor regularity of $\partial\Omega$, the integral representation (3.54) can be written

$$1 - E(z,w) = \frac{1}{4\pi^2} \int_{\partial\Omega} \int_{\partial\Omega} H(u,v) \frac{\bar{u}du}{u-z} \frac{vd\bar{v}}{\bar{v}-\bar{w}}, \tag{3.55}$$

and assuming in addition that $\partial\Omega$ admits a Schwarz function,

$$1 - E(z,w) = \frac{1}{4\pi^2} \int_{\partial\Omega} \int_{\partial\Omega} H(u,v) \frac{S(u)du}{u-z} \frac{\overline{S(v)}d\bar{v}}{\bar{v}-\bar{w}}. \tag{3.56}$$

Since, in equation (1.48), $S_+(z)$ is analytic in Ω, it follows that $S(z)$ is analytic as far into Ω as $S_-(z)$ has an analytic continuation. Recall also, (1.50), that $S_-(z) = C_\Omega(z)$ for $z \in \mathbb{C} \setminus \overline{\Omega}$ and that therefore $S_-(z)$ represents the analytic continuation of $C_\Omega(z)$ into Ω, as far as it exists. Using the above formulas, and the analyticity/antianalyticity of $H(u,v)$, one therefore arrives at the following result.

Theorem 3.3.2. *Assume that the exterior Cauchy transform $S_-(z)$ has an analytic continuation across $\partial\Omega$ to some compact set $K \subset \Omega$, equivalently that the Schwarz function $S(z)$ is analytic in $\Omega \setminus K$. Then also $E(z,w)$ has an analytic/antianalytic continuation, in both variables, across $\partial\Omega$, to $K \times K$.*

The theorem is valid for any bounded domain Ω, but without assuming some light regularity of $\partial\Omega$, as above, the proof becomes much more technical (see [491], [240]).

One major importance of the above theorem is that it directly shows that any boundary which admits analytic continuation of the exterior Cauchy transform has to be analytic. In fact, under the assumptions in the theorem, denote by $F(z,w)$ the analytic/antianalytic continuation of $E(z,w)$ assured by the theorem, so that $F(z,w)$ is analytic in z, antianalytic in w, whenever $z,w \in \mathbb{C}\setminus K$, and $F(z,w) = E(z,w)$ whenever $z,w \in \mathbb{C}\setminus\Omega$.

From the definition of $E(z,w)$ we see that, on the diagonal,

$$E(z,z) = \exp\left[-\frac{1}{\pi}\int_\Omega \frac{d\sigma_\zeta}{|\zeta - z|^2}\right] > 0$$

for $z \in \mathbb{C}\setminus\overline{\Omega}$. When $z \in \partial\Omega$ the integral above diverges (equals $+\infty$), and hence $E(z,z) = 0$, except if Ω is very thin at z. The latter does not occur if the assumptions in Theorem 3.3.2 are satisfied (see [240]), hence it follows that $F(z,z)$, which is real-analytic in $\mathbb{C}\setminus K$, satisfies $F(z,z) > 0$ outside Ω, $F(z,z) = 0$ on $\partial\Omega$.

Corollary 3.3.1. *Under the assumptions of the theorem, the boundary $\partial\Omega$ is a real-analytic variety (possibly having some singular points), being defined in terms of the analytic continuation $F(z,w)$ of $E(z,w)$ by the equation*

$$\partial\Omega: \quad F(z,z) = 0.$$

The corollary applies to weak solutions of the Hele-Shaw problem because the exterior Cauchy transform of $\Omega(t)$ ($t > 0$) then does have an analytic continuation to $K = \overline{\Omega(0)}$. Thus the corollary provides another avenue to Theorem 3.1.2.

Now we wish to spell out in detail what $E(z,w)$ and $F(z,w)$ look like in case Ω is a quadrature domain. Assuming that the quadrature identity (3.45) holds, the representation (3.54) reduces, when z,w are outside Ω, to a finite sum, in other words $1 - E(z,w)$ becomes a rational function (outside Ω). This rational function then is $1 - F(z,w)$. The quadrature nodes $z_1,\ldots,z_m$ will be its poles, and setting $P_n(z) = \prod_{k=1}^m (z - z_k)^{n_k}$ one finds the exact expression

$$F(z,w) = \frac{Q(z,\bar{w})}{P_n(z)\overline{P_n(w)}},$$

where $Q(z,w) = \sum_{k,j=0}^n a_{kj}z^k w^j$ is the same polynomial as in (3.48) (if both are normalized so that $a_{nn} = 1$). In particular, the equation in Corollary 3.3.1 becomes effectively $Q(z,\bar{z}) = 0$, i.e., the same as equation (3.48), as expected.

Taking into account the positive semidefiniteness of $1 - E(z,w)$, and the behavior at infinity, one concludes that the structure of $Q(z,w)$ is such that $1 - F(z,w)$ has the decomposition

$$1 - F(z,w) = \sum_{k=0}^{n-1} \frac{P_k(z)\,\overline{P_k(w)}}{P_n(z)\,\overline{P_n(w)}},$$

where each $P_k(z)$ is a polynomial of degree k (exactly). It follows that the equation for the boundary can be written in the more precise form

$$\partial\Omega: \quad |P_n(z)|^2 = \sum_{k=0}^{n-1} |P_k(z)|^2. \tag{3.57}$$

Here $P_n(z)$ is as before, and $P_{n-1}(z)$ is (essentially) the numerator in the exterior Cauchy transform (1.50), specifically

$$S_-(z) = \sqrt{\frac{|\Omega|}{\pi}} \frac{P_{n-1}(z)}{P_n(z)} = \sum_{k=1}^{m} \sum_{j=0}^{n_k-1} \frac{j! c_{kj}}{(z - z_k)^{j+1}},$$

simultaneously the singular part of the Schwarz function $S(z)$. These statements follow from formulas in Section 3.3.1.

The remaining polynomials, $P_k(z)$ for $0 \le k \le n-2$, are only implictly related to the quadrature date, and are difficult to get hold of in general. The difference in information content between the usual Cauchy transform $C_\Omega(z)$ and the double Cauchy transform $C_\Omega(z, w)$, for z and w outside the domain, or between $S_-(z)$ and $F(z, w)$, is that the first contains explicit information only of the quadrature data, or the two polynomials $P_n(z)$, P_{n-1}, whereas the latter contains complete information of the domain (all the polynomials $P_k(z)$, or $Q(z, w)$).

Example 3.3.1. In the case of the disk, $\Omega = \mathbb{D}(0, r)$, we have

$$1 - F(z, w) = \frac{r^2}{z\bar{w}}, \quad S_-(z) = \frac{r^2}{z}, \quad S_+(z) = 0.$$

Example 3.3.2. For the two point quadrature domain Ω defined by

$$\mathrm{Bal}\,(\pi r^2(\delta_{-1} + \delta_{+1}), 1) = \chi_\Omega,$$

where $r > 1$, we have

$$Q(z, w) = z^2 w^2 - z^2 - w^2 - 2r^2 zw,$$

$$P_2(z) = z^2 - 1, \quad P_1(z) = \sqrt{2}rz, \quad P_0(z) = 1,$$

$$1 - F(z, w) = \frac{1 + 2r^2 z\bar{w}}{(z^2 - 1)(\bar{w}^2 - 1)},$$

$$S_-(z) = \frac{2r^2 z}{z^2 - 1} = \frac{r^2}{z+1} + \frac{r^2}{z-1}, \quad S(z) = \frac{z}{z^2 - 1}\left(r^2 + \sqrt{z^2 + r^4 - 1}\right).$$

The above domain Ω has been called [142], [359] the "smash sum" of the disks $\mathbb{D}(\pm 1, r)$, because it can also be obtained as

$$\mathrm{Bal}\,(\chi_{\mathbb{D}(-1,r)} + \chi_{\mathbb{D}(+1,r)}, 1) = \chi_\Omega.$$

This follows from general properties of balayage, for example (3.25). In [359] it is shown rigorously that it is also the continuum limit of the appropriate internal DLA process (and some other processes).

3.4 Weak solutions for cusp formation

3.4.1 Local approximation of the complex curve near a 3/2 cusp

In this section, we present a weak solution for the critical case of cusp formation. The solution is weak both in the sense explained above, and in the sense of hyperbolic partial differential equations, which may be allowed to develop finite discontinuities across curves in $\mathbb{C}$, generically known as *shocks*.

For the purpose of this discussion, the domain is the exterior domain, assumed to develop a 3/2 cusp at a critical time t_c. Given the manifest self-similarity of the boundary when approaching a cusp singularity, the spectral curve is approximated with a generic elliptic curve, with four branch points set for convenience at $\infty, e_{1,2,3}$, such that $e_1 = \bar{e}_2, e_1 + e_2 + e_3 = 0$. Moreover, the Laplacian growth time is measured from the critical value t_c, i.e., with respect to $T \equiv t - t_c$, the curve will become fully self-similar both in the spatial and time variables.

Since the flow near a singular finger tip does not depend on details of a boundary away from the tip of the finger, it is sufficient to consider a "finger" with a symmetry axis. In Cartesian coordinates aligned with a finger axis, the equation for the boundary is given by a curve (in $\mathbb{R}^2$) $\pm y(x)$. The *height* function $Y(X, T)$ is defined as an analytic function on a finite part of the complex plane $X = x + iy$ cut along the symmetry axis of the finger, such that the boundary values are real and equal to the graph of the finger $Y(X)_{X=x\pm i0} = \pm y(x, T)$. Then the Darcy law (cf. (1.39), (1.45), (1.54)) can be expressed through the height function:

$$\partial_T Y = -\partial_X \phi, \qquad (3.58)$$

where (up to constant factors) the analytic function $\phi(X) = \psi + ip$ is the complex potential of the flow, ψ is the stream function, and p is the pressure.

The height function $Y(X, T)$ defines a complex curve (or Riemann surface) evolving in time. Since at real $X = x$ the height $Y = \pm y(x)$ is real, the curve has the property $\overline{Y(X)} = Y(\overline{X})$.

The flow potential $\phi(X)$ has an important interpretation. Darcy's law states that ϕ is a local parameter of the curve. This means that the potential $\phi(X)$ is an analytic function. Locally it can be inverted, so that X, and other physical quantities, say, the height function, become functions of ϕ. Then Darcy's law (3.58) states that functions $(Y(\phi), X(\phi))$ treated as functions of the potential, are regular and single-valued. In the case of critical flow, when (Y, X) is hyperelliptic, the variable ϕ covers the physical plane twice: ϕ and $-\phi$ correspond to the same X, such that the double-valued function $Y(X)$ becomes a single-valued function $Y(\phi)$ on the double covering.

The fact that the potential is a local parameter of the curve is the essence of the Darcy law. The law can be formulated entirely in terms of the curve: the curve (Y, X), with a local parameter ϕ, moves in time such that the differential $id\mathcal{U} = Y \, dX + \phi \, dT$ is closed. In this form Darcy's law becomes identical to Whitham equations describing slow modulation of fast oscillating non-linear waves [590, 191, 325]. This fact was recognized in [326].

One consequence is the form taken by the Polubarinova–Galin equation ([199], and Sections 1.4.2 and 1.6 above): treating Y and X as functions of the complex potential ϕ, the Darcy law reads

$$\{X, Y\} = 1, \tag{3.59}$$

where the brackets are defined via $\{X, Y\} = \partial_T X \partial_\phi Y - \partial_\phi X \partial_T Y$ and time derivative is taken at fixed ϕ.

In the following, we will describe the flow before and up to cusp formation through the generating function

$$\mathcal{U}(X) = -i \int_e^X Y \, dX, \tag{3.60}$$

where the integration starts at the tip of the finger e and goes through the fluid. Then the Darcy law reads

$$\partial_T \mathcal{U} = i\phi. \tag{3.61}$$

3.4.2   Integral form of the Darcy law

Let us integrate (3.58) over a closed path B in the fluid:

$$\frac{d}{dT} \mathrm{Im} \oint_B d\mathcal{U} = \oint_B \mathbf{V} \times d\boldsymbol{\ell} = \oint_B d\psi, \tag{3.62}$$

$$\frac{d}{dT} \mathrm{Re} \oint_B d\mathcal{U} = -\oint_B \mathbf{V} \cdot d\boldsymbol{\ell} = -\oint_B dp. \tag{3.63}$$

The imaginary part measures a flux of fluid through the closed path. If the path goes around a drain (infinity), the integral $\mathrm{Im} \oint_B d\mathcal{U}$ counts a mass of fluid drained up to time T, equal to QT. In a canonical anti-clockwise orientation this reads

$$i \frac{d}{dT} \oint_\infty d\mathcal{U} = Q > 0. \tag{3.64}$$

The real part measures circulation along the path. It follows from the differential form of the Darcy law (3.58), valid before a singularity, that the fluid is irrotational $\nabla \times \mathbf{V} = 0$ and therefore $\frac{d}{dt} \mathrm{Re} \oint_B d\mathcal{U} = 0$. Therefore $\mathrm{Re} \oint_B d\mathcal{U} = 0$ is a constant. Provided that the constant does not depend on the choice of the path, the constant can be determined by taking the integral around a drain where $\oint_\infty d\mathcal{U}$ is purely imaginary as in (3.64). Thus $\mathrm{Re} \oint_B d\mathcal{U} = 0$ on any contour in the fluid.

If there are shocks, the height function and $d\mathcal{U}$ jump through a shock. The condition $\mathrm{Re} \oint_B d\mathcal{U} = 0$ for a path crossing a shock must be understood as a sum of the integral over a path in the fluid plus an oriented jump of $d\mathcal{U}$ at points where the path crosses a shock. A proper way to write this condition is to extend the integration over the Riemann surface (Y, X) where the height function and the differential $d\mathcal{U}$ are smooth. The physical plane where the height function suffers a discontinuity consists of patches of different sheets of the curve. Then the above condition reads

$$\text{Re} \oint_B d\mathcal{U} = 0, \quad \text{any cycle on the curve.} \tag{3.65}$$

Curves with condition (3.65) are rather restrictive. We call them Krichever–Boutroux curves, since they were first employed a century ago by Boutroux in his study of isomonodromy invariance of Painlevé transcendents, and later by Krichever in a far-reaching generalization.

Equations (3.64), (3.65) combined give a compact and complete formulation of the Hele-Shaw problem: find an evolution of a real Krichever–Boutoux curve (condition (3.64)), of a given degree, with respect to its residue (time T) at a marked point (condition (3.64)).

Contrary to the differential form of the Darcy law, equations (3.64), (3.65) extend beyond singularities. They admit a weak solution with shocks. Shocks are curved lines where patches of the complex curve covering the physical plane are joined together. At these lines, the height function jumps. Condition (3.65) being specified for a contour going along both sides of a shock, reads

$$\text{Re discc}\,\mathcal{U}|_{\text{shocks}} = 0. \tag{3.66}$$

These conditions determine position and motion of shocks, as we show in the next section.

3.4.3 Formation of the 3/2-singularity: self-similar elliptic curve

In this case the finger parametrization is especially simple: Y^2 is a polynomial of the third degree – an elliptic curve having one branch point at infinity. Fixing a scale and the origin, the curve reads:

$$-Y^2 = 4X^3 - g_2 X - g_3 = 4(X - e)(X - e_2)(X - e_1), \tag{3.67}$$

where g_2 and g_3 are real time-dependent coefficients. One of the branch points of the curve (3.67) may always be chosen real. We denote it as e. It is the tip of a finger, where $y(e) = 0$. The other two are conjugated $e_1 = \bar{e}_2$ (condition (3.65) excluding a possibility of real values for e_1 and e_2 – in this case disc $d\mathcal{U}$ through the branch cut connecting e_1 and e_2 is real and does not change sign, so that Re $\oint_{e_1,e_2} d\mathcal{U}$ cannot vanish). This is a real elliptic curve with a negative discriminant $g_2^3 - 27g_3^2 < 0$. The coefficient g_2 is determined by the drain rate. Equation (3.64) gives $Q \sim \dot{g}_2$. We set the rate Q such that

$$g_2 = -12T. \tag{3.68}$$

In this setting the time is counted from a cusp event $-Y^2 = 4X^3$ occurring at a critical time $T = 0$. The real coefficient g_3 is determined by the condition (3.65).

This is the simplest curve arising in singular flow. Its distinctive property is self-similarity. Thus it follows from (3.64), (3.65) that

$$Y(X,T) = |T|^{3/4} Y\left(|T|^{-1/2}X, 1\right),   \tag{3.69}$$

where $Y(X,1)$ is a unique universal function with no free parameters.

The scaling property alone is sufficient to express $\mathcal{U}$ through the height function and potential:

$$\mathcal{U}(X) = -i \int_e^X Y\, dX = -\frac{2i}{5}(XY - 2T\phi).   \tag{3.70}$$

We briefly describe this curve in the following subsection.

3.4.4   Smooth flow and a degenerate curve

At $T < 0$, before the singularity, the flow is smooth, there are no discontinuities. This follows also from condition (3.65). As $g_2 > 0$ the degenerate elliptic curve has vanishing discriminant, $g_2^3 = 27g_3^2$. The only unknown coefficient is thus determined, $g_3 = -8(-T)^{3/2}$. Degeneration of the curve means that the two branch points coincide to a real double point: $e_1 = e_2 = +\sqrt{-T} > 0$. Then $e = -2e_1 = -2\sqrt{-T}$. The degenerate curve

$$Y^2 = -4\left(X - e\right)\left(X + \frac{e}{2}\right)^2, \quad e = -2\sqrt{-T}, \, T < 0   \tag{3.71}$$

represents a torus with one pinched cycle. The real period of the curve is infinite, the imaginary half-period is $\int_e^{-\infty} \frac{dX}{Y} = \pi i\sqrt{12}(-T)^{-1/4}$. This is a general situation – a continuous flow corresponds to a degenerate curve where all cycles but one are pinched (the number of non-degenerate cycles is equal to the connectivity of the fluid boundary [326]). The condition that all cycles located in the bulk of the fluid are pinched follows from incompressibility of the fluid. Only in this case the fluid potential ϕ is analytic.

Alternatively, we notice two holomorphic functions $Y(\phi)$ and $X(\phi)$ with an asymptote $Y^2 \sim X^3$ at large X are polynomials of the third and second degree, respectively. The Darcy law in the form of (3.59) determines the polynomials

$$X = -\left(\frac{\phi}{6}\right)^2 + e(T), \quad Y = 2\left(\frac{\phi}{6}\right)^3 - 3e(T)\left(\frac{\phi}{6}\right),   \tag{3.72}$$

and yields to a differential equation for time-dependence of the coefficients

$$e\partial_T e = -2,   \tag{3.73}$$

which prompts (3.71). The real solution exists only at $T < 0$.

3.5 Generalizations

3.5.1 Multidimensional Hele-Shaw flow and other generalizations

The governing law for the evolution of the fluid boundary in a Hele-Shaw cell makes sense, in its mathematical form, in any number of dimensions. In three space dimensions there are then other physical models. Some examples are porous medium flow [439, 440], electro-chemical deposition [571], dendritic growth, growth processes in biology [46, 388, 421].

The classical formulation of the zero surface tension version in any number of dimensions is that a family of bounded domains $\Omega(t)$, smooth both in time and space, makes up a Hele-Shaw evolution with a point source of unit strength at a if the boundary $\partial\Omega(t)$ moves in the normal direction with speed equal to the normal derivative of the Green function for the Laplacian, $G_{\Omega(t)}(x, a)$. A convenient way of formulating this moving boundary condition is to say that for any smooth test function φ the identity

$$\frac{d}{dt} \int_{\Omega(t)} \varphi(x) dx = - \int_{\partial\Omega(t)} \varphi(x) \frac{\partial G_{\Omega(t)}(x, a)}{\partial n} ds \qquad (3.74)$$

is to hold, where ds is surface measure on $\partial\Omega(t)$ and the Green function is normalized so that $-\Delta G_{\Omega(t)}(\cdot, a) = \delta_a$ in the sense of distributions (cf. [554]). Certainly, conformal mapping and other complex variable methods do not apply in the higher-dimensional case, but the potential theory tools work practically unchanged. For example, the construction and basic properties of weak solutions work in the same way with only minor modifications (see, e.g., [40], [139], [236], [243]). When it comes to geometry and regularity of the free boundary (Sections 3.4 and 4.6) the situation in higher dimensions is also basically the same as in two dimensions, but the theory is somewhat less complete. For example, there is no such precise statement as Theorem 3.4.1 in higher dimensions, but it is known, in the well-posed (injection) case, that $\partial\Omega(t) \setminus \overline{\Omega(0)}$ is a (smooth) real-analytic manifold except possibly for a small exceptional set, and that outside the convex hull of $\overline{\Omega(0)}$ there are no exceptional points at all. These statements are based on the known regularity results for the obstacle problem [76], [77], [80], [79], [193], [431] combined with geometric results (relevant parts of Theorem 4.7.2 remain true [243]).

Strong solutions to the higher-dimensional Hele-Shaw problem, with and without surface tension, have been studied in a number of papers, for example [30], [179], [180], [554]. Compare also [100], [154]. What one can basically prove is the existence and uniqueness, locally in time in both time directions, of strong solutions under the assumption that the initial domain has a smooth real-analytic boundary. Under certain assumptions, like strong starlikeness of the initial domain, it is possible even to prove the existence globally in time in the well-posed time direction (injection) [248]. Another such example is found in [387]. Corner development in higher dimensions has been studied by S. Gardiner and T. Sjödin [202].

A generalization of the Hele-Shaw problem in another direction is to replace the Laplace operator by a more general elliptic operator in divergence form. This allows for considering irregular and anisotropic media in for example a porous medium interpretation of the model. A selection of papers going in this direction is [41], [207], [364], [378], [465], [305]. In the latter paper the terminology elliptic growth is used. See also [484], where techniques of several complex variables are used. A moving boundary problem, related to quadrature surfaces (see [515], [246]) in the same way as Hele-Shaw flow problems are related to quadrature domains, has been studied by M. Onodera [416].

Closely related to weak solutions and the obstacle problem is the treatment, in two dimensions, of Coulomb gas ensembles, or "Quantum Hele-Shaw", by H. Hedenmalm and N. Makarov in [262], which in particular establishes a rigorous connection between the Laplacian growth problem and random matrix theory (see also Chapter 6 below). This, and many related papers, actually treat the exterior problem, i.e., the case that the fluid region $\Omega(t)$ contains a neighborhood of infinity. If $D(t)$ denotes the complement of $\Omega(t)$, a compact set ("droplet"), the balayage formulation then can be taken to be

$$\text{Bal}\,(t\delta_\infty - \chi_{D(0)}, 0) = -\chi_{D(t)}, \qquad (3.75)$$

where $0 \le t < |D(0)|$. This equation is obtained from the usual one, say (3.24) (which does not make sense when $\Omega(0)$ has infinite area), by subtracting the constant one from both members and replacing δ_0 with δ_∞. Here δ_∞ formally means a unit source at infinity, which becomes meaningful in the above equation if it is represented by a uniform mass distribution of unit mass on a large circle $|z| = R$. Or one simply interprets (3.75) on the Riemann sphere, compare Section 3.5.2 below.

The exterior problem may be named "motion by equilibrium measure", and when integrated up by the Baiocchi transform (3.1) to a weak solution, this equilibrium measure becomes a *weighted equilibrium measure*, see [487] for this notion. The relationship between random matrix theory, weighted equilibrium measures and quadrature domains is amply discussed, in terms of an illustrative example, in [32]. The relationship between weighted equilibrium measures and partial balayage has been made fully explicit in [483], where the example in [32] is used as a starting point. In the terminology of the above references, equation (3.75) expresses that $\chi_{D(t)}$ is the weighted equilibrium measure in the presence of an external background potential Q such that $-\Delta Q = t\delta_\infty - \chi_{D(0)}$. In order to produce $D(t)$ there is actually a great flexibility in choosing an external potential $Q = Q_t$ such that $\text{Bal}\,(-\Delta Q, 0) = -\chi_{D(t)}$. One may for example start by defining $Q(z) = \frac{1}{4}|z|^2$ in $D(0)$ and then extending it appropriately outside $D(0)$ with the asymptotic behaviour $Q(z) = \frac{|D(0)|-t}{2\pi} \log|z| + O(1)$ as $|z| \to \infty$. Compare discussions in Chapter 6.

The suction version of the motion by equilibrium measure is the continuous deterministic counterpart of the standard version of DLA (with Brownian motion particles emitted at infinity) [583], [583], [304].

3.5.2 Hele-Shaw flow on Riemannian manifolds

There are no difficulties involved in generalizing the theory of weak solutions and partial balayage to Riemannian manifolds of any dimension. The resulting equations in a local coordinate will in fact be similar to the ones obtained for elliptic operators in divergence form. To briefly sketch the formalism, let M be an oriented Riemannian mainfold of dimension n. Local coordinates on M are denoted $(x^1, x^2, \ldots, x^n)$. We adopt the usual notations of differential geometry and tensor analysis, in particular those of [192]. Thus, e.g., the Riemannian metric is described by an expression (for a covariant 2-tensor)

$$ds^2 = g_{ij} dx^i \otimes dx^j$$

(summation understood, $1 \le i, j \le n$), where (g_{ij}) is positive definite (as a matrix) at each point. The inverse (as a matrix) is the corresponding contravariant tensor, with components g^{ij}:

$$g_{ij} g^{jk} = \delta_j^k,$$

where δ_j^k is the Kronecker delta (or identity matrix). The volume form on M is

$$\text{vol}^n = \sqrt{g}\, dx^1 \wedge \cdots \wedge dx^n = \sqrt{g}\, dx,$$

where $g = \det{(g_{ij})}$.

We use Δ to denote the ordinary Laplacian operator with respect to the local coordinates chosen. There is also the invariant Laplacian, Δ_{inv}, most conveniently defined (on functions) by the requirement

$$(\Delta_{\text{inv}} f)\, \text{vol}^n = d^*df,$$

where * is the *Hodge star*, taking p-forms to $(n-p)$-forms. This gives

$$\Delta_{\text{inv}} f = \frac{1}{\sqrt{g}} \frac{\partial}{\partial x^j} \left(\sqrt{g} g^{ij} \frac{\partial f}{\partial x^i} \right).$$

The dynamical law (3.74) for Laplacian growth with a point source of unit strength at a point a on the manifold M can be written as

$$\frac{d}{dt} \int_{\Omega(t)} \varphi\, \text{vol}^n = - \int_{\partial\Omega(t)} \varphi\, {}^*dG_{\Omega(t)}(\cdot, a), \qquad (3.76)$$

to hold for all smooth functions φ with compact support in M. The Green function $G_\Omega(\cdot, a)$ for a relatively compact subdomain $\Omega \subset M$ is determined by

$$-d^*dG_\Omega(\cdot, a) = \varepsilon_a \quad \text{in } \Omega,$$
$$G_\Omega(\cdot, a) = 0 \quad \text{on } \partial\Omega.$$

Here ε_a denotes the Dirac measure (point mass at a) considered as an n-form (more precisely, an n-form current), acting on test functions φ by $\langle \varepsilon_a, \varphi \rangle = \varphi(a)$. The Green function exists under very weak assumptions, but the formulation of the law (3.76) requires some smoothness of $\partial \Omega(t)$.

Like in the planar case one passes from the above classical formulation of Laplacian growth to a weak, or balayage, formulation: any smooth evolution $t \mapsto \Omega(t) \subset\subset M$ satisfying the dynamics (3.76) can be produced from the initial domain $\Omega(0)$ by partial balayage according to

$$\mathrm{Bal}\,(\chi_{\Omega(0)} \mathrm{vol}^n + t\varepsilon_a, \mathrm{vol}^n) = \chi_{\Omega(t)} \mathrm{vol}^n. \qquad (3.77)$$

Here partial balayage on M acts on measures, considered as n-forms, on M. The precise definition is perfectly natural: for μ and λ general n-forms satisfying some mild regularity requirements (including μ having compact support),

$$\mathrm{Bal}\,(\mu, \lambda) = \mu + d^* du,$$

where u is the smallest nonnegative function on M satisfying

$$\mu + d^* du \le \lambda \quad \text{on } M.$$

Alternatively, $\mu \mapsto \mathrm{Bal}\,(\mu, \lambda)$ can be defined as the orthogonal projection onto the set of n-forms $\le \lambda$ in a suitable Hilbert space having the energy as the squared norm (in case M is a bounded subdomain of $\mathbb{R}^n$ this will be the Sobolev space $H^{-1}(M)$).

The equation (3.77) spells out to

$$\chi_{\Omega(0)} \mathrm{vol}^n + t\varepsilon_a + d^* du = \chi_{\Omega(t)} \mathrm{vol}^n,$$

where $u \ge 0$ in all M, $u = 0$ on $M \setminus \Omega(t)$ (compare (3.4)–(3.6)). On "dividing" by vol^n (in any fixed system of coordinates) this becomes

$$\chi_{\Omega(0)} + t\delta_a + \Delta_{\mathrm{inv}} u = \chi_{\Omega(t)},$$

where δ_a is the Dirac distribution as a 0-form, so that

$$\varepsilon_a = \delta_a \, \mathrm{vol}^n.$$

Let us see what the above looks like in the case of a metric of the form

$$ds^2 = \rho^2((dx^1)^2 + \cdots + (dx^n)^2) \qquad (3.78)$$

for some function $\rho > 0$. Then $\sqrt{g} = \rho^n$ and the invariant Laplacian is

$$\Delta_{\mathrm{inv}} f = \sum_{i=1}^n \frac{1}{\rho^n} \frac{\partial}{\partial x^i} \left(\rho^{n-2} \frac{\partial f}{\partial x^i} \right).$$

We see that the two-dimensional case is particularly nice: when $n = 2$, $\Delta_{\text{inv}} f = \frac{1}{\rho^2} \Delta f$, so that the equation for u becomes, after multiplication by ρ^2,

$$\rho^2 \chi_{\Omega(0)} + t\rho(a)^2 \delta_a + \Delta u = \rho^2 \chi_{\Omega(t)}.$$

Thus, in two dimensions and in terms of any local coordinate for which the metric has the form (3.78), the Hele-Shaw flow can be thought of simply as weighted Hele-Shaw flow for the ordinary Laplacian, with the weight ρ^2 and with the Dirac measure scaled by a factor $\rho(a)^2$. Compare [261]. Expressed in terms of the balayage operator in the plane, and within the range of the local coordinate in question, we have

$$\text{Bal}\left(\rho^2 \chi_{\Omega(0)} + t\rho(a)^2 \delta_a, \rho^2\right) = \rho^2 \chi_{\Omega(t)}.$$

Continuing in two dimensions, assume that $M \subset \mathbb{C}$ with metric $ds^2 = \rho^2(dx^2 + dy^2)$. The Gaussian curvature of the metric is

$$\kappa = -\Delta_{\text{inv}} \log \rho = -\frac{\Delta \log \rho}{\rho^2}.$$

Some beautiful results are obtained by starting with an initially empty set, i.e., on considering

$$\text{Bal}\left(t\rho(a)^2 \delta_a, \rho^2\right) = \rho^2 \chi_{\Omega(t)}.$$

A first observation is that if κ is constant, then the domains $\Omega(t)$ will all be circular disks with respect to the Euclidean metric in the plane (however with the point a usually off-center) and at the same time geodesic disks, centered at a with respect to the Riemannian metric. This is a simple consequence of the obvious fact that Hele-Shaw evolution on a Riemannian manifold is invariant under rigid (with respect to the metric) transformations on the manifold. If for example the constant curvature is negative then we may assume that M is part of the unit disk $\mathbb{D}$ equipped with the Poincaré metric

$$ds = \rho_{\text{Poinc}}(z)|dz| = \frac{|dz|}{1 - |z|^2}.$$

By a rigid transformation, i.e., in this case, a Möbius transformation mapping $\mathbb{D}$ onto itself, any point $a \in \mathbb{D}$ can be moved to the origin, and Hele-Shaw flow started at the origin will certainly consist of expanding disks. When these fluid domains are moved back to the original position around a by the inverse Möbius transformation they will remain circular disks, however not with center a. But considered as hyperbolic disks, i.e., geodesic disks with respect to the Poincaré metric, their center will be the injection point a. Similiar discussions apply in the case of a metric with constant positive (or zero) curvature.

Some deeper results, due to H. Hedenmalm, S. Shimorin, A. Olofsson [261], [263] say that if $M \subset \mathbb{D}$ is equipped with any real-analytic metric for which $\kappa \leq 0$ (M hyperbolic), or even just $2\Delta \log \rho + \Delta \log \rho_{\text{Poinc}} \geq 0$ (called "weakly hyperbolic"), then the domains $\Omega(t)$ are simply connected as long as they remain compactly contained in M.

If M is not a planar domain and has nontrivial topology in itself, then the fluid domains $\Omega(t)$ may very well change topology, even in the case of zero or negative curvature. In fact, it is just to think of a cylinder $M = \mathbb{C}/\mathbb{Z}$ or a torus $M = \mathbb{C}/(\mathbb{Z} + i\mathbb{Z})$ with the flat metric (from $\mathbb{C}$). Hele-Shaw evolution with a point source on such a surface is equivalent to a Hele-Shaw evolution on $\mathbb{C}$ with a periodic distribution of sources, and the fluid domains in this picture will initially consist of infinitely many disjoint disks, which however at a later instant will be glued together. Such periodic problems are treated by, for example, S. Richardson [472], D. Crowdy [111], A. Silva, G. Vasconcelos [524].

Further results and examples, on Hele-Shaw evolutions on closed surfaces in $\mathbb{R}^3$, are given in [560].

Chapter 4

Geometric Properties

In this chapter we discuss geometric properties of general Hele-Shaw flows. Special classes of univalent functions that admit explicit geometric interpretations are considered. In particular, we ask the following question: which geometrical properties are preserved during the time evolution of the moving boundary? We also discuss the geometry of weak solutions.

4.1 Distance to the boundary

In this brief section, we use some observations from [269] to estimate the minimal distance from the source to the free boundary. We consider injection $(Q > 0)$ into a bounded fluid domain $\Omega(t)$ parameterized by a univalent function $f : \mathbb{D} \to \Omega(t)$ with Taylor expansion

$$f(\zeta, t) = a_1(t)\zeta + a_2(t)\zeta^2 + \cdots \quad (a_1(t) > 0).$$

From the Löwner–Kufarev type equation (1.23)–(1.24) satisfied by f we obtain

$$\frac{\dot{a}_1(t)}{a_1(t)} = \frac{Q}{4\pi^2} \int\limits_0^{2\pi} \frac{1}{|f'(e^{i\theta}, t)|^2} d\theta.$$

Here the right-hand member can be estimated by means of Jensen's inequality and the mean-value property for harmonic functions:

$$\frac{1}{2\pi} \int\limits_0^{2\pi} \frac{1}{|f'(e^{i\theta}, t)|^2} d\theta \geq \exp\left[\frac{1}{2\pi} \int\limits_0^{2\pi} \log \frac{1}{|f'(e^{i\theta}, t)|^2} d\theta\right] = \frac{1}{|f'(0, t)|^2} = \frac{1}{a_1(t)^2}.$$

Thus $\dot{a}_1(t)/a_1(t) \geq Q/(2\pi a_1(t)^2)$, or

$$\frac{d}{dt}\left(a_1(t)^2 - \frac{Qt}{\pi}\right) \geq 0. \tag{4.1}$$

We conclude that $a_1(t)^2 - \frac{Qt}{\pi}$, and in particular $a_1(t)$ itself, is monotone increasing and that hence

$$a_1(t)^2 \geq a_1(0)^2 + \frac{Qt}{\pi} \quad (t > 0). \tag{4.2}$$

See also [336]. The Koebe one-quarter theorem (see, e.g., [213]) now yields the inequality

$$\mathrm{dist}(\partial\Omega(t), 0) \geq \frac{1}{4}\sqrt{a_1(0)^2 + \frac{Qt}{\pi}}.$$

A more general result will be given at the end of this chapter.

4.2   A topological result

The injection version of the Hele-Shaw problem is in principle well posed, but it may still happen that a solution, when considered as a strong solution and hence not allowed to change topology, breaks down if the initial geometry is sufficiently entangled. As we shall see in Theoerem 4.5.2, the initial domain being starshaped with respect to the injection point is a good enough condition for this not to happen, and the strong solution then goes on forever. On the other hand, we have seen in Section 2.1.3 that, in the non-starshaped case, univalence may be lost even in the case of degree-three polynomial solutions. Indeed, referring to the example in that section, if the initial domain is given by parameter values satisfying $\frac{16}{15}\sqrt[4]{\frac{3}{5}} < A_2 < \frac{2\sqrt{2}}{3}$, $A_3 = \frac{1}{3}$, then soon after the two nearby cusps on the initial boundary have been resolved a double point will be created, which makes the solution to break down as a univalent solution.

The Polubarinova–Galin equation itself is actually not affected by this loss of univalence as long as the derivative of the mapping function does not vanish on the closed unit disk. So as a solution to this equation the evolution may go on, but the mapping functions will no longer be univalent, only locally univalent. By f being *locally univalent* we shall mean here that $f' \neq 0$ on the closed unit disk $\overline{\mathbb{D}}$. What however the next theorem says is that the situation cannot go on like this forever, sooner or later the solution will lose also its local univalence, i.e., a zero of the derivative of the mapping function will reach the unit circle. To put it another way, any global (in time) locally univalent solution of the Polubarinova–Galin equation is actually univalent all the time.

Theorem 4.2.1 ([239]). *Let $f(\zeta, t)$ be a global $(0 \leq t < \infty)$ locally univalent solution of the Polubarinova–Galin equation (1.22) driven by injection with rate $Q > 0$. Then $f(\zeta, t)$ is actually univalent, even on the closed unit disk, for all $0 \leq t < \infty$.*

Proof. The proof is based on a result from the Löwner theory, namely Theorem 6.2 in [446]. Recall that in terms of the Taylor expansion

$$f(\zeta, t) = \sum_{j=1}^{\infty} a_j(t)\zeta^j$$

of the mapping function, the zeroth-order moment is given by

$$M_0(t) = \sum_{j=1}^{\infty} j|a_j(t)|^2.$$

Let

$$N_0(t) = \sum_{j=2}^{\infty} j|a_j(t)|^2 \tag{4.3}$$

denote what remains after the first term. Then

$$N_0(t) = M_0(t) - a_1(t)^2 = M_0(0) + \frac{Qt}{\pi} - a_1(t)^2.$$

Since $a_1(t)^2 - \frac{Qt}{\pi}$ is an increasing function of t by (4.1), $N_0(t)$ is decreasing, and in particular

$$0 \leq N_0(t) \leq N_0(0). \tag{4.4}$$

It also follows that

$$\frac{a_1(t)}{\sqrt{M_0(t)}} = \frac{1}{\sqrt{1 + \frac{N_0(t)}{a_1(t)^2}}} \nearrow 1 \quad \text{as } t \to \infty, \tag{4.5}$$

and that

$$|a_k(t)| \leq \sqrt{\frac{N_0(0)}{k}} \quad \text{for } k \geq 2. \tag{4.6}$$

For (say) $|\zeta| \leq \frac{1}{2}$ and $t \geq 0$ we have, using (4.4),

$$|f(\zeta,t)| \leq a_1(t) + \sum_{j=2}^{\infty} |a_j(t)| 2^{-j}$$

$$\leq a_1(t) + \left(\sum_{j=2}^{\infty} j|a_j(t)|^2 \right)^{1/2} \left(\sum_{j=2}^{\infty} \frac{1}{j} 2^{-2j} \right)^{1/2}$$

$$\leq a_1(t) + \sqrt{N_0(t)} \leq a_1(t) + \sqrt{N_0(0)}.$$

Thus $|f(\zeta,t)| \leq Ca_1(t)$ $(0 \leq t < \infty)$ in a neighborhood of the origin. It is known that such an estimate holding (with $a_1(t) \to \infty$ as $t \to \infty$) together with f satisfying a Löwner–Kufarev equation (1.24) for some P with $\mathrm{Re}\,P > 0$ in $\mathbb{D}$, is enough to conclude that $f(\zeta,t)$ is univalent for all $t \geq 0$, see Theorem 6.2 in [446]. In our case it follows that $f(\zeta,t)$ is actually univalent in the closed unit disk, because by assumption it is locally univalent there, and if for two different points $\zeta_1, \zeta_2 \in \partial\mathbb{D}$ we had $f(\zeta_1,t) = f(\zeta_2,t)$ for some t, then univalence in the open disk $\mathbb{D}$ would be lost in the next instance (the boundary is really propagating with positive speed under our assumptions). □

4.3   Special classes of univalent functions

Let us define some special classes of univalent functions which will parameterize
our phase domains.

A domain $\Omega \subset \mathbb{C}$, $0 \in \Omega$ is said to be *starlike* (with respect to the origin)
if each ray starting at the origin intersects Ω in a connected set. If a function
$f(\zeta)$ maps $\mathbb{D}$ onto a domain which is starlike, $f(0) = 0$, then we say that $f(\zeta)$
is a *starlike function*. If a function $f(\zeta)$ maps $\mathbb{D}$ onto a domain which is convex,
$f(0) = 0$, then we say that $f(\zeta)$ is a *convex function*. We denote the class of starlike
functions by S^* and the class of convex functions by C. A necessary and sufficient
condition for a function $f(\zeta)$, $\zeta \in \mathbb{D}$, $f(0) = 0$ to be starlike is that the inequality

$$\operatorname{Re} \frac{\zeta f'(\zeta)}{f(\zeta)} > 0, \quad \zeta \in \mathbb{D} \tag{4.7}$$

holds. Similarly, a necessary and sufficient condition for a function f to be convex
is the inequality

$$\operatorname{Re} \left(1 + \frac{\zeta f''(\zeta)}{f'(\zeta)} \right) > 0, \quad \zeta \in \mathbb{D}. \tag{4.8}$$

These standard characterizations can be found, e.g., in [17, 156, 216, 446].

A simple way to generalize the class S^* is to introduce a class of so-called
starlike functions of order α, $0 < \alpha \leq 1$, obtained by replacing 0 in the right-hand
side of (4.7) by the constant α. Let us denote it by S_α^*. It is known that any
convex function is in $S_{1/2}^*$ (see [379]). Unfortunately, the classes S_α^* do not admit
any clear geometric interpretation. A more reasonable generalization was given by
Brannan, Kirwan [72] and Stankiewicz [531]. A function $f : \mathbb{D} \to \mathbb{C}$, $f(0) = 0$ is
said to be *strongly starlike of order* α in $\mathbb{D}$, $0 < \alpha \leq 1$, if for all $\zeta \in \mathbb{D}$

$$\left| \arg \frac{\zeta f'(\zeta)}{f(\zeta)} \right| < \alpha \frac{\pi}{2}. \tag{4.9}$$

The set of all such functions is denoted by $S^*(\alpha)$. This class of functions has a
better geometric description: every *level line* $L_r = \{f(re^{i\theta}), \theta \in [0, 2\pi)\}$, $f \in$
$S^*(\alpha)$ is reachable from outside by the radial angle $\pi(1 - \alpha)$ (see Figure 4.1).

The inequalities (4.7)–(4.9) also give sufficient conditions for an analytic function f to be univalent (see [27] for a collection of sufficient conditions of univalence). Kaplan [295] proved that if $f(\zeta)$ and $g(\zeta)$ are analytic in $\mathbb{D}$, $g \in C$, and

$$\operatorname{Re} \frac{f'(\zeta)}{g'(\zeta)} > 0, \quad \zeta \in \mathbb{D},$$

then f is univalent in $\mathbb{D}$. Kaplan gave the name *close-to-convex* to univalent functions f that satisfy the above condition. The close-to-convex functions have a nice
geometric characterization: every level line L_r of a close-to-convex function f has

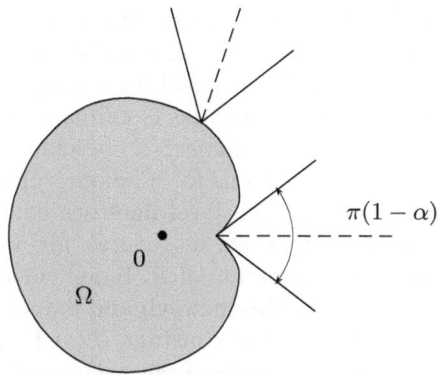

Fig. 4.1: Strongly starlike functions of order α $(S^*(\alpha))$

no "large hairpin" turns, that is there are no sections of the curve L_r in which the tangent vector turns backward through an angle greater than π.

We say that a simply connected hyperbolic domain Ω is *convex in the direction of the real axis* $\mathbb{R}$ if each line parallel to $\mathbb{R}$ either misses Ω, or the intersection with Ω is a connected set. The study of this class goes back to Fejér [185] and Robertson [477]. If a function $f(\zeta)$ maps $\mathbb{D}$ onto a domain which is convex in the direction of the real axis, $f(0) = 0$, then we say that $f(\zeta)$ is a *convex function in the direction of the real axis* and denote the class of such function by $C_{\mathbb{R}}$ (see Figure 4.2). The criterion that characterizes these functions is as follows: the unit circle

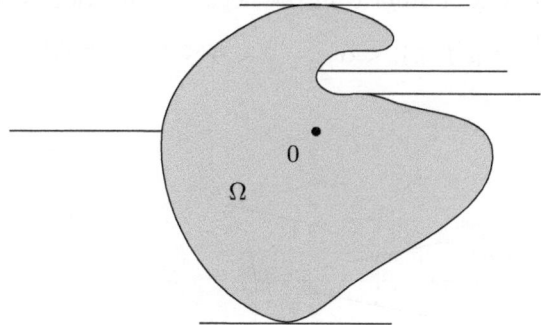

Fig. 4.2: Convex functions in the direction of the real axis $(C_{\mathbb{R}})$

can be divided into two disjoint arcs I and J, $I = \{\zeta = e^{i\theta}, \ \theta \in [0, \varphi] \cup [\psi, 2\pi]\}$, $J = \{\zeta = e^{i\theta}, \ \theta \in [\varphi, \psi]\}$, such that

$$\operatorname{Re} \zeta f'(\zeta) \geq 0, \quad \text{for} \quad \zeta \in I,$$
$$\operatorname{Re} \zeta f'(\zeta) \leq 0, \quad \text{for} \quad \zeta \in J.$$

The harmonic function Re $\zeta f'(\zeta)$ changes its sign in $\mathbb{D}$. Therefore, the level lines L_r for $0 < r < 1$ need not be convex in the direction of $\mathbb{R}$. Hengartner and Schober [267] proved this in 1973. Their proof used a slightly modified argument. An example of such a function was given by Goodman and Saff [215]. Fejér and Szegő in 1951 [186] proved that if the domain $f(\mathbb{D})$ is symmetric with respect to the imaginary axis, then the above conditions for a holomorphic function f, $f(0) = 0$, are sufficient for its univalence and the level lines are convex in the direction of $\mathbb{R}$. Prokhorov [453] proved in 1988 that, in general, for any $r \in (0, \sqrt{2} - 1)$ the level lines are still convex in the direction of $\mathbb{R}$. Independently this result (even a more general one) was obtained by Ruscheweyh and Salinas [486]. An example by Goodman and Saff [215] shows that the constant $\sqrt{2} - 1$ can not be improved.

Now we discuss conformal maps from the right half-plane. A simply-connected domain Ω with a boundary that contains more than two points in the extended complex plane $\overline{\mathbb{C}}$ is said to be *convex in the negative direction* of the real axis $\mathbb{R}^-$ if its complement can be covered by a family of non-intersecting parallel rays starting at the same direction of $\mathbb{R}^-$. A holomorphic univalent map $f(\zeta)$, $\zeta \in \mathbb{H}^+$, $\mathbb{H}^+ = \{\zeta : \text{Re}\,\zeta > 0\}$, is said to be convex in the negative direction if $f(\mathbb{H}^+)$ is as above. We denote this class by $H_{\mathbb{R}}^-$. This is somehow an analogue of the class of starlike functions for the half-plane. The criterion for this property is

$$\text{Re}\ f'(\zeta) > 0, \quad \zeta = \xi + i\eta \in \mathbb{H}^+.$$

We define a subclass $H_{\mathbb{R}}^-(\alpha)$ of $H_{\mathbb{R}}^-$ of functions whose level lines $L_a = \{f(a + i\eta), \ \eta \in (-\infty, \infty)\}$ are reachable by the angles $\pi(1 - \alpha)$ with their bisectors co-directed with $\mathbb{R}^-$. We call these functions *convex of order α in the negative direction* (see Figure 4.3). The necessary and sufficient condition for a holomorphic function f to be convex of order α in the negative direction is that

$$\left| \arg f'(\zeta) \right| < \alpha \frac{\pi}{2}, \quad 0 < \alpha \le 1, \quad \zeta \in \mathbb{H}^+. \tag{4.10}$$

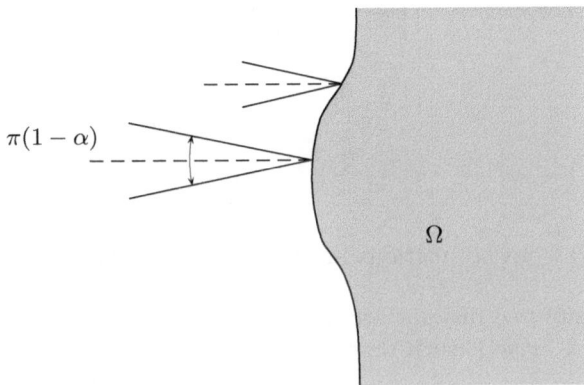

Fig. 4.3: Convex functions of order α in the negative direction $(H_{\mathbb{R}}^-(\alpha))$

4.4   Hereditary shape of phase domains

In this section we shall find some geometric properties which are preserved during the time evolution of the moving boundary.

4.4.1   Bounded dynamics

Simple examples show that virtually no geometric properties are preserved in the case of suction, $Q < 0$. So we henceworth assume that $Q > 0$.

Starlike dynamics

Let us start with starlike dynamics. We suppose that the initial function f_0 is analytic in the closure of $\mathbb{D}$ to guarantee the local in time existence of solutions (see Section 1.4.3). The following theorem was proved in [272] (see also [561]). Here we use a slightly modified argument.

Theorem 4.4.1. *Let $Q > 0$, $f_0 \in S^*$, and be analytic and univalent in a neighbourhood of $\bar{U}$. Then the family of domains $\Omega(t)$ (in the sequel, the family of univalent functions $f(\zeta, t)$) remain in S^* as long as the solution to the Polubarinova–Galin equation exists.*

Proof. If we consider f in the closure of $\mathbb{D}$, then the inequality sign in (4.7) is to be replaced by $(\geq)$ where equality can be attained only for $|\zeta| = 1$.

The proof is based on consideration of a critical map $f \in S^*$, such that the image of $\mathbb{D}$ under the map $\zeta f'(\zeta, t)/f(\zeta, t)$, $|\zeta| \leq 1$ touches the imaginary axis, say there exist $t' \geq 0$ and $\zeta_0 = e^{i\theta_0}$, such that

$$\arg \frac{\zeta_0 f'(\zeta_0, t')}{f(\zeta_0, t')} = \frac{\pi}{2} \quad (\text{or } -\tfrac{\pi}{2}), \tag{4.11}$$

and for any $\varepsilon > 0$ there are $t > t'$ and $\theta \in (\theta_0 - \varepsilon, \theta_0 + \varepsilon)$ such that

$$\arg \frac{e^{i\theta} f'(e^{i\theta}, t)}{f(e^{i\theta}, t)} \geq \frac{\pi}{2} \quad (\text{or } \leq -\tfrac{\pi}{2}). \tag{4.12}$$

For definiteness we consider the sign $(+)$ in (4.11). Without loss of generality, let us assume $t' = 0$. Since $f'(e^{i\theta}, t) \neq 0$, our assumption about the sign in (4.11) yields that

$$\operatorname{Im} \frac{\zeta_0 f'(\zeta_0, 0)}{f(\zeta_0, 0)} > 0, \tag{4.13}$$

(the negative case is considered similarly).

Since ζ_0 is a critical point and the image of the unit disk $\mathbb{D}$ under the mapping $\frac{\zeta f'(\zeta,0)}{f(\zeta,0)}$ touches the imaginary axis at the point $\zeta_0 = e^{i\theta_0}$, we deduce that

$$\frac{\partial}{\partial\theta}\arg\frac{e^{i\theta}f'(e^{i\theta},0)}{f(e^{i\theta},0)}\bigg|_{\theta=\theta_0} = 0,$$

$$\frac{\partial}{\partial r}\arg\frac{re^{i\theta_0}f'(re^{i\theta_0},0)}{f(re^{i\theta},0)}\bigg|_{r=1} \geq 0.$$

From this we calculate

$$\mathrm{Re}\left[1 + \frac{\zeta_0 f''(\zeta_0,0)}{f'(\zeta_0,0)} - \frac{\zeta_0 f'(\zeta_0,0)}{f(\zeta_0,0)}\right] = 0, \tag{4.14}$$

$$\mathrm{Im}\left[1 + \frac{\zeta_0 f''(\zeta_0,0)}{f'(\zeta_0,0)} - \frac{\zeta_0 f'(\zeta_0,0)}{f(\zeta_0,0)}\right] \geq 0. \tag{4.15}$$

By straightforward calculations one derives

$$\frac{\partial}{\partial t}\arg\frac{\zeta f'(\zeta,t)}{f(\zeta,t)} = \mathrm{Im}\frac{\partial}{\partial t}\log\frac{f'(\zeta,t)}{f(\zeta,t)} = \mathrm{Im}\left(\frac{\frac{\partial}{\partial t}f'(\zeta,t)}{f'(\zeta,t)} - \frac{\frac{\partial}{\partial t}f(\zeta,t)}{f(\zeta,t)}\right). \tag{4.16}$$

We now differentiate the equation (1.22) with respect to θ,

$$\mathrm{Im}\left(\overline{f'(\zeta,t)}\frac{\partial}{\partial t}f'(\zeta,t) - \overline{\zeta f'(\zeta,t)}\dot{f}(\zeta,t) - \overline{\zeta^2 f''(\zeta,t)}\dot{f}(\zeta,t)\right) = 0, \tag{4.17}$$

for $|\zeta| = 1$. This equality is equivalent to the following:

$$|f'(\zeta,t)|^2\mathrm{Im}\left(\frac{\frac{\partial}{\partial t}f'(\zeta,t)}{f'(\zeta,t)} - \frac{\frac{\partial}{\partial t}f(\zeta,t)}{f(\zeta,t)}\right)$$

$$= \mathrm{Im}(\overline{\zeta f'(\zeta,t)}\dot{f}(\zeta,t))\left(\frac{\zeta f''(\zeta,t)}{f'(\zeta,t)} - \frac{\zeta f'(\zeta,t)}{f(\zeta,t)} + 1\right).$$

Substituting (1.22) and (4.14) in the latter expression we finally have

$$\frac{\partial}{\partial t}\arg\frac{\zeta f'(\zeta,t)}{f(\zeta,t)}\bigg|_{\zeta=e^{i\theta_0},t=0}$$

$$= -\frac{Q}{2\pi|f'(e^{i\theta_0},0)|^2}\mathrm{Im}\left(\frac{e^{i\theta_0}f'(e^{i\theta_0},0)}{f(e^{i\theta_0},0)} + \frac{e^{i\theta_0}f''(e^{i\theta_0},0)}{f'(e^{i\theta_0},0)}\right).$$

The right-hand side of this equality is strictly negative because of (4.13), (4.15). Therefore,

$$\arg\frac{e^{i\theta}f'(e^{i\theta},t)}{f(e^{i\theta},t)} < \frac{\pi}{2}$$

for $t > 0$ (close to 0) in some neighbourhood of θ_0. This contradicts the assumption that $\Omega(t)$ fails to be starlike for some $t > 0$ and ends the proof for the class S^*. $\square$

The property of preservation of starlikeness is especially interesting in view
of Novikov's theorem [412], that says: if two bounded domains are starlike with
respect to a common point and have the same exterior gravity potential, then they
coincide.

We continue with strongly starlike functions of order α. We shall prove that
starting with a bounded phase domain Ω_0 which is strongly starlike of order α and
bounded by an analytic curve we obtain a subordination chain of domains $\Omega(t)$
(and functions $f(\zeta, t)$) under injection at the origin which remain strongly starlike
of order $\alpha(t)$ with a decreasing order $\alpha(t)$. The following monotonicity theorem is
found in [247]. A similar result has recently and independently been obtained in
[337]

Theorem 4.4.2. *Let $f_0 \in S^*(\alpha)$, $\alpha \in (0, 1]$, be analytic and univalent in a neighbourhood of $\bar{\mathbb{D}}$. Then the strong solution $f(\zeta, t)$ to the Polubarinova–Galin equation
(1.22) forms a subordination chain (see the definition in Section 8.1) of strongly
starlike functions of order $\alpha(t)$ with a strictly decreasing $\alpha(t)$ during the time of
existence, $\alpha(0) = \alpha$.*

Proof. Let t_0 be such that the strong solution $f(\zeta, t)$ exists during the time $t \in
[0, t_0)$, $t_0 > 0$. Since all functions $f(\zeta, t)$ have analytic univalent extension into
a neighbourhood of $\bar{\mathbb{D}}$ during the time of the existence of the strong solution to
(1.22), their derivatives $f'(\zeta, t)$ are continuous and do not vanish in $\bar{\mathbb{D}}$. Moreover,
$f(\zeta, t)$ are starlike in $\mathbb{D}$ (see Theorem 4.4.1). Therefore, there exists $\alpha(t)$, $0 <
\alpha(t) \leq 1$, such that $f(\zeta, t) \in S^*(\alpha(t))$ and $f(\zeta, t) \notin S^*(\alpha(t) - \varepsilon)$ for any $\varepsilon > 0$.

Let us fix $t' \in [0, t_0)$ and consider the set A of all points ζ, $|\zeta| = 1$ for which
$|\arg \frac{\zeta f'(\zeta, t')}{f(\zeta, t')}| = \alpha \pi / 2$. First, we deal with the subset A^+ of A where

$$\arg \frac{\zeta f'(\zeta, t')}{f(\zeta, t')} = \frac{\alpha \pi}{2}. \tag{4.18}$$

The sets A^+ and $A^- = A \setminus A^+$ are closed and do not intersect. One of the sets
A^+ and A^- is allowed to be empty. Without loss of generality we suppose that
$A^+ \neq \emptyset$. For any point $\zeta \in A^+$, we have

$$\operatorname{Im} \frac{\zeta f'(\zeta, t')}{f(\zeta, t')} > 0. \tag{4.19}$$

The argument $\arg \frac{\zeta f'(\zeta, t')}{f(\zeta, t')}$ attains its maximum on $\zeta \in \partial \mathbb{D}$ at the points of
A^+. Therefore,

$$\frac{\partial}{\partial \theta} \arg \frac{e^{i\theta} f'(e^{i\theta}, t')}{f(e^{i\theta}, t')} = 0, \quad \zeta = e^{i\theta} \in A^+.$$

The argument $\arg \frac{re^{i\theta} f'(re^{i\theta}, t')}{f(re^{i\theta}, t')}$, $e^{i\theta} \in A^+$ attains its maximum on $r \in [0, 1]$ at
$r = 1$. Hence,

$$\frac{\partial}{\partial r} \arg \frac{re^{i\theta} f'(re^{i\theta}, t')}{f(re^{i\theta}, t')} \bigg|_{r=1} \geq 0.$$

We calculate

$$\text{Re} \left[1 + \frac{\zeta f''(\zeta, t')}{f'(\zeta, t')} - \frac{\zeta f'(\zeta, t')}{f(\zeta, t')} \right] = 0, \tag{4.20}$$

$$\text{Im} \left[1 + \frac{\zeta f''(\zeta, t')}{f'(\zeta, t')} - \frac{\zeta f'(\zeta, t')}{f(\zeta, t')} \right] \geq 0, \tag{4.21}$$

where $\zeta \in A^+$.

Let us represent the derivative

$$\frac{\partial}{\partial t} \arg \frac{\zeta f'(\zeta, t)}{f(\zeta, t)} = \text{Im} \frac{\partial}{\partial t} \log \frac{f'(\zeta, t)}{f(\zeta, t)} = \text{Im} \left(\frac{\frac{\partial}{\partial t} f'(\zeta, t)}{f'(\zeta, t)} - \frac{\frac{\partial}{\partial t} f(\zeta, t)}{f(\zeta, t)} \right). \tag{4.22}$$

Now we differentiate the Polubarinova–Galin equation (1.22) with respect to θ as

$$\text{Im} \left(\overline{f'(\zeta, t)} \frac{\partial}{\partial t} f'(\zeta, t) - \overline{\zeta f'(\zeta, t)} \dot{f}(\zeta, t) - \overline{\zeta^2 f''(\zeta, t)} \dot{f}(\zeta, t) \right) = 0, \quad \zeta = e^{i\theta}. \tag{4.23}$$

This equality is equivalent to

$$|f'(\zeta, t)|^2 \text{Im} \left(\frac{\frac{\partial}{\partial t} f'(\zeta, t)}{f'(\zeta, t)} - \frac{\frac{\partial}{\partial t} f(\zeta, t)}{f(\zeta, t)} \right)$$

$$= \text{Im} \left[\overline{\zeta f'(\zeta, t)} \dot{f}(\zeta, t) \left(\overline{\left(\frac{\zeta f''(\zeta, t)}{f'(\zeta, t)} \right)} - \frac{\zeta f'(\zeta, t)}{f(\zeta, t)} + 1 \right) \right].$$

Substituting (1.22) and (4.20) in the latter expression we have

$$\frac{\partial}{\partial t} \arg \frac{\zeta f'(\zeta, t)}{f(\zeta, t)} \bigg|_{\zeta \in A^+, t=t'} = - \frac{Q}{|f'(\zeta, t')|^2} \text{Im} \left(\frac{\zeta f'(\zeta, t')}{f(\zeta, t')} + \frac{\zeta f''(\zeta, t')}{f'(\zeta, t')} \right).$$

The right-hand side of this equality is continuous on A^+ and strictly negative because of (4.19), (4.21). Therefore,

$$\max_{\zeta \in A^+} \frac{\partial}{\partial t} \arg \frac{\zeta f'(\zeta, t)}{f(\zeta, t)} \bigg|_{t=t'} = -\delta < 0.$$

There exists a neighbourhood $A^+(\delta)$ on the unit circle of A^+ such that $A^+(\delta)$ and A^- do not intersect and

$$\frac{\partial}{\partial t} \arg \frac{\zeta f'(\zeta, t)}{f(\zeta, t)} \bigg|_{\zeta \in A^+(\delta), t=t'} < -\frac{\delta}{2}.$$

There is a positive number σ such that

$$\max_{\zeta \in \partial \mathbb{D} \backslash A^+(\delta)} \arg \frac{\zeta f'(\zeta, t)}{f(\zeta, t)} \bigg|_{t=t'} = \frac{\alpha \pi}{2} - \sigma.$$

We choose such $s > 0$ that

(i) $t' + s < t_0$;

(ii) $\dfrac{\partial}{\partial t} \arg \dfrac{\zeta f'(\zeta, t)}{f(\zeta, t)} \Big|_{\zeta \in A^+(\delta)} < 0, \quad t \in [t', t' + s]$;

(iii) $\displaystyle\max_{\zeta \in \partial \mathbb{D} \setminus A^+(\delta)} \arg \dfrac{\zeta f'(\zeta, t)}{f(\zeta, t)} \leq \dfrac{\alpha \pi}{2} - \dfrac{\sigma}{2}, \quad t \in [t', t' + s]$.

The condition (ii) implies that

$$\arg \frac{\zeta f'(\zeta, t)}{f(\zeta, t)} < \frac{\alpha \pi}{2}, \quad t \in (t', t' + s], \ \zeta \in A^+(\delta).$$

Thus, the condition (iii) yields

$$\alpha^+(t) := \max_{\zeta \in \partial \mathbb{D}} \arg \frac{\zeta f'(\zeta, t)}{f(\zeta, t)} < \frac{\alpha \pi}{2} = \alpha(t'), \quad \text{for all } t \in (t', t' + s].$$

This means that $\alpha^+(t)$ is strictly decreasing in $[0, t_0)$.

If the set $A^- \neq \emptyset$, then we can define the function

$$\alpha^-(t) := -\min_{\zeta \in \partial \mathbb{D}} \arg \frac{\zeta f'(\zeta, t)}{f(\zeta, t)}.$$

Similar argumentation shows that $\alpha^-(t)$ is strictly decreasing.

If $A^- = \emptyset$ (or $A^+ = \emptyset$), then $\alpha(t) = \alpha^+(t)$ (or $= \alpha^-(t)$) for $t \in [t', t' + s]$, s sufficiently small.

We set the function $\alpha(t) = \max\{\alpha^+(t), \alpha^-(t)\}$ in the case $A^+ \neq \emptyset$ and $A^- \neq \emptyset$. The so-defined function $\alpha(t)$ is strictly decreasing, that ends the proof. $\square$

Directional convex dynamics

We proceed with the class $C_{\mathbb{R}}$ and the dynamics under injection $(Q > 0)$.

Theorem 4.4.3. *If the initial domain $\Omega(0)$ is convex in the direction of $\mathbb{R}$, then the family of domains $\Omega(t)$ (in the sequel, the family of univalent functions $f(\zeta, t)$) preserves this property as long as the classical solution of the Hele-Shaw problem exists and the level lines of the function $f(\zeta, t)$ also are convex in the direction of $\mathbb{R}$ in a neighbourhood $|\zeta| \in (1 - \varepsilon, 1]$.*

Remark 4.4.1. The last requirement is fulfilled always if the initial domain Ω_0 is symmetric with respect to the imaginary axis (Fejér and Szegő [186]).

Proof. Let us again consider a critical map $f \in C_{\mathbb{R}}$, such that the image of $\mathbb{D}$ under the map $\zeta f'(\zeta, t)$, $|\zeta| \leq 1$ touches the imaginary axis. In other words, there exists $\zeta_0 = e^{i\theta_0}$, which satisfies the equality $\arg \zeta_0 f'(\zeta_0, 0) = \frac{\pi}{2}$ (or $-\frac{\pi}{2}$)

at the initial instant $t = 0$ and for any $\varepsilon > 0$ there is such $t > 0$ and $\theta \in (\theta_0 - \varepsilon, \theta_0 + \varepsilon)$ that $\arg e^{i\theta} f'(e^{i\theta}, t) \geq \frac{\pi}{2}$ (or $\leq -\frac{\pi}{2}$). Of course, $\arg e^{i\varphi} f'(e^{i\varphi}, 0) = \frac{\pi}{2}$ and $\arg e^{i\psi} f'(e^{i\psi}, 0) = -\frac{\pi}{2}$. In this case the curve $e^{i\theta} f'(e^{i\theta}, t)$ intersects with the imaginary axis at the points φ, ψ. Then, $\frac{\partial}{\partial \theta} \arg e^{i\theta} f'(e^{i\theta}, t) \geq 0$, or $\mathrm{Re}\,(1 + \zeta f''(\zeta, 0)/f'(\zeta, 0)) > 0$ for $\zeta = re^{i\theta}$ for all $r \sim 1$, $r \neq 1$, and $\theta \sim \varphi$ or ψ. Thus, the level lines of the function f are of positive curvature, therefore, due to the argument of continuity, they are still of positive curvature locally in time $t > 0$ and reachable by horizontal rays. So we suppose that $\arg \zeta_0 \equiv \theta_0 \neq \varphi$, or ψ on the smooth boundary of $f(\mathbb{D}, 0) = \Omega_0$. Let us assume $\theta_0 \in [0, \varphi) \cup (\psi, 2\pi]$. For other locations of θ_0 the proof is similar. For definiteness, we put

$$\arg \zeta_0 f'(\zeta_0, 0) = \frac{\pi}{2}. \tag{4.24}$$

Since $f'(e^{i\theta}, t) \neq 0$, we have that $\mathrm{Im}\,\zeta_0 f'(\zeta_0, 0) > 0$. Since ζ_0 is a critical point and the image of the unit disk $\mathbb{D}$ under the mapping $\zeta f'(\zeta, 0)$ touches the imaginary axis at the point $\zeta_0 = e^{i\theta_0}$, we deduce that

$$\frac{\partial}{\partial \theta} \arg e^{i\theta} f'(e^{i\theta}, 0)\big|_{\theta = \theta_0} = 0,$$

$$\frac{\partial}{\partial r} \arg e^{i\theta_0} f'(re^{i\theta_0}, 0)\big|_{r=1} \geq 0.$$

We calculate

$$\mathrm{Re}\left[1 + \frac{\zeta_0 f''(\zeta_0, 0)}{f'(\zeta_0, 0)}\right] = 0, \tag{4.25}$$

$$\mathrm{Im}\left[1 + \frac{\zeta_0 f''(\zeta_0, 0)}{f'(\zeta_0, 0)}\right] \geq 0. \tag{4.26}$$

Let us show that in (4.26) the equality sign is never attained. If it were so, we would conclude that $(\zeta f'(\zeta, 0))'\big|_{\zeta = \zeta_0} = 0$ because of (4.24). This means that the function $\mathrm{Re}\,\zeta f'(\zeta, 0)$ admits both signs in a neighbourhood of ζ_0 in $\mathbb{D}$. This contradicts the condition that the level lines preserve the convexity property in the direction of $\mathbb{R}$. Therefore, there is a strict inequality in (4.26).

Then we derive

$$\frac{\partial}{\partial t} \arg \zeta f'(\zeta, t) = \mathrm{Im}\,\frac{\partial}{\partial t} \log f'(\zeta, t) = \mathrm{Im}\,\frac{\frac{\partial}{\partial t} f'(\zeta, t)}{f'(\zeta, t)}. \tag{4.27}$$

Now we use the result of differentiating (4.17) and come to the equality

$$|f'(\zeta, t)|^2 \mathrm{Im}\,\frac{\frac{\partial}{\partial t} f'(\zeta, t)}{f'(\zeta, t)} = \mathrm{Im}\,(\overline{\zeta f'(\zeta, t)}\,\dot{f}(\zeta, t))\left(\frac{\overline{\zeta f''(\zeta, t)}}{f'(\zeta, t)} - 1\right).$$

Substituting (4.37) and (4.25) in the latter expression we finally have

$$\frac{\partial}{\partial t} \arg \zeta f'(\zeta, t)\Big|_{\zeta = e^{i\theta_0}} = -\frac{Q}{2\pi |f'(e^{i\theta_0}, 0)|^2}\mathrm{Im}\left(1 + \frac{e^{i\theta_0} f''(e^{i\theta_0}, 0)}{f'(e^{i\theta_0}, 0)}\right).$$

The right-hand side of this equality is strictly negative because of (4.26) and the remark thereafter. Hence, $\arg e^{i\theta} f'(e^{i\theta}, t) < \frac{\pi}{2}$ for $t > 0$ close to 0 in a neighbourhood of θ_0. This contradicts the assumption that ζ_0 is a critical point and ends the proof for the class $C_{\mathbb{R}}$. $\square$

Close-to-convex dynamics

Of course, a function from $C_{\mathbb{R}}$ is close-to-convex. But in general, a result analogous to the previous theorem for close-to-convex functions is not true [272], because the level lines of a close-to-convex function continues to be close-to-convex but may fail to be $C_{\mathbb{R}}$. Here we give an example to prove that the solutions of the inner problem do not preserve the property of the initial flow domain to be close-to-convex.

It is known [295] that close-to-convexity is equivalent to the following analytic assertion: for any θ_1, θ_2 such that $0 < \theta_2 - \theta_1 < 2\pi$, the inequality

$$\int_{\theta_1}^{\theta_2} \left[1 + \operatorname{Re} \frac{re^{i\theta} f''(re^{i\theta})}{f'(re^{i\theta})} \right] d\theta \geq -\pi, \quad 0 < r \leq 1,$$

holds (the equality sign is possible when $r = 1$). Let

$$H(\theta_1, \theta_2, f(\zeta, t)) = \int_{\theta_1}^{\theta_2} \left[1 + \operatorname{Re} \left(\zeta \frac{\partial}{\partial \zeta} \log f'(\zeta, t) \right) \right] d\theta, \quad \zeta = re^{i\theta}.$$

Let $f(\zeta, t')$ be a critical close-to-convex mapping, i.e., there are θ_1 and θ_2, such that $0 < \theta_2 - \theta_1 < 2\pi$ and $H(\theta_1, \theta_2, f(e^{i\theta}, t')) = -\pi$. Without loss of generality we assume $t' = 0$. Therefore, the integrals $J_1(\theta) = H(\theta, \theta_2, f_0(e^{i\theta}))$ and $J_2(\theta) = H(\theta_1, \theta, f_0(e^{i\theta}))$, as differentiable functions of the first and the second argument of H respectively, have the local maxima $J_1(\theta)$ at the point θ_1 and $J_2(\theta)$ at the point θ_2, i.e., $J_1'(\theta_1) = 0$ and $J_2'(\theta_2) = 0$. This means

$$\operatorname{Re} \left(e^{i\theta_1} \frac{f_0''(e^{i\theta_1})}{f_0'(e^{i\theta_1})} \right) = \operatorname{Re} \left(e^{i\theta_2} \frac{f_0''(e^{i\theta_2})}{f_0'(e^{i\theta_2})} \right) = -1. \tag{4.28}$$

The function $J_3(r) = H(\theta_1, \theta_2, f_0(re^{i\theta}))$ locally decreases in $r \in (0, 1]$ in the neighbourhood of 1^-, and J_3 is differentiable in this semi-interval. Hence,

$$J_3'(1^-) = \operatorname{Im} \int_{\theta_1}^{\theta_2} d\left(\frac{e^{i\theta} f_0''(e^{i\theta})}{f_0'(e^{i\theta})} \right) = \operatorname{Im} \left(\frac{e^{i\theta_2} f_0''(e^{i\theta_2})}{f_0'(e^{i\theta_2})} - \frac{e^{i\theta_1} f_0''(e^{i\theta_1})}{f_0'(e^{i\theta_1})} \right) \leq 0.$$

$$\tag{4.29}$$

Checking the sign of

$$\frac{\partial H(\theta_1, \theta_2, f_0(e^{i\theta}))}{\partial t}$$

at the point $t = 0^+$ we come to the decision about close-to-convexity. If it is negative, then there is a neighbourhood $(0, \varepsilon)$ where $H(\theta_1, \theta_2, f(\zeta, t)) < -\pi$, that contradicts the condition of close-to-convexity.

As in preceding subsections we deduce from the Polubarinova–Galin equation

$$\frac{\partial}{\partial t} H(\theta_1, \theta_2, f(e^{i\theta}, t)) \bigg|_{t=0} = \frac{Q}{|f'(e^{i\theta}, 0)|^2} \operatorname{Im} \frac{e^{i\theta} f''(e^{i\theta}, 0)}{f'(e^{i\theta}, 0)} \bigg]_{\theta_1}^{\theta_2}. \qquad (4.30)$$

We have $Q > 0$ for injection and consider an example critical map

$$f_0(\zeta) = \int\limits_0^\zeta \exp\left[-\frac{1}{2\pi} \int\limits_0^{2\pi} \left(\gamma(\theta) - \theta - \frac{\pi}{2}\right) S(\theta, \zeta) d\theta\right] d\zeta,$$

where $S(\theta, \zeta)$ is the Schwarz–Poisson kernel

$$S(\theta, \zeta) = \frac{e^{i\theta} + \zeta}{e^{i\theta} - \zeta},$$

$\gamma(\theta) = \frac{3}{2}\pi(1 + \sin(\frac{\alpha}{\pi}(\theta - \pi)))$, $\alpha = \pi - \arcsin\frac{2}{3}$, $\theta_2 = -\theta_1 = \pi - \frac{\pi^2}{2\alpha}$. This function satisfies the condition $f_0(\bar\zeta) = \overline{f_0(\zeta)}$, hence $|f_0'(e^{i\theta_1})| = |f_0'(e^{i\theta_2})|$. By (4.29) we obtain that the right-hand side in (4.30) is not positive. So it suffices to prove that

$$\operatorname{Im} \frac{e^{i\theta} f_0''(e^{i\theta})}{f_0'(e^{i\theta})} \bigg|_{\theta_1}^{\theta_2} \neq 0.$$

Integrating by parts we get

$$\operatorname{Im} \frac{e^{i\theta} f_0''(e^{i\theta})}{f_0'(e^{i\theta})} \bigg|_{\theta_1}^{\theta_2} = \frac{-3\alpha^2}{2\pi^2} \int\limits_0^{2\pi} \sin(\frac{\alpha}{\pi}(\theta - \pi)) \log\left[1 + \cos\left(\theta + \frac{\pi^2}{2\alpha}\right)\right] d\theta. \qquad (4.31)$$

From the obvious inequality $\frac{\pi}{2} < \frac{\pi^2}{2\alpha} < \pi$ it easily follows that the right-hand side of (4.31) is strictly negative and remains negative for $f(\zeta, t)$ in some time interval $t \in [0, \varepsilon)$.

To complete the proof we show that $f_0(\zeta)$ is close-to-convex and univalent verifying the condition $\operatorname{Re} f_0'(\zeta) \geq 0$. This condition is equivalent to the inequality

$$-\frac{\pi}{2} \leq \operatorname{Re} \frac{1}{2\pi} \int\limits_0^{2\pi} \left(\gamma(\theta) - \theta - \frac{\pi}{2}\right) \frac{e^{i\theta} + \zeta}{e^{i\theta} - \zeta} d\theta \leq \frac{\pi}{2}.$$

The right-hand side inequality is equivalent to $\int_0^{2\pi} (\gamma(\theta) - \theta - \pi) P(\zeta, \theta) d\theta \leq 0$, where $P(\zeta, \theta) \equiv \operatorname{Re} S(\zeta, \theta)$ is the Poisson kernel. The sign is obviously verified. The left-hand side inequality can be considered analogously.

4.4.2 Dynamics with small surface tension

As we mentioned in Section 1.4.4, in most practical experiments zero surface tension process never occurs in the three-dimensional case. A 2D approximation of the 3D effect is given by introducing surface tension in the 2D case. At the same time the non-zero surface tension model regularizes the ill-posed problem.

We recall from Section 1.4.4 that the governing equations for the non-zero surface tension model are

$$-\Delta p = Q\delta_0 \quad \text{in} \quad \Omega(t), \tag{4.32}$$

$$p = \gamma\kappa \quad \text{on} \quad \Gamma(t), \tag{4.33}$$

$$V_n = -\frac{\partial p}{\partial n} \quad \text{on} \quad \Gamma(t), \tag{4.34}$$

where κ is the curvature of the boundary and γ is surface tension. The problem of the existence of the solution in the non-zero surface tension case was discussed in Section 1.4.4.

Now we obtain an equation for the free boundary using an auxiliary parametric univalent map. To derive it let us consider the complex potential $W(z,t)$, Re $W = p$. For each fixed t this is an analytic function defined in $\Omega(t)$ which solves the Dirichlet problem (4.32)–(4.33) in the sense that its real part induces the same distribution as the solution of the problem (4.32)–(4.33). In the neighbourhood of the origin we have the expansion

$$W(z,t) = -\frac{Q}{2\pi} \log z + w_0(z,t), \tag{4.35}$$

where $w_0(z,t)$ is an analytic regular function in $\Omega(t)$.

To derive the equation for the free boundary $\Gamma(t)$ we use the same arguments as in Sections 1.4.1, 1.4.2 and consider the Riemann conformal univalent map $f(\zeta,t)$ from the unit disk $\mathbb{D}$ into the phase plane $f : \mathbb{D} \to \Omega(t)$, $f(0,t) = 0$, $f'(0,t) > 0$. Then the moving boundary is parameterized by $\Gamma(t) = \{f(e^{i\theta},t), \theta \in [0,2\pi)\}$. The normal velocity V_n of $\Gamma(t)$ in the outward direction is given by $V_n = -\partial p/\partial n$. The normal outer vector is given by the formula

$$\mathbf{n} = \zeta\frac{f'}{|f'|}, \quad \zeta \in \partial\mathbb{D}.$$

Therefore, the normal velocity is obtained as

$$V_n = \mathbf{V} \cdot \mathbf{n} = -\text{Re}\left(\frac{\partial W}{\partial z}\zeta\frac{f'}{|f'|}\right).$$

The superposition $(W \circ f)(\zeta,t)$ is an analytic function in the unit disk. Since the Laplacian is invariant under a conformal map, the solution to the Dirichlet problem (4.32)–(4.33) is given in terms of the ζ-plane as

$$(W \circ f)(\zeta, t) = -\frac{Q}{2\pi} \log \zeta + \frac{\gamma}{2\pi} \int_0^{2\pi} \kappa(e^{i\theta}, t) \frac{e^{i\theta} + \zeta}{e^{i\theta} - \zeta} d\theta + iC, \qquad (4.36)$$

where

$$\kappa(e^{i\theta}, t) = \frac{\mathrm{Re}\,\left(1 + e^{i\theta} f''(e^{i\theta}, t)/f'(e^{i\theta}, t)\right)}{|f'(e^{i\theta}, t)|}, \quad \theta \in [0, 2\pi).$$

We calculate

$$\frac{\partial \kappa(e^{i\theta}, t)}{\partial \theta} = \frac{-\mathrm{Im}\, e^{2i\theta} S_f(e^{i\theta}, t)}{|f'(e^{i\theta}, t)|},$$

with the Schwarzian derivative (see, e.g., [156], [446], [447])

$$S_f(\zeta) = \frac{\partial}{\partial \zeta} \left(\frac{f''(\zeta, t)}{f'(\zeta, t)}\right) - \frac{1}{2} \left(\frac{f''(\zeta, t)}{f'(\zeta, t)}\right)^2.$$

Differentiating (4.36) we deduce that

$$\zeta \frac{\partial W}{\partial z} f'(\zeta, t) = -\frac{Q}{2\pi} + \frac{\gamma}{\pi} \int_0^{2\pi} \frac{\kappa(e^{i\theta}) \zeta e^{i\theta}}{(e^{i\theta} - \zeta)^2} d\theta, \quad \zeta \in \mathbb{D}.$$

Integrating by parts we obtain

$$\zeta \frac{\partial W}{\partial z} f'(\zeta, t) = -\frac{Q}{2\pi} - \frac{\gamma}{2\pi i} \int_0^{2\pi} \frac{\mathrm{Im}\, e^{2i\theta} S_f(e^{i\theta}, t)}{|f'(e^{i\theta}, t)|} \frac{e^{i\theta} + \zeta}{e^{i\theta} - \zeta} d\theta.$$

On the other hand, we have $V_n = \mathrm{Re}\, \dot{f}\, \overline{\zeta f'}/|f'|$, and applying the Sokhotskiĭ–Plemelj formulae [399] we, finally, get

$$\mathrm{Re}\, \dot{f}(\zeta, t) \overline{\zeta f'(\zeta, t)} = \frac{Q}{2\pi} - \gamma H \left[i\mathrm{Im}\, \frac{\zeta^2 S_f(\zeta, t)}{|f'(\zeta, t)|}\right](\theta), \qquad (4.37)$$

$\zeta = e^{i\theta}$, where the Hilbert transform in (4.37) is of the form

$$\frac{1}{\pi} \mathbf{p.v.}_\theta \int_0^{2\pi} \frac{\psi(e^{i\theta'}) d\theta'}{1 - e^{i(\theta - \theta')}} = H[\psi](\theta).$$

For $\gamma = 0$ we just have equation (1.22).

From (4.37) one can derive a Löwner–Kufarev type equation by the Schwarz–Poisson formula:

$$\dot{f} = -\zeta f' \frac{1}{2\pi} \int_0^{2\pi} \frac{1}{|f'(e^{i\theta}, t)|^2} \left(-\frac{Q}{2\pi} + \gamma H \left[i\mathrm{Im}\, \frac{e^{2i\theta} S_f(e^{i\theta}, t)}{|f'(e^{i\theta}, t)|}\right](\theta)\right) \frac{e^{i\theta} + \zeta}{e^{i\theta} - \zeta} d\theta,$$

$$(4.38)$$

where $\zeta \in \mathbb{D}$.

An interesting question appears when $\gamma \to 0$. It turns out that the solution in the limiting γ-surface-tension case need not always be the corresponding zero surface tension solution (see the discussion in [522, 547, 554]). This means that starting with a domain $\Omega(0) = \Omega(0, \gamma)$ we come to the domain $\Omega(t, \gamma)$ at an instant t using surface tension γ and to the domain $\Omega(t)$ at the same instant t in the zero surface tension model. Then the domain $\lim_{\gamma \to 0} \Omega(t, \gamma) = \Omega(t, 0)$ is not necessarily the same as $\Omega(t)$ (see numerical evidence in [459]). If the boundary Γ is highly curved, then the condition (4.33) must be used even though γ is small. Obviously, the non-zero surface tension model never develops cusps. Thus, solutions and geometric behaviour of the free boundary for small γ are of particular interest.

4.4.3   Geometric properties in the presence of surface tension

We need the following technical lemma.

Lemma 4.4.1. *For the function $f : \mathbb{D} \to \mathbb{C}$ which parameterizes the phase domain $\Omega(t)$ we have the equality*

$$\frac{\partial}{\partial \theta} H\left[\frac{ie^{2i\theta}\,\mathrm{Im}\,S_f(e^{i\theta})}{|f'(e^{i\theta})|}\right](\theta) = -H[iA](\theta)$$

with

$$A(\zeta) = \frac{\mathrm{Re}\left(2\zeta^2 S_f(\zeta) + \zeta\left[\left(\frac{f''(\zeta)}{f'(\zeta)}\right)'' - \frac{f''(\zeta)}{f'(\zeta)}\left(\frac{f''(\zeta)}{f'(\zeta)}\right)'\right]\right) + \mathrm{Im}\,\frac{\zeta f''(\zeta)}{f'(\zeta)}\,\mathrm{Im}\,\zeta^2 S_f(\zeta)}{|f'(\zeta)|}.$$

Proof. Let

$$\Phi(\zeta) = \frac{i}{\pi}\int\limits_0^{2\pi} \frac{\mathrm{Im}\,e^{2i\theta'}\,S_f(e^{i\theta'})}{|f'(e^{i\theta'}, t)|}\,\frac{e^{i\theta'}}{e^{i\theta'} - \zeta}\,d\theta', \quad \zeta \in \mathbb{D}.$$

Then, by the Sokhotskiĭ–Plemelj formulae we deduce that

$$\lim_{\zeta \to (1-0)e^{i\theta}} \Phi(\zeta) = i\frac{\mathrm{Im}\,e^{2i\theta'}\,S_f(e^{i\theta'})}{|f'(e^{i\theta'}, t)|} - H\left[i\frac{\mathrm{Im}\,e^{2i\theta'}\,S_f(e^{i\theta'})}{|f'(e^{i\theta'}, t)|}\right](\theta).$$

We note that the second term in the above relation is real, which can be easily seen by the definition of Φ and the Schwarz integral formula. The left-hand side in (4.37) is differentiable in θ and that is why one can calculate the derivative in question as the limit

$$\frac{\partial}{\partial \theta} H\left[i\frac{\mathrm{Im}\,e^{2i\theta'}\,S_f(e^{i\theta'})}{|f'(e^{i\theta'}, t)|}\right](\theta) = \mathrm{Im}\lim_{\zeta \to (1-)e^{i\theta}} \zeta\Phi'(\zeta).$$

We calculate

$$\Phi'(\zeta) = \frac{i}{\pi} \int\limits_0^{2\pi} \frac{\operatorname{Im} e^{2i\theta'} S_f(e^{i\theta'})}{|f'(e^{i\theta'}, t)|} \frac{e^{i\theta'}}{(e^{i\theta'} - \zeta)^2} d\theta'.$$

Integration by parts leads to the equality

$$\zeta \Phi'(\zeta) = \frac{1}{\pi} \int\limits_0^{2\pi} A(e^{i\theta'}) \frac{\zeta \, d\theta'}{e^{i\theta'} - \zeta}.$$

Thus, we apply the Sokhotskiĭ–Plemelj formulae once again and get the assertion of the Lemma 4.4.1. □

Theorem 4.4.4. *Let $Q > 0$ and the surface tension γ be sufficiently small. If the initial domain $\Omega(0)$ is strongly starlike of order α, then there exists $t = t(\gamma) \leq t_0$, such that the family of domains $\Omega(t)$ (in the sequel, the family of univalent functions $f(\zeta, t)$) preserves this property during the time $t \in [0, t(\gamma)]$.*

Remark 4.4.2. For $\gamma = 0$ we have the result of Theorem 4.3.2.

Proof. If we consider f in the closure of $\mathbb{D}$, then the inequality sign in (4.9) can be replaced by ($\leq$) where equality can be attained for $|\zeta| = 1$.

We suppose that there exists a critical map $f \in S^*(\alpha)$ of exact order α, that is the image of $\mathbb{D}$ under the map $\zeta f'(\zeta, t)/f(\zeta, t)$, $|\zeta| \leq 1$ touches the boundary rays $l^\pm$ of the angle $\arg w \in [-\alpha\frac{\pi}{2}, \alpha\frac{\pi}{2}]$. Say there exist such $t' \geq 0$ and $\zeta_0 = e^{i\theta_0}$, that

$$\arg \frac{\zeta_0 f'(\zeta_0, t')}{f(\zeta_0, t')} = \alpha \frac{\pi}{2} \quad (\text{or } -\alpha\frac{\pi}{2}), \tag{4.39}$$

and for any $\varepsilon > 0$ there are $t > t'$ and $\theta \in (\theta_0 - \varepsilon, \theta_0 + \varepsilon)$ such that

$$\arg \frac{e^{i\theta} f'(e^{i\theta}, t)}{f(e^{i\theta}, t)} \geq \alpha \frac{\pi}{2} \quad (\text{or } \leq -\alpha\frac{\pi}{2}).$$

For definiteness we put the sign $(+)$ in (4.39). Without loss of generality, let us assume $t' = 0$. Since $f'(e^{i\theta}, t) \neq 0$, our assumption about the sign in (4.39) yields that

$$\operatorname{Im} \frac{\zeta_0 f'(\zeta_0, 0)}{f(\zeta_0, 0)} > 0, \tag{4.40}$$

(the negative case is considered similarly).

Since ζ_0 is a critical point and the image of the unit disk $\mathbb{D}$ under the mapping $\frac{\zeta f'(\zeta, 0)}{f(\zeta, 0)}$ touches the ray l^+ at the point $\zeta_0 = e^{i\theta_0}$, we deduce that

$$\frac{\partial}{\partial \theta} \arg \frac{e^{i\theta} f'(e^{i\theta}, 0)}{f(e^{i\theta}, 0)} \bigg|_{\theta = \theta_0} = 0,$$

$$\frac{\partial}{\partial r} \arg \frac{r e^{i\theta_0} f'(r e^{i\theta_0}, 0)}{f(r e^{i\theta}, 0)} \bigg|_{r=1} \geq 0.$$

We calculate

$$\text{Re}\left[1+\frac{\zeta_0 f''(\zeta_0,0)}{f'(\zeta_0,0)}-\frac{\zeta_0 f'(\zeta_0,0)}{f(\zeta_0,0)}\right]=0, \tag{4.41}$$

$$\text{Im}\left[1+\frac{\zeta_0 f''(\zeta_0,0)}{f'(\zeta_0,0)}-\frac{\zeta_0 f'(\zeta_0,0)}{f(\zeta_0,0)}\right]\geq 0. \tag{4.42}$$

One can derive

$$\frac{\partial}{\partial t}\arg\frac{\zeta f'(\zeta,t)}{f(\zeta,t)}=\text{Im}\,\frac{\partial}{\partial t}\log\frac{f'(\zeta,t)}{f(\zeta,t)}=\text{Im}\left(\frac{\frac{\partial}{\partial t}f'(\zeta,t)}{f'(\zeta,t)}-\frac{\frac{\partial}{\partial t}f(\zeta,t)}{f(\zeta,t)}\right). \tag{4.43}$$

We now differentiate the equation (4.37) with respect to θ using Lemma 4.4.1,

$$\text{Im}\left(\overline{f'(\zeta,t)}\frac{\partial}{\partial t}f'(\zeta,t)-\overline{\zeta f'(\zeta,t)}\dot{f}(\zeta,t)-\overline{\zeta^2 f''(\zeta,t)}\dot{f}(\zeta,t)\right)=-\gamma H[iA](\theta), \tag{4.44}$$

for $\zeta=e^{i\theta}$. This equality is equivalent to the following:

$$|f'(\zeta,t)|^2\text{Im}\left(\frac{\frac{\partial}{\partial t}f'(\zeta,t)}{f'(\zeta,t)}-\frac{\frac{\partial}{\partial t}f(\zeta,t)}{f(\zeta,t)}\right)$$

$$=\text{Im}\,(\overline{\zeta f'(\zeta,t)}\dot{f}(\zeta,t))\left(\frac{\zeta f''(\zeta,t)}{f'(\zeta,t)}-\frac{\zeta f'(\zeta,t)}{f(\zeta,t)}+1\right)-\gamma H[iA](\theta).$$

Substituting (4.37) and (4.41) in the latter expression we finally have

$$\frac{\partial}{\partial t}\arg\frac{\zeta f'(\zeta,t)}{f(\zeta,t)}\bigg|_{\zeta=e^{i\theta_0},t=0}$$

$$=\frac{1}{|f'(e^{i\theta_0},0)|^2}\left(-\frac{Q}{2\pi}+\gamma H\left[i\text{Im}\,\frac{e^{2i\theta}S_f(e^{i\theta},t)}{|f'(e^{i\theta},t)|}\right](\theta_0)\right) \tag{4.45}$$

$$\times\text{Im}\left(\frac{e^{i\theta_0}f'(e^{i\theta_0},0)}{f(e^{i\theta_0},0)}+\frac{e^{i\theta_0}f''(e^{i\theta_0},0)}{f'(e^{i\theta_0},0)}\right)-\gamma\frac{H[iA(e^{i\theta})](\theta_0)}{|f'(e^{i\theta_0},0)|^2}.$$

The right-hand side of this equality is strictly negative for small γ because of (4.40), (4.42). Therefore,

$$\arg\frac{e^{i\theta}f'(e^{i\theta},t)}{f(e^{i\theta},t)}<\alpha\frac{\pi}{2}$$

for $t>0$ (close to 0) in some neighbourhood of θ_0. This contradicts the assumption that $\Omega(t)$ fails to be starlike for $t>0$ and ends the proof for the class $S^*(\alpha)$. □

Problem 4.4.1. Let $\Omega(0)$ be starlike of order α (not strongly), $f(\zeta,0)\in S^*_\alpha$, and $Q>0$. Do the domains of the Hele-Shaw evolution $\Omega(t)$ remain starlike of the same order?

Problem 4.4.2. Let $\Omega(0)$ be convex. It is known that for $Q > 0$ the Hele-Shaw evolution $\Omega(t)$ is not necessarily convex. Are there conditions that guarantee convexity of $\Omega(t)$ all the time?

Problem 4.4.3. If $Q = 0$ and $\gamma \neq 0$, then the driving mechanism is only surface tension. Let $\Omega(0)$ be convex. Are the domains $\Omega(t)$ convex for all t?

4.4.4   Unbounded regions with bounded complement

We now consider a Hele-Shaw cell in which the fluid occupies a full neighbourhood of infinity, so that the complementary set is a finite bubble. Injection or suction is supposed to take place at the point of infinity. This model has various applications in the boundary problems of gas mechanics, problems of metal or polymer swamping, etc., where the air viscosity is neglected. More about this problem is found in [175, 279].

As usual we denote by $\Omega(t)$ the fluid domain, i.e., in this case the unbounded complement of the bubble in consideration (see Figure 4.4). Let p be the pressure in the domain $\Omega(t)$ occupied by the fluid. We construct the complex potential $W(z,t)$, $\mathrm{Re}\, W = p$. For each fixed t this is an analytic function defined in $\Omega(t)$ which solves the problem

$$-\Delta p = 0 \qquad \text{in} \quad \Omega(t), \tag{4.46}$$

$$p = 0 \qquad \text{on} \quad \Gamma(t), \tag{4.47}$$

$$V_n = -\frac{\partial p}{\partial n} \qquad \text{on} \quad \Gamma(t), \tag{4.48}$$

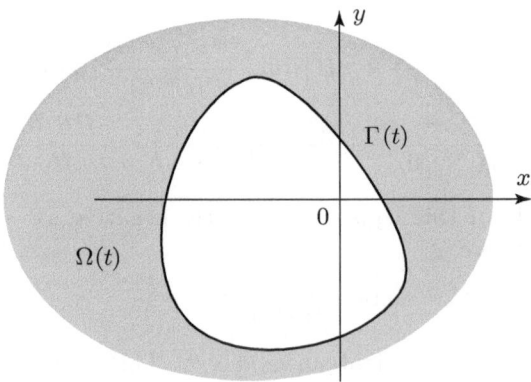

Fig. 4.4: $\Omega(t)$ is the complement to a bounded simply
connected bubble with the boundary $\Gamma(t)$ and
the sink/source at infinity

normalized about infinity by

$$p \sim \frac{Q}{2\pi} \log |z|, \quad \text{as } |z| \to \infty,$$

where Q is the rate of bubble release caused by air extraction, $Q > 0$ in the case of contracting bubble, $Q < 0$ otherwise. Thus p is Q times the Green function of $\Omega(t)$ with pole at infinity, and V_n equals Q times the density of the equilibrium measure for the bubble $\mathbb{C} \setminus \Omega(t)$.

Mathematical treatment for the case of a contracting bubble was presented in [175]. In particular, the problem of the limiting configuration was solved. It was proved that the moving boundary tends to a finite number of points which give the minimum to a certain potential. There an interesting problem was posed: to describe domains whose dynamics presents only one limiting point. Howison [279] proved that a contracting elliptic bubble has a homothetic dynamics to a point (in particular, this is obvious for a circular one). Entov, Ètingof [175] (see also [560]) have shown that a contracting bubble which is convex at the initial instant preserves this property up to the moment when its boundary reduces to a point. These domains are called "simple" in [175]. Further results on bubble contraction are given in [177].

Let us set up in the table some information about the dynamics for bounded and unbounded domains for the inner and outer stable (well-posed) problems known at the moment using special univalent functions. Here 'yes' means that the property is preserved whereas 'no' means that is not. For the outer problem we consider the complement to $\Omega(t)$.

Class of univalent functions	Inner problem	Outer problem
starlike (or strongly starlike)	yes	no
convex	no	yes
close-to-convex	no	no
convex in the direction of $\mathbb{R}$ (with the condition for level lines)	yes	no

For the ill-posed case fewer results are known. Injected air forms a bubble which grows as time increases. It was shown [279] that three kinds of behavior can occur. Firstly, the solution may cease to exist in finite time; secondly, the solution may exist for all time and the free boundary may have one or more limit points as t tends to infinity; and thirdly, the bubble may exist for all time and fill the whole space as t tends to infinity. As remarked earlier (Section 3.3.3) the latter case occurs if and only if the initial bubble is elliptical, because $\Omega(0)$ has to be a null quadrature domain. The multidimensional case was treated in [140], [194]. See also [296] and [298].

4.4.5 Unbounded regions with the boundary extending to infinity

This model corresponds to the moving fluid front which for definiteness we suppose
to be located to the right. More precisely, we denote by $\Omega(t)$ a simply connected
domain in the phase z-plane occupied by the moving fluid and its moving boundary
$\Gamma(t) = \partial\Omega(t)$ contains the point at infinity. With $z = x + iy$ one can construct a
parametrization $\Gamma(t)$ by the equation $\phi(x, y, t) \equiv \phi(z, t) = 0$, so that $\phi(\infty, t) \equiv$
0. Assuming a natural normalization for $\Gamma(t)$ close to ∞, we require that $\Gamma(t)$
is a vertical straight line near infinity (see Figure 4.5). The initial situation is

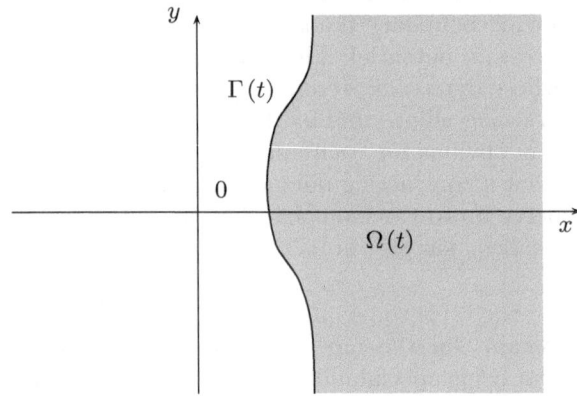

Fig. 4.5: $\Omega(t)$ is an infinite domain with the boundary
$\Gamma(t)$ extending to infinity and the sink/source
at the infinity

represented at instant $t = 0$ as $\Omega(0) = \Omega_0$, and the boundary $\partial\Omega_0 = \Gamma(0) \equiv \Gamma_0$
is defined parameterically by an implicit function $\phi(x, y, 0) = 0$. We construct
the complex potential $W(z, t)$, $\mathrm{Re}\, W = p$, where p is, as usual, a pressure field
in $\Omega(t)$. For each fixed t the potential W is an analytic function defined in $\Omega(t)$
which solves the problem

$$-\Delta p = 0 \qquad \text{in} \quad \Omega(t), \tag{4.49}$$

$$p = 0 \qquad \text{on} \quad \Gamma(t), \tag{4.50}$$

$$V_n = -\frac{\partial p}{\partial n} \qquad \text{on} \quad \Gamma(t). \tag{4.51}$$

We assume that the velocity tends to a constant value Q as $x \to \infty$, that is
negative when fluid is removed to the right and positive otherwise. In terms of the
potential p we have $p(x, y, t)/x \to Q$ as $x \to \infty$ for any t fixed. The problem of
existence was discussed in Subsection 1.4.4. It is noteworthy that for this case the
local solvability and uniqueness was proved by Kimura [309] in presence of surface
tension.

We consider the auxiliary parametric complex ζ-plane, $\zeta = \xi + i\eta$. The Riemann Mapping Theorem yields that there exists a conformal univalent map $f(\zeta, t)$ of the right half-plane $\mathbb{H}^+ = \{\zeta : \mathrm{Re}\, \zeta > 0\}$ into the phase plane $f : \mathbb{H}^+ \to \Omega(t)$. The half-plane $\mathbb{H}^+$ is a natural parametric domain for Ω. The function $f(\zeta, 0) = f_0(\zeta)$ produces a parametrization of Γ_0. The smoothness of the boundary $\Gamma(t)$ and its behavior in the neighbourhood of ∞ allows us to assume the normalization $f(\zeta, \cdot) = a\zeta + a_0 + \frac{a_{-1}}{\zeta} + \cdots$, $\zeta \sim \infty$, $a > 0$, i.e., the function f has an analytic continuation on the imaginary axis $\partial\mathbb{H}^+$ near ∞. Thus, the moving boundary is parameterized by $\Gamma(t) = f(\partial\mathbb{H}^+, t)$. The normal velocity V_n of $\Gamma(t)$ in the outward direction is given by $V_n = -\partial p/\partial n$. The normal exterior vector is given by the formula

$$\mathbf{n} = -\frac{\partial f}{\partial \zeta}\left|\frac{\partial f}{\partial \zeta}\right|^{-1}, \quad \zeta \in \partial\mathbb{H}^+.$$

The harmonic function p is a linear one. The normalization about infinity implies that $W \circ f = Q\zeta$ and the Polubarinova–Galin equation is of the form

$$\mathrm{Re}\left(\dot{f}(\zeta, t)\overline{f'(\zeta, t)}\right) = -Q, \quad \mathrm{Re}\, \zeta = 0. \tag{4.52}$$

An application of the Schwarz integral formula enables us to deduce a Löwner–Kufarev type equation in the right half-plane

$$\frac{\partial f}{\partial t} = \frac{1}{\pi}\frac{\partial f}{\partial \zeta}\int\limits_{-\infty}^{\infty} \frac{Q}{|f'(i\eta', t)|^2}\frac{d\eta'}{i\eta' - \zeta}, \quad \zeta \in \mathbb{H}^+, \tag{4.53}$$

with the initial condition $f(\zeta, 0) = f_0(\zeta)$. Taking into account surface tension this equation becomes

$$\mathrm{Re}\left(\dot{f}(\zeta, t)\overline{f'(\zeta, t)}\right) = -Q + \gamma H\left[\frac{i\mathrm{Im}\, S_f}{|f'|}\right](\eta), \quad \mathrm{Re}\, \zeta = 0,$$

with the Hilbert transform defined as

$$\frac{1}{\pi i}\mathbf{p.v.}_\eta\int\limits_{-\infty}^{\infty} \frac{\psi(i\eta')d\eta'}{\eta' - \eta} = H[\psi](\eta).$$

Hereditary properties

We are going to prove that if the initial interface possesses the property of being convex of order α in the negative direction ($H_{\mathbb{R}}^-(\alpha)$), then the free boundary remains convex in the negative direction of the same order in so far as the solution to the Hele-Shaw problem exists in the case $Q > 0$ (the liquid moves to the left). An important remark is that the level lines of a function from $H_{\mathbb{R}}^-(\alpha)$ remain convex in the negative direction.

We are going to prove the following statement.

Theorem 4.4.5. *Let $Q > 0$ and $\Omega(0)$ be a domain convex in the negative direction of order (α). Let the solutions to the equation (4.52) exist during the time $t \in [0, t_0]$. Then, for all $t \in [0, t_0]$ the family of functions $f(\zeta, t)$ and the family of domains $\Omega(t)$ preserve the same property of convexity.*

Proof. Let us again suppose the contrary. In other words, there exists $\zeta_0 = e^{i\theta_0}$ that satisfies the equality

$$\arg f'(\zeta_0, 0) = \alpha \frac{\pi}{2} \quad (\text{or } -\alpha \frac{\pi}{2}) \tag{4.54}$$

at the initial instant $t = 0$ and for any $\varepsilon > 0$ there is such $t > 0$ and $\theta \in (\theta_0 - \varepsilon, \theta_0 + \varepsilon)$ that $\arg f'(e^{i\theta}, t) \geq \alpha \frac{\pi}{2}$ (or $\leq -\alpha \frac{\pi}{2}$). For definiteness we put the sign $(+)$ in (4.54). This means that the mapping $f(\zeta, 0)$ is critical for the property of convexity in the negative direction. Since the free boundary $\Gamma(t)$ tends to a vertical line as $\eta \to \pm\infty$, we can consider finite critical points $\zeta_0 = i\eta_0$ and the set of such critical points lies in the compact subset on the imaginary axis.

$\Gamma(t)$ is analytic, the function f is analytically extendable onto the imaginary axis, and the derivative $f'(\zeta_0, t) \neq 0$. Suppose that $\operatorname{Im} f'(\zeta_0, 0) > 0$ (for $\operatorname{Im} f'(\zeta_0, 0) < 0$ the proof is similar). Now we show that $f''(\zeta_0, 0) \neq 0$. If not, then the point ζ_0 is a branch point of the function $f'(\zeta, 0)$ and in a neighbourhood of this point in $\mathbb{H}^+$ the quantity $\arg f'(\zeta, 0) - \alpha\pi/2$ admits both positive and negative values. This contradicts the assumption that the function $f(\zeta, 0)$ is convex in the negative direction.

The image of the right half-plane $\mathbb{H}^+$ under the map $f'(\zeta, 0)$ touches the ray $\arg w = \alpha \frac{\pi}{2}$ at the point $f'(\zeta_0, 0)$. Thus, the following statements are true

$$\frac{\partial}{\partial \eta} \arg f'(i\eta, 0)\Big|_{\eta=\eta_0} = 0, \quad \frac{\partial}{\partial \xi} \arg f'(\xi + i\eta_0, 0)\Big|_{\xi=0} \leq 0.$$

Calculation of the left-hand sides of these formulae leads to

$$\operatorname{Re} \frac{f''(i\eta_0, 0)}{f'(i\eta_0, 0)} = 0, \quad \operatorname{Im} \frac{f''(i\eta_0, 0)}{f'(i\eta_0, 0)} < 0. \tag{4.55}$$

We differentiate (4.52) with respect to η and at the point η_0 using (4.55) we obtain

$$|f'(i\eta, 0)|^2 \operatorname{Im} \frac{\dot{f}'(i\eta, 0)}{f'(i\eta, 0)} = \operatorname{Im} \dot{f}(i\eta, 0)\overline{f''(i\eta, 0)}$$

$$= \operatorname{Im} \dot{f}(i\eta, 0)\overline{f'(i\eta, 0)}\frac{\overline{f''(i\eta, 0)}}{\overline{f'(i\eta, 0)}}. \tag{4.56}$$

From (4.55) and (4.56) it follows that

$$\frac{\partial}{\partial t} \arg f'(i\eta_0, t)\Big|_{t=0} = -Q\operatorname{Im} \frac{f''(i\eta_0, 0)}{f'(i\eta_0, 0)}. \tag{4.57}$$

The inequality in (4.55) with $Q > 0$ implies that the right-hand side of (4.57) is strictly negative for $t > 0$ close to 0 and the inequality $\arg f'(i\eta_0, t) < \alpha\frac{\pi}{2}$ holds in some neighbourhood of $i\eta_0$. This contradicts the assumption that $\Omega(t)$ fails to be convex in the negative direction for $t \geq 0$ and ends the proof. $\square$

Finally, in this section we should say that there are several other processes that involve planar dynamics in Hele-Shaw cells. Let us refer the reader, for example, to the papers [60, 182, 277, 282, 423], [452] and the references therein. We mention here a 600-paper bibliography of free and moving boundary problems for Hele-Shaw and Stokes flow since 1898 up to 1998 collected by Gillow and Howison [208]. Let us mention also a recent work [261] where the authors study the Hele-Shaw flow on hyperbolic surfaces.

Most of the results presented in this section are found in [247, 272, 454, 564, 561, 563].

4.5   Infinite life-span of starlike dynamics

In this section we prove precisely that, starting with a starlike bounded analytic phase domain Ω_0, the Hele-Shaw chain of subordinating domains $\Omega(t)$, $\Omega_0 = \Omega(0)$ exists for all time under injection at the point of starlikeness. Suppose that at the initial time the phase domain Ω_0 occupied by the fluid is simply connected and bounded by a smooth analytic curve Γ_0.

In Section 4.4 we proved (Theorem 4.4.1) that, starting with a phase domain Ω_0 which is strongly starlike of order α and bounded by an analytic curve, we obtain a subordination chain of domains $\Omega(t)$ (and functions $f(\zeta, t)$) strongly starlike of order $\alpha(t)$ with a decreasing order $\alpha(t)$.

In this section we shall first prove that if the strong solution to (1.22) exists during the time interval $[0, t_0)$, then the limiting function $\lim\limits_{t \to t_0 - 0} f(\zeta, t) \equiv f(\zeta, t_0)$ is analytic in some neighbourhood of the unit disk $\mathbb{D}$. Here the limit is taken with respect to the uniform convergence on compacts of the unit disk $\mathbb{D}$. It exists because $f(\zeta, t)$ is a subordination chain and due to the Carathéodory Kernel Theorem. Then, we shall give the main result about the infinite life-span, see also [247].

Theorem 4.5.1. *Let the strong solution to* (1.22) *with* $Q > 0$ *exist during the time interval* $[0, t_0)$, $0 < t_0 < \infty$, $\Omega(t) = f(\mathbb{D}, t)$, *and let the initial function* $f(\zeta, 0)$ *be analytic and univalent in a neighbourhood of the closure of the unit disk* $\mathbb{D}$. *Then the function* $f(\zeta, t)$ *is analytic in* $\mathbb{D}_{R(t)}$, *where the radius of analyticity* $R(t) > 1$ *is a nondecreasing function in* $t \in [0, t_0]$. *The function* $f(\zeta, t)$ *is univalent in* $\mathbb{D}$, *and possibly* $f(\zeta, t_0)$ *has a vanishing derivative at some points of the unit circle* $\partial\mathbb{D}$ *or is not univalent on* $\partial\mathbb{D}$. *It follows that* $\Omega(t_0) \equiv f(\mathbb{D}, t_0)$ *is a simply connected domain with an analytic boundary* $\partial\Omega(t_0)$ *with possible analytic singularities in the form of finitely many cusps and double points. In the case there are no singularities the strong solution can be extended to some time interval* $[0, t_0 + \varepsilon)$.

Proof. By the Carathéodory Kernel Theorem the domain

$$\Omega(t_0) = \bigcup_{t \in [0,t_0)} \Omega(t)$$

is just the same as in the formulation of the theorem and $\Omega(t_0)$ is a simply connected domain. It follows from Proposition 3.3.1 that $\Omega(t_0)$ is also the same as the domain at time t_0 for the weak solution.

We note also that since the normal velocity on the boundary never vanishes, we have strict monotonicity of the subordination chain of domains:

$$\overline{\Omega(s)} \subset \Omega(t) \quad \text{for } s < t \text{ and } s, t \in (0, t_0). \tag{4.58}$$

Letting $t \to t_0$ we see that (4.58) holds for $t = t_0$, i.e., for $\Omega(t_0)$ as well.

The strong solution exists in the time interval $t \in [0, t_0)$ and coincides with the weak one. Therefore, the statements about $f(\zeta, t_0)$ and $\partial\Omega(t_0)$ follow directly from Theorem 3.1.2.

Let us prove existence of an extension of the solution to the time interval $[0, t_0 + \varepsilon)$ when there are no singularities on $\partial\Omega(t_0)$. Construct the subordination chain of mappings $f_2(\zeta, t)$ satisfying the Polubarinova–Galin equation (1.22) with the initial data $f_2(\zeta, 0) \equiv f(\zeta, t_0)$. The strong solution exists and is unique locally in time, say $t \in [0, \varepsilon)$. Moreover, we have $\lim_{t \to t_0 - 0} f(\zeta, t) = \lim_{t \to 0+0} f_2(\zeta, t) = f(\zeta, t_0)$ and $\lim_{t \to t_0 - 0} f'(\zeta, t) = \lim_{t \to 0+0} f_2'(\zeta, t) = f'(\zeta, t_0)$ locally uniformly in $\mathbb{D}_{1+\eta}$. We recall equation (1.23):

$$\dot{f}(\zeta, t) = \zeta f'(\zeta, t) \frac{Q}{4\pi^2} \int_0^{2\pi} \frac{1}{|f'(e^{i\theta}, t)|^2} \frac{e^{i\theta} + \zeta}{e^{i\theta} - \zeta} d\theta, \quad t \in [0, t_0), \ |\zeta| < 1.$$

A similar equation is valid for the chain $f_2(\zeta, t)$ in the time interval $[0, \varepsilon)$. Taking the limit in the above equation as $t \to t_0 - 0$ we observe that there exists the one-sided limit $\dot{f}(\zeta, t_0 - 0)$. Similarly, there exists the one-sided limit $\dot{f}_2(\zeta, 0^+)$ and they are equal. Let us define $f(\zeta, t) \equiv f_2(\zeta, t - t_0)$ in the interval $t \in [t_0, t_0 + \varepsilon)$. Above observations yield that the so extended function is continuous in the interval $t \in [0, t_0 + \varepsilon)$, analytic, univalent and starlike in some neighbourhood of $\overline{\mathbb{D}}$. Moreover, it is differentiable at the point $t = t_0$, and being extended onto the unit circle, satisfies equation (1.22). Thus, it is a unique strong solution in the interval $t \in [0, t_0 + \varepsilon)$. This finishes the proof of the theorem. $\qquad\square$

Lemma 4.5.1. *Under the assumptions of the previous theorem, if Ω_0 is starlike ($f_0 \in S^*$) then the limiting domain $\Omega(t_0)$ has no singularities on the boundary.*

Proof. The function $f(\zeta, t)$ belongs to the class $S^*(\alpha(t))$ with $\alpha(t) < 1$ for any $t \in (0, t_0)$ due to Theorem 4.4.2, where $\alpha(t)$ strictly decreases with respect to t.

Define the limiting function $f(\zeta, t_0) = \lim\limits_{t \to t_0 - 0} f(\zeta, t)$, where the limit is taken locally uniformly in $\mathbb{D}$. The function $f(\zeta, t_0)$ is univalent, strongly starlike of order $\alpha(t_0) = \lim\limits_{t \to t_0 - 0} \alpha(t) < 1$. According to the geometric characterization of the class $S^*(\alpha(t_0))$, the boundary of the domain $\Omega(t_0) = f(\mathbb{D}, t_0)$ is reachable by the radial external angles $\pi(1 - \alpha(t_0))$, which implies that there is no cusp or a double point on the boundary of $\Omega(t_0)$. This completes the proof. $\qquad\square$

Theorem 4.5.2. *Starting with a starlike phase domain Ω_0 with an analytic boundary the life-span of the strong Hele-Shaw starlike dynamics $\Omega(t)$ is infinite.*

Proof. Indeed, if the strong solution exists during the finite interval $t \in [0, t_0)$ and does not in $t \in [t_0, t_0 + \varepsilon)$ for any $\varepsilon > 0$, then this contradicts Theorem 4.5.1 and Lemma 4.5.1. $\qquad\square$

Problem 4.5.1. Of course, any result on life-span of the strong solution would be welcomed.

- Are there other geometric conditions that guarantee infinite life-span?
- Given some geometric characteristics of the initial domain can we estimate life-span of the strong solution?
- Can we estimate life-span of polynomial solutions of degree n depending on n?

4.6 Solidification and melting in potential flows

Another free boundary problem which we consider in this book, is the problem of pattern formation in a forced hydrodynamic flow. The Ivantsov problem of dendritic solidification [285] and the Saffman–Taylor problem of viscous fingering [488] present a basis for a mathematical treatment of two-dimensional solidification/melting in a forced potential flow. Such a problem arises, for example, in models of artificial freezing and thawing of flows in porous media (see [20, 131, 211, 217, 318, 376, 377]). The behavior of a solution to our problem have common features with solutions to the one-phase zero surface tension Hele-Shaw problem, melting corresponds to the stable case of the injection into the Hele-Shaw cell, and crystallization to the unstable case of suction. Mathematically, the problem that appears for the complex potential of the unfrozen flow is governed by Darcy's law which takes into account additional equations for the temperature field. One of the typical features of this problem is that there is, in general, no unique solution. At the same time the existence can be proved in a usual way.

Let us formulate the governing equations. In the exterior part $\Omega(t)$ of the crystal cross-section we introduce the complex coordinate $z = x + iy$ and the complex flow potential $W = \varphi + i\psi$, where φ is the velocity potential and ψ is the stream function. We consider a dimensionless model such that $\varphi = p$ refers to the pressure and gravity is neglected. Let us denote the temperature field by

$\theta(z) \equiv \theta(x,y)$ in $\Omega(t)$ and suppose that the phase transition is taken under the temperature $\theta = 0$ on $\Gamma(t) = \partial\Omega(t)$. A suitable scaling leads to the condition that $\theta(x,y) = \pm 1$ if $x \to -\infty$, where we suppose that the fluid moves to the right and $(+)$ corresponds to melting and $(-)$ to crystallization with supercooling. Within strong Hele-Shaw assumptions, the mathematical model is described by the following equations [20, 131, 121, 217, 318]:

$$\begin{cases} \nabla \cdot \mathbf{V} = 0, \ \mathbf{V} = \nabla\varphi, \quad Pe\,\mathbf{V} \cdot \nabla\theta = \Delta\theta, \ z \in \Omega(t); \\ \theta = 0, \ V_n = -\frac{\partial\theta}{\partial n}, \qquad \frac{\partial\varphi}{\partial n} = 0, \quad z \in \Gamma(t). \end{cases} \qquad (4.59)$$

In this system the Péclet number Pe is a measure of the intensity of heat transfer by convection compared with conduction. We note that this model is time-reversible [121]. In fact, reversing the sign of the temperature changes only the sign condition for $\lim_{x\to-\infty} \theta$ and for the kinematic boundary condition which are both reversed.

4.6.1   Close-to-parabolic semi-infinite crystal

Let us specify the shape of the initial crystal. In this subsection we suppose that the initial melting crystal is approximately parabolic when $x, y \in \Gamma(t)$, $x \to \infty$. Let us note that if $\Gamma(0)$ admits such a normalization, then $\Gamma(t)$ is of the same normalization for $t > 0$. We add to (4.59) the initial conditions

$$\lim_{x\to-\infty} \theta = 1, \quad \lim_{y\to\pm\infty} \frac{\partial\theta}{\partial y} = 0.$$

The Boussinesq transformation [66] applied to the convective heat transfer equation (4.59) leads to uncoupling of the problem and permits us to apply analytic univalent functions. Uncoupling means that the initial problem (4.59) may be split into two independent tasks ([318, 376, 377]), the first of which is the problem of heat exchange, the second refers to the free boundary nature of the problem. In fact, the Boussinesq transformation is equivalent to the existence of a conformal univalent map from the phase domain $\Omega(t)$ onto the plane of the complex potential $W = \varphi + i\psi$. Under this transformation the boundary of the crystal cross-section is mapped onto the slit directed along the positive real axis $\psi = 0$, $\varphi > 0$ in the W-plane. Thus, the problem admits the form:

$$Pe\,\frac{\partial\theta}{\partial\varphi} = \Delta\theta, \quad W \in D, \qquad (4.60)$$

where $D = \mathbb{C} \setminus [0, \infty)$. The boundary conditions are

$$\lim_{\varphi\to-\infty} \theta = 1, \quad \lim_{\psi\to\pm\infty} \frac{\partial\theta}{\partial\psi} = 0, \quad \theta = 0, \quad W \in \partial D. \qquad (4.61)$$

Now let us introduce the auxiliary parametric complex ζ-plane, $\zeta = \xi + i\eta$. The Riemann Mapping Theorem yields that there exists a conformal univalent map $f(\zeta, t)$ of the left half-plane $\mathbb{H}^- = \{\zeta : \operatorname{Re} \zeta < 0\}$ onto the phase domain $f : \mathbb{H}^- \to \Omega(t)$. The parabolic shape of $\Gamma(t)$ implies the normalization $f(\zeta, \cdot) = -\zeta^2 + a_1 \zeta + a_0 + a_{-1}/\zeta + \cdots$. Fortunately, for the problem (4.60)–(4.61) the method of separating variables is applicable. First we introduce the map $\mathbb{H}^- \to D$ given by $W = -\zeta^2$. Then we are looking for a similarity solution $\theta = g(\xi)$. Elementary calculation leads to the relations

$$\left| \frac{\partial \theta}{\partial \psi} \right| = \frac{\sigma}{\sqrt{\varphi}} \quad \text{on the slit } \partial D,$$

where $\sigma = \sqrt{Pe/\pi}$, and

$$\left| \frac{\partial \theta}{\partial n} \right| = \frac{\sigma}{\sqrt{\varphi}} |W'(z)|.$$

The unit normal outer vector to the moving interface is $\mathbf{n} = f'(i\eta, t)/|f'(i\eta, t)|$, and the normal velocity, hence, is of the form

$$V_n = \operatorname{Re} \dot{f}(i\eta, t) \overline{\frac{f'(i\eta, t)}{|f'(i\eta, t)|}}, \quad \eta \in (-\infty, \infty).$$

Besides, we have

$$\pm i \left| \frac{\partial \theta}{\partial \psi} \right| \overline{W'(z)} = -V_n \hat{n}, \quad \psi = \pm 0, \ \varphi > 0.$$

Changing variables, $W = -\zeta^2$ we come to the Polubarinova–Galin type equation for the free boundary:

$$\operatorname{Re} \dot{f}(i\eta, t) \overline{f'(i\eta, t)} = -2\sigma, \quad \xi = 0, \eta \in (-\infty, \infty). \tag{4.62}$$

Using the same argumentation as in the above subsections we prove the following statement

Theorem 4.6.1. *If the initial crystal interface possesses the property to be convex of order α in the negative direction $(H_{\mathbb{R}}^-(\alpha))$, then the free boundary remains convex in the negative direction of the same order as long as the solution to the problem (4.60)–(4.61) (or (4.62)) exists ($\sigma > 0$ and the liquid moves to the right).*

The result for $\alpha = 1$ is obtained in [319].

4.7 Geometry of weak solutions

Some of the results on geometry of solutions to Hele-Shaw flow problems are most easily discussed in terms of weak solutions, in fact, they will really be results on the geometry of domains obtained by partial balayage (see Section 3.2). We recall that the weak solution $\Omega(t)$ of the one point injection Hele-Shaw problem with

$Q = 1$ is expressed as
$$\mathrm{Bal}\,(\chi_{\Omega(0)} + t\delta_0, 1) = \chi_{\Omega(t)}.$$
in terms of balayage. What will count in the results on geometry is just the support, or even the convex hull of the support, of the measure $\mu = \chi_{\Omega(0)} + t\delta_0$.

4.7.1   Starlikeness of the weak solution

We have already proved via conformal mappings that starting with a domain $\Omega(0)$ which is starlike with respect to the origin, the injection at the origin gives a strong solution with the starlikeness preserved. In the weak formulation the preservation of starlikeness is even easier to show. The following result is originally due to Di Benedetto and Friedman [139].

Theorem 4.7.1. *Let* $\mathrm{Bal}\,(\chi_{\Omega(0)} + t\delta_0) = \chi_{\Omega(t)}$, *where* $\Omega(0)$ *is starlike with respect to the origin. Then also* $\Omega(t)$ *is starlike with respect to the origin for* $t > 0$.

Proof. We write
$$\chi_{\Omega(0)} + t\delta_0 + \Delta u = \chi_{\Omega(t)}$$
with $u \geq 0$, $u = 0$ outside $\Omega(t)$, as usual. Then in terms of polar coordinates
$$\frac{1}{r}\frac{\partial}{\partial r}\left(r\frac{\partial u}{\partial r}\right) + \frac{1}{r^2}\frac{\partial^2 u}{\partial \theta^2} = 1 - \chi_{\Omega(0)}$$
in $\Omega(t) \setminus \{0\}$. Multiplying by r^2 and then applying $\frac{1}{r}\frac{\partial}{\partial r}$ to both parts gives
$$\frac{1}{r}\frac{\partial}{\partial r}\left(r\frac{\partial}{\partial r}\left(r\frac{\partial u}{\partial r}\right)\right) + \frac{1}{r^2}\frac{\partial^2}{\partial \theta^2}\left(r\frac{\partial u}{\partial r}\right) = \frac{1}{r}\frac{\partial}{\partial r}\left(r^2(1 - \chi_{\Omega(0)})\right).$$

Here the left-hand member is $\Delta(r\frac{\partial u}{\partial r})$ and the right-hand member is non-negative due to the starlikeness of $\Omega(0)$. Thus $r\frac{\partial u}{\partial r}$ is subharmonic in $\Omega(t)\setminus\{0\}$, and $r\frac{\partial u}{\partial r} = 0$ on $\partial\Omega(t)$.

Near the origin, $u(z) = -\frac{t}{2\pi}\log|z| + O(1)$, hence $r\frac{\partial u}{\partial r} = -\frac{t}{2\pi} + O(|z|)$, as $z \to 0$. It follows that $r\frac{\partial u}{\partial r}$ is regular harmonic in all $\Omega(t)$, and strictly negative at the origin. Thus $r\frac{\partial u}{\partial r} < 0$ in $\Omega(t)$ by the maximum priciple, from which the starlikeness follows. $\square$

4.7.2   The inner normal theorem

Next we show that, for any initial domain $\Omega(0)$, $\Omega(t)$ has very good properties outside the convex hull of $\Omega(0)$, e.g., that there are natural bounds on the curvature of $\partial\Omega(t)$. So we consider a weak solution
$$\chi_{\Omega(t)} = \mathrm{Bal}\,(\chi_{\Omega(0)} + t\delta_0, 1),$$

or more generally,

$$\chi_{\Omega(t)} = \text{Bal}\,(\chi_{\Omega(0)} + \nu(t), 1),$$

for any measure $\nu(t) \geq 0$ which vanishes outside $\Omega(0)$. Let $K = \text{conv}\,\overline{\Omega(0)}$ be the closed convex hull of $\Omega(0)$.

Theorem 4.7.2 ([243, 244, 245]). *Under the above assumptions $\Omega = \Omega(t)$ has the following properties:*

(i) *$\partial\Omega \setminus K$ is smooth analytic;*

(ii) *for any $z \in \partial\Omega \setminus K$ the inward normal ray N_z from z intersects K (if $\Omega(0)$ is connected, then it has to intersect $\overline{\Omega(0)}$ itself);*

(iii) *the normal rays N_z in (ii) do not intersect each other before they reach K;*

(iv) *Ω can be expressed as a union of disks with centers on $K \cap \Omega$:*

$$\Omega = \bigcup_{a \in K \cap \Omega} \mathbb{D}(a, r(a))$$

for suitable $r(a) > 0$.

Proof. Set $\mu = \chi_{\Omega(0)} + \nu(t)$ and write

$$\chi_\Omega = \text{Bal}\,(\mu, 1) = \mu + \Delta u,$$

where u is the smallest function satisfying $u \geq 0$, $\Delta u \leq 1 - \mu$.

We first assume that K lies in the lower half-plane $K \subset \{y \leq 0\}$, i.e., that $\Omega(0) \subset \{y < 0\}$, and we shall study the geometry of

$$\Omega^+ = \Omega \cap \{y > 0\}.$$

Let u^* be the reflection of u with respect to the real axis, i.e., $u^*(x+iy) = u(x-iy)$, and set

$$v = u - \inf(u, u^*) = (u - u^*)_+.$$

Since $\Delta u \leq 1 - \mu \leq 1$ everywhere, we have $\Delta u^* \leq 1$. Therefore, $\Delta \inf(u, u^*) \leq 1$ everywhere. Indeed, the infimum of two superharmonic functions is again superharmonic, so

$$\Delta \inf(u, u^*) - 1 = \Delta\left(\inf(u, u^*) - \frac{1}{4}|z|^2\right) = \Delta \inf\left(u - \frac{1}{4}|z|^2, u^* - \frac{1}{4}|z|^2\right) \leq 0.$$

Since $\Delta u = 1$ in Ω^+ it follows that $\Delta v \geq 0$ in Ω^+. Moreover, $v = 0$ on $\partial(\Omega^+)$.

The maximum principle now shows that $v \leq 0$ in Ω^+. This means that $u \leq u^*$ in Ω^+, i.e., that u is smaller (or at least not larger) at any point in the upper half-plane, than in the reflected point in the lower half-plane. On the real axis this gives

$$\frac{\partial u}{\partial y}(x, 0) \leq 0, \tag{4.63}$$

and in general it shows that the reflection of Ω^+ in the real axis is contained in Ω:

$$(\Omega^+)^* \subset \Omega. \tag{4.64}$$

Now since $\nabla u = 0$ on $(\partial\Omega)^+$ and $\Delta(\frac{\partial u}{\partial y}) = \frac{\partial}{\partial y}\Delta u = 0$ in Ω^+ we can apply the maximum principle again, now to $\frac{\partial u}{\partial y}$, to obtain that $\frac{\partial u}{\partial y} \leq 0$ in Ω^+. This inequality is everywhere strict because if we had equality at some point, then it would follow that $u = 0$ in a whole component of Ω^+; and this is impossible because $1 = \chi_\Omega = \mu + \Delta u = \Delta u$ in Ω^+.

The conclusion now is that $u(x+iy)$ is a strictly decreasing function of $y > 0$ in Ω^+. Therefore, since $u = 0$ outside Ω, every vertical line L in the upper half-plane intersects Ω^+ in at most one segment ($L \setminus \Omega^+$ is connected). It follows that $(\partial\Omega)^+$ is a graph of a function, say

$$(\partial\Omega)^+ = \{z = x + iy, \, y = g(x)\}.$$

The domain of definition of the function g may consist of more than one interval. It follows from the general regularity theory (e.g., [77, 492, 193, 431]) that g is real-analytic.

Next, with K still in the lower half-plane we shall obtain a similar convexity statement, but for semicircles instead of vertical lines. Let (r, θ) be the polar coordinates. In the proof of Theorem 4.7.1 we studied

$$r\frac{\partial u}{\partial r} = x\frac{\partial u}{\partial x} + y\frac{\partial u}{\partial y},$$

here we shall study

$$\frac{\partial u}{\partial \theta} = -y\frac{\partial u}{\partial x} + x\frac{\partial u}{\partial y}, \tag{4.65}$$

in Ω^+. Since $\Delta u = 1$ in Ω^+ and the coefficients in

$$\Delta u = \frac{1}{r}\frac{\partial}{\partial r}\left(r\frac{\partial u}{\partial r}\right) + \frac{1}{r^2}\frac{\partial^2 u}{\partial \theta^2}$$

do not depend on θ, $\frac{\partial u}{\partial \theta}$ is harmonic,

$$\Delta\frac{\partial u}{\partial \theta} = \frac{\partial}{\partial \theta}\Delta u = 0 \quad \text{in } \Omega^+.$$

As to the boundary values of $\frac{\partial u}{\partial \theta}$ on $\partial(\Omega^+)$, we have

$$\frac{\partial u}{\partial \theta} = 0 \quad \text{on } (\partial\Omega)^+.$$

By (4.63), (4.65) we have that

$$\frac{\partial u}{\partial \theta} \leq 0 \quad \text{for } x > 0, \qquad \frac{\partial u}{\partial \theta} \geq 0 \quad \text{for } x < 0,$$

on the real axis.

Now consider a circular arc

$$C_R = \{z = re^{i\theta}, \, r = R, 0 < \theta < \pi\}$$

in the upper half-plane. We shall prove that $C_R \backslash \Omega^+$ consists of at most one segment (more precisely, that it is connected), and we shall argue by contradiction.

So suppose $C_R \backslash \Omega^+$ has at least two components. Then there are points $z_1 = Re^{i\theta_1}$ and $z_2 = Re^{i\theta_2}$ with $0 < \theta_1 < \theta_2 < \pi$, such that $z_1, z_2 \in (\partial\Omega)^+$, and $z = Re^{i\theta} \in \Omega^+$ for all $\theta_1 < \theta < \theta_2$. Since $u(z_1) = u(z_2) = 0$ and $u(z) > 0$, we have, integrating along C_R,

$$\int_{z_1}^{z} \frac{\partial u}{\partial \theta} d\theta = u(z) - u(z_1) > 0, \qquad \int_{z}^{z_2} \frac{\partial u}{\partial \theta} d\theta = u(z_2) - u(z) < 0.$$

Therefore, there must be points $z = Re^{i\theta}$ with $\theta_1 < \theta < \theta_2$ arbitrarily close to θ_1 for which $\frac{\partial u}{\partial \theta} > 0$. Similarly, there must be points $z = Re^{i\theta}$ with $\theta_1 < \theta < \theta_2$ arbitrarily close to θ_2 for which $\frac{\partial u}{\partial \theta} < 0$.

Now we apply the maximum principle. Every component of $\{\frac{\partial u}{\partial \theta} > 0\}$ must reach some part of the negative real axis, because we know that $\frac{\partial u}{\partial \theta} \leq 0$ on all other possible parts of the boundary of that component. Similarly, every component of $\{\frac{\partial u}{\partial \theta} < 0\}$ must reach some part of the positive real axis. But it is obviously topologically impossible to have components of $\{\frac{\partial u}{\partial \theta} > 0\}$ stretching from points arbitrarily close to z_1 (the rightmost end point of the described component of C_R) to the negative real axis, and simultaneously, components of $\{\frac{\partial u}{\partial \theta} > 0\}$ stretching from points arbitrary close to z_2 to the positive real axis.

This contradiction shows that $C_R \backslash \Omega^+$ actually is connected. The same reasoning applies with the center of the polar coordinates at any point of the real axis. Thus for any semicircle C in the upper half-plane with the center on the real axis, $C \backslash \Omega^+$ is connected. Together with the first part of the proof, saying that $L \backslash \Omega^+$ is connected for any vertical semiline L, what we have proved can be expressed by saying that the complement of Ω^+ in the upper half-plane is convex with respect to the Poincaré metric in the upper half-plane.

Next, like for Euclidean convexity, $C^+ \backslash \Omega^+$ being convex (in the Poincaré metric) implies that it is the intersection of Poincaré half-planes, or that, turning to the complements, Ω^+ is the union of such. This means that

$$\Omega^+ = \bigcup_{a \in \mathbb{R}} \mathbb{D}(a, r(a)) \tag{4.66}$$

for suitable radii $r(a) \geq 0$.

Let $z \in (\partial\Omega)^+$ and let N_z be the inward normal ray at z. Since $(\partial\Omega)^+$ is a graph of a function, N_z intersects the real axis at a point $p(z)$, which we call the *foot point* of the normal. In terms of the equation $y = g(x)$ for $(\partial\Omega)^+$ we have

$z = x + ig(x)$ and one easily computes $p(z) = x + g(x)g'(x)$. The fact (4.66) that Ω^+ is a union of semidisks $(D(a, r(a)))^+$ implies that actually

$$(\mathbb{D}(p(z), |z - p(z)|))^+ \subset \Omega^+. \tag{4.67}$$

Let $z_1 \neq z_2$. Assume that two normals N_{z_1} and N_{z_2} intersect in $\mathbb{C}^+$, say $w \in N_{z_1} \cap N_{z_2}$, where $w \in \mathbb{C}^+$. We may assume that $|z_1 - w| \leq |z_2 - w|$, for example. Then z_1 is contained in the closure of the disk with the center at w and the radius $|z_2 - w|$, and therefore, in the interior of the larger disk

$$\mathbb{D}(p(z_2), |z_2 - p(z_2)|).$$

But $z_1 \in (\partial\Omega)^+$, so this disk must contain points outside $\overline{\Omega^+}$ as well. This contradicts (4.67), and we conclude that N_{z_1} and N_{z_2} actually cannot intersect in $\mathbb{C}^+$.

One easily sees that the inner ball property (4.66), or the fact that the inner normals do not intersect in $\mathbb{C}^+$, is equivalent to the foot point $p(z) = x + g(x)g'(x)$ being a monotone increasing function of x.

So far we have assumed that $K \subset \{y < 0\}$. Adapting the results we have obtained to all half-planes containing K easily gives the statements of the theorem. We give some details below.

(i) We already know that $\partial\Omega \setminus K$ is analytic but with possible singularities. But since near any point of $\partial\Omega \setminus K$, $\partial\Omega$ will be a graph seen from several different angles (choosing different half-planes containing K) there can be no singular points.

(ii) If the inward normal N_z at a point $z \in \partial\Omega \setminus K$ did not intersect K we could find a half-plane $H \supset K$ which does not meet N_z, contradicting that $\partial\Omega \setminus H$ is a graph when seen from H.

(iii) Similarly, if $w \in N_{z_1} \cap N_{z_2}$, $w \in \Omega \setminus K$, $z_1, z_2 \in \partial\Omega \setminus K$, $z_1 \neq z_2$, we choose a half-plane $H \supset K$, such that $w \notin \bar{H}$. By (ii), both N_{z_1} and N_{z_2} intersect K. This shows that $z_1 \notin \bar{H}$, $z_2 \notin \bar{H}$, and we are back in the situation with $H = \{y < 0\}$, i.e., we have a contradiction. Thus N_{z_1} and N_{z_2} do not meet before they reach K.

(iv) From what we already did it follows that

$$\Omega \setminus K = \left(\bigcup_{a \in \partial K} \mathbb{D}(a, r(a)) \right) \setminus K$$

for suitable $r(a) \geq 0$. Also, by (4.64),

$$\bigcup_{a \in \partial K} \mathbb{D}(a, r(a)) \subset \Omega.$$

Obviously, $r(a) = 0$ for $a \in \partial K \setminus \Omega$. The points from $\Omega \cap K$ can be trivially covered by the disks $\mathbb{D}(a, r(a)) \subset \Omega$ with $a \in \Omega \cap K$. Now (iv) follows, and the proof of the theorem is complete. $\qquad\square$

4.7.3   Distance to the boundary (revisited)

Now we discuss the distances from points in the initial domain $\Omega(0)$ to points on the boundary of $\Omega(t)$. The following theorem is due to Sakai [495].

Theorem 4.7.3. *Let $\mu \geq 0$ be a measure with a support in the disk $\mathbb{D}_R$, $R > 0$. Define $r(\mu)$ by*

$$\pi(r(\mu))^2 = \int d\mu,$$

and let Ω be the saturated set (3.21) for $\mathrm{Bal}\,(\mu, 1)$. Then

$$\Omega \subset \mathbb{D}(0, r(\mu) + R),$$

and if $r(\mu) \geq 2R$ we moreover have

$$\mathbb{D}(0, r(\mu) - R) \subset \Omega.$$

Proof. First we recall from (3.28) that

$$\mathrm{Bal}\,(\mu, 1) = \chi_\Omega + \mu\chi_{\mathbb{C}\backslash\Omega}.$$

From this it follows that

$$\int\limits_{\mathbb{D}_R} \varphi d\mu \leq \int\limits_{\Omega} \varphi d\sigma + \int\limits_{\mathbb{C}\backslash\Omega} \varphi d\mu$$

for all functions φ in $\mathbb{C}$ which are integrable and subharmonic in Ω. In particular, taking $\varphi = \pm 1$, $|\Omega| \leq \int d\mu$.

The upper bound $\Omega \subset \mathbb{D}(0, r(\mu) + R)$ is actually a direct consequence of the Inner Normal Theorem 4.7.2. By that theorem Ω is a union of disks with centers in $\overline{\mathbb{D}(0, R)}$, so if Ω contained points outside $\mathbb{D}(0, r(\mu) + R)$, then it would contain a disk of radius greater than $r(\mu)$, which is impossible because $|\Omega| \leq \int d\mu = |\mathbb{D}(0, r(\mu))|$.

Next assume $r(\mu) \geq 2R$ and let $z \notin \Omega$. We shall show that $|z| \geq r(\mu) - R$, hence that $\mathbb{D}(0, r(\mu) - R) \subset \Omega$. By a rotation we may assume that $z = \rho \geq 0$.

If $0 < \rho < R$ we choose

$$\varphi(\zeta) = G(\zeta, \rho) + c,$$

where $G(\zeta, \rho) = \frac{1}{2\pi} \log |\frac{R^2 - \rho\zeta}{R(\zeta - \rho)}|$ and $c > 0$. Since $\rho \notin \Omega$, φ is subharmonic in Ω. The level lines of $G(\cdot, \rho)$ and φ are circles, so

$$\{\zeta \in \mathbb{C} : \varphi(\zeta) > 0\} = \{\zeta \in \mathbb{C} : G(\zeta, \rho) > -c\} = \mathbb{D}(a, t)$$

for some disk $\mathbb{D}(a, t)$, which contains $\mathbb{D}(0, R)$ because $c > 0$.

Now choose c so that

$$\int_{\mathbb{D}(a,t)} G(\zeta,\rho)\, d\sigma = 0.$$

By a straightforward computation (see Section 6 in [495]) this gives $R < t < 2R$. Thus, using that $\mathbb{D}(a,t)$ is exactly the set where φ is positive we get

$$c\pi r(\mu)^2 = c\int_{\mathbb{D}(0,R)} d\mu \leq \int_{\mathbb{D}(0,R)} (G(\zeta,\rho)+c)\, d\mu(\zeta) = \int_{\mathbb{D}(0,R)} \varphi\, d\mu$$

$$\leq \int_{\Omega} \varphi\, d\sigma + \int_{\mathbb{C}\backslash\Omega} \varphi d\mu \leq \int_{\mathbb{D}(a,t)} \varphi\, d\sigma = \int_{\mathbb{D}(a,t)} c\, d\sigma = c\pi t^2.$$

Hence $r(\mu) \leq t < 2R$, contrary to our assumption in the beginning. Thus there is no $0 < \rho < R$ with $\rho \notin \Omega$. So $\mathbb{D}(0,R) \subset \Omega$ and $\mathrm{Bal}\,(\mu,1) = \chi_\Omega$.

If $\rho \geq R$ then we take instead φ to be the Poisson kernel for the disk $\mathbb{D}(-R, R+\rho)$ and with the pole at ρ:

$$\varphi(\zeta) = \frac{(R+\rho)^2 - |\zeta + R|^2}{|(R+\rho) - (\zeta + R)|^2} = \frac{(R+\rho)^2 - |\zeta + R|^2}{|\zeta - \rho|^2}.$$

Since $\rho \notin \Omega$, φ is subharmonic in Ω. Also in this case the level lines of φ are circles. The circle where $\varphi = 1$ passes through the center $-R$ of $\mathbb{D}(-R, R+\rho)$ and through the pole $\zeta = \rho$, hence $\varphi \geq 1$ inside that circle, in particular $\varphi \geq 1$ in $\mathbb{D}(0,R)$. Moreover, $\mathbb{D}(-R, R+\rho)$ is exactly the set where $\varphi \geq 0$. All this, combined with the mean value property of φ in $\mathbb{D}(-R, R+\rho)$, gives that

$$\pi r(\mu)^2 \int_{\mathbb{D}(0,R)} d\mu \leq \int_{\mathbb{D}(0,R)} \varphi\, d\mu \leq \int_{\Omega} \varphi\, d\sigma \leq \int_{\mathbb{D}(-R,R+\rho)} \varphi\, d\sigma$$

$$= |\mathbb{D}(-R, R+\rho)|\, \varphi(-R) = \pi(R+\rho)^2.$$

Thus $\rho \leq R - r(\mu)$ as required. $\square$

Corollary 4.7.1. *In the one point injection Hele-Shaw case* $\mu = \chi_{\Omega(0)} + t\delta_a$. *This gives, if* $a \in \mathbb{D}(0,R)$, $\Omega(0) \subset \mathbb{D}(0,R)$, *that*

$$\Omega(t) \subset \mathbb{D}(0, \sqrt{(|\Omega(0)| + t)/\pi} + R).$$

If, in addition, $t \geq 4\pi R^2 - |\Omega(0)|$, *then*

$$\mathbb{D}(0, \sqrt{(|\Omega(0)| + t)/\pi} - R) \subset \Omega(t).$$

Chapter 5

Capacities and Isoperimetric Inequalities

Isoperimetric inequalities have been known since antiquity. The simplest version of an isoperimetric theorem reads in two equivalent forms:

- Among all planar shapes with the same perimeter the circle has the largest area.

- Among all planar shapes with the same area the circle has the shortest perimeter.

This is the solution of what is sometimes known as Dido's problem because of the story that Queen Dido of Tyre bargained for some land bounded on one side by the (straight) Mediterranean coast and agreed to pay a fixed sum for as much land as could be enclosed by a bull's hide. Both statements can be expressed in a more algebraic form which indeed underlines the fact that they are equivalent. Denote the perimeter and area of a planar shape by L and A, respectively. Then, $4\pi A \leq L^2$. The equality only holds for a circle. In higher-dimensional spaces, for example, if S is a surface area while V a volume of a three-dimensional body, then $36\pi V^2 \leq S^3$ (see, e.g., [104], [153]).

Pappus of Alexandria (ca 300 A.D.) wrote: *bees, then, know just this fact which is useful to them, that the hexagon is greater than the square and the triangle and will hold more honey for the same expenditure of material in constructing each. But we, claiming a greater share of wisdom than the bees, will investigate a somewhat wider problem, namely that, of all equilateral and equiangular plane figures having the same perimeter, that which has the greater number of angles is always greater, and the greatest of them all is the circle having its perimeter equal to them.*

Probably, the most representative work on isoperimetric inequalities in various aspects of mathematical physics is the famous monograph [441] written by **George Pólya** (1887–1985) and **Gabor Szegő** (1895–1985) (see also [395]). By an isoperimetric inequality we mean an inequality that links a measure of volume with a measure of its boundary. We shall be concerned mainly with the following related question: how is the area of the phase domain controlled by the capacity of its boundary (or conformal radius of the domain)?

5.1 Conformal invariants and capacities

We start by giving some background on quantities we are going to use in isoperimetric inequalities. These quantities are moduli, reduced moduli and capacities.

5.1.1 Modulus of a family of curves

The notion of the *modulus of a family of curves* goes back to early works of Grötzsch [229, 230]. Later, Ahlfors and Beurling [8, 10] introduced the notion of *extremal length* (the reciprocal of the *modulus*) which stimulated the active development of the method of extremal lengths. Major contributions to the subject were made by Jenkins [287], [288], Strebel [533] and Ohtsuka [420] who connected the modulus problem with the problem of the extremal partitioning of a Riemann surface and proved the existence of the extremal metric by Schiffer's variations.

Let Ω be a domain in $\mathbb{C}$ and $\rho(z)$ be a real-valued, Borel measurable, non-negative function in $L^2(\Omega)$, which we use to define a differential metric $\rho(z)|dz|$ on Ω.

Let γ be a locally rectifiable curve in Ω. The integral

$$l_\rho(\gamma) := \int_\gamma \rho(z)|dz|, \tag{5.1}$$

is said to be the *ρ-length* of γ. If $\rho(z) \equiv 1$ in Ω, then the 1-length of any rectifiable $\gamma \subset \Omega$ coincides with its Euclidian length. The integral

$$A_\rho(\Omega) := \int_\Omega \rho^2 \, d\sigma \tag{5.2}$$

is called the *ρ-area* of Ω. Recall that $d\sigma = dxdy$.

Let Γ be a family of curves γ in Ω. Denote by

$$L_\rho(\Gamma) := \inf_{\gamma \in \Gamma} l_\rho(\gamma)$$

the ρ-length of the family Γ. Then, the quantity

$$m(\Omega, \Gamma) = \inf_\rho \frac{A_\rho(\Omega)}{L_\rho^2(\Gamma)}$$

is said to be the *modulus* of the family Γ in Ω where the infimum is taken over all metrics ρ in Ω.

Another equivalent and suitable (in a view of further applications) definition of the modulus can be formulated as follows. Denote by P the family of all *admissible* (for Γ) metrics in Ω, that is, metrics $\rho \in P$ that satisfy the additional condition $l_\rho(\gamma) \geq 1$ for all $\gamma \in \Gamma$. If $P \neq \emptyset$, then we can define the modulus as

$$m(\Omega, \Gamma) = \inf_{\rho \in P} A_\rho(\Omega).$$

If there is a metric ρ^*, such that $m(\Omega, \Gamma) = A_{\rho^*}(\Omega)$, then this metric is called *extremal*.

Two main properties of the modulus are its conformal invariance and the uniqueness of the extremal metric (if it exists). More precisely, let Γ be a family of curves in a domain $\Omega \in \overline{\mathbb{C}}$, and let $w = f(z)$ be a conformal map of Ω onto $\widetilde{\Omega} \in \overline{\mathbb{C}}$. If $\widetilde{\Gamma} := f(\Gamma)$, then

$$m(\Omega, \Gamma) = m(\widetilde{\Omega}, \widetilde{\Gamma}).$$

Let ρ_1 and ρ_2 be two extremal metrics for the modulus $m(\Omega, \Gamma)$. Then, $\rho^* := \rho_1 = \rho_2$ almost everywhere. Moreover, $L_{\rho^*}(\Gamma) = 1$.

The property of monotonicity reads as follows. If $\Gamma_1 \subset \Gamma_2$ in Ω, then $m(\Omega, \Gamma_1) \leq m(\Omega, \Gamma_2)$.

Example 5.1.1. Let Ω be a rectangle $\{z = x + iy : 0 < x < a, 0 < y < b\}$ and Γ be the family of curves in Ω that connect the opposite horizontal sides of Ω. Then, $m(\Omega, \Gamma) = a/b$.

Example 5.1.2. Let Ω be an annulus $\{z = re^{i\theta} : 1 < r < R, 0 < \theta \leq 2\pi\}$ and Γ be the family of curves in Ω that separate the opposite boundary components of Ω. Then, $m(\Omega, \Gamma) = \frac{1}{2\pi} \log R$.

Example 5.1.3. Let Ω be an annulus $\{z = re^{i\theta} : 1 < r < R, 0 < \theta \leq 2\pi\}$ and Γ be the family of curves in Ω that connect the two boundary components of Ω. Then, $m(\Omega, \Gamma) = \frac{2\pi}{\log R}$.

For more information see, e.g., [8, 288, 420, 562].

5.1.2   Reduced modulus and capacity

Let $\Omega \subset \overline{\mathbb{C}}$ be a simply connected hyperbolic domain, $a \in \Omega$, $|a| < \infty$. We consider the doubly connected domain $\Omega_\varepsilon = \Omega \setminus \mathbb{D}(a, \varepsilon)$ for a sufficiently small $\varepsilon > 0$. The quantity

$$M(\Omega, a) := \lim_{\varepsilon \to 0} \left(M(\Omega_\varepsilon) + \frac{1}{2\pi} \log \varepsilon \right)$$

is said to be the *reduced modulus* of the circular domain Ω with respect to the point a, where $M(\Omega_\varepsilon)$ is the modulus of the doubly connected domain Ω_ε with respect to the family of curves that separate its boundary components.

Let a simply connected hyperbolic domain Ω have the conformal radius $R(\Omega, a)$ with respect to a fixed point $a \in \Omega$. Then, the quantity $M(\Omega, a)$ exists, is finite, and is equal to $\frac{1}{2\pi}\log R(\Omega, a)$, see [288, 562]. An immediate corollary when $|a| < \infty$ says that if $f(z)$ is a conformal map of Ω, such that $|f(a)| < \infty$, then $M(f(\Omega), f(a)) = M(\Omega, a) + \frac{1}{2\pi}\log|f'(a)|$.

Now, we define the reduced modulus $M(\Omega, \infty)$ of a simply connected domain Ω, $\infty \in \Omega$ with respect to infinity as the reduced modulus of the image of Ω under the map $1/z$ with respect to the origin,

$$M(\Omega, \infty) = -\frac{1}{2\pi}\log R(\Omega, \infty).$$

So, if Ω is a simply connected hyperbolic domain, $a \in \Omega$, $|a| < \infty$, and $f(z) = a_{-1}/(z-a) + a_0 + a_1(z-a) + \cdots$ is a conformal map from Ω, then $M(f(\Omega), \infty) = M(\Omega, a) - \frac{1}{2\pi}\log|a_{-1}|$.

We give all further definitions only for compact sets, however, there are generalizations to the Borel sets as well. Denote by $\text{Lip}(\Omega)$ the class of functions $u : \Omega \to \mathbb{R}$ satisfying the Lipschitz condition in Ω, i.e., for every function $u \in \text{Lip}(\Omega)$ there is a constant c such that for any two points $z_1, z_2 \in \Omega$ the inequality $|u(z_1) - u(z_2)| \leq c|z_1 - z_2|$ holds. In the case $\infty \in \Omega$ the continuity of u at ∞ is required. Functions from $\text{Lip}(\overline{\mathbb{C}})$ are absolutely continuous on lines which are parallel to the axes and the integral

$$I(u) := \int_{\overline{\mathbb{C}}} |\nabla u|^2 d\sigma$$

exists.

An ordered pair of disjoint compact sets K_1, K_2 is called a *condenser* $C = \{K_1, K_2\}$ with the *field* $\overline{\mathbb{C}} \setminus \{K_1 \cup K_2\}$. The *capacity of a condenser* C is the quantity

$$\text{cap } C := \inf I(u)$$

as u ranges over the class $\text{Lip}(\overline{\mathbb{C}})$ and $0 \leq u(z) \leq 1$ whenever $z \in \overline{\mathbb{C}}$, $u(z) \equiv 0$ in K_1, $u(z) \equiv 1$ in K_2.

A condenser C is said to be *admissible* if there exists a continuous real-valued in $\overline{\mathbb{C}}$ function $\omega(z)$, $0 \leq \omega(z) \leq 1$ which is harmonic in $\overline{\mathbb{C}} \setminus \{K_1 \cup K_2\}$ and $\omega(z) = 0$ for $z \in K_1$, $\omega(z) = 1$ for $z \in K_2$. This function is said to be a *potential* . The Dirichlet principle yields that in the definition of capacity equality appears only in the case of an admissible condenser and $u(z) \equiv \omega(z)$ almost everywhere for the potential function ω. Then by Green's formula

$$\text{cap } C = \int_{\overline{\mathbb{C}} \setminus \{K_1 \cup K_2\}} |\nabla\omega|^2 d\sigma = \int_{\partial K_2} \left|\frac{\partial\omega}{\partial n}\right| ds.$$

Obviously, the capacity is a conformal invariant, that is, if C_f is a condenser $\overline{\mathbb{C}} \setminus f(\overline{\mathbb{C}} \setminus \{K_1 \cup K_2\})$ for a conformal map f in $\overline{\mathbb{C}} \setminus \{K_1 \cup K_2\}$, then $\text{cap } C = \text{cap } C_f$.

If K_1 and K_2 are two disjoint continua, then we can construct the conformal map $w = f(z)$ of the doubly connected domain $\overline{\mathbb{C}} \setminus \{K_1 \cup K_2\}$ onto an annulus $1 < |w| < R$ and the potential function for the condenser $C = \{K_1, K_2\}$ is

$$\omega(z) = \frac{\log \frac{R}{|f(z)|}}{\log R}, \quad z \in \overline{\mathbb{C}} \setminus \{K_1 \cup K_2\},$$

$\omega(z) \equiv 0$ in K_1, $\omega(z) \equiv 1$ in K_2. Therefore, cap $C = 2\pi / \log R$.

Let $C = \{K_1, K_2\}$ and $C_k = \{K_1^k, K_2^k\}$, $k = 1, \dots, n$ be such condensers that all C_k have non-intersecting fields and

$$K_1 \subset \bigcap_{k=1}^{n} K_1^k, \quad K_2 \subset \bigcap_{k=1}^{n} K_2^k.$$

From the definition of capacity and from the Dirichlet principle one can derive the inequality (Grötzsch Lemma)

$$\frac{1}{\text{cap } C} \geq \sum_{k=1}^{n} \frac{1}{\text{cap } C_k}. \tag{5.3}$$

(possibly with equality, see, e.g., [149]).

Let K be a compact set in $\mathbb{C}$. We consider condensers of special type $C_R = \{|z| \geq R, K\}$ for R large. If $C_{R_1, R_2} = \{|z| \leq R_1\} \cup \{|z| \geq R_2\}$ for $R_1 < R_2$, then the inequality (5.3) implies

$$\frac{1}{\text{cap } C_{R_2}} \geq \frac{1}{\text{cap } C_{R_1}} + \frac{1}{2\pi} \log \frac{R_2}{R_1}.$$

Therefore, the function $\frac{1}{\text{cap } C_R} - \frac{1}{2\pi} \log R$ increases with increasing R and the limit

$$\text{cap } K = \lim_{R \to \infty} R \exp\left(-\frac{2\pi}{\text{cap } C_R} \right) \tag{5.4}$$

exists and is said to be the *logarithmic capacity* of the compact set $K \subset \mathbb{C}$. Equality (5.4) is also known as Pfluger's theorem (see, e.g., [446], Theorem 9.17).

Next we briefly summarize the definition and some properties of the logarithmic capacity of a compact set $K \subset \mathbb{C}$ following Fekete. For $n = 2, 3, \dots$ we consider

$$\Delta_n(K) = \max_{z_1, \dots, z_n \in K} \prod_{1 \leq k < j \leq n} |z_k - z_j|.$$

The maximum exists and is attained for so-called Fekete points $z_{nk} \in \partial K$, $k = 1, \dots, n$. The value Δ_n is equal to the Vandermonde determinant

$$\Delta_n(K) = \left| \det_{k=1,\dots,n} (1 \; z_{nk} \; \dots \; z_{nk}^{n-1}) \right|.$$

Then, the limit

$$\text{cap } K = \lim_{n \to \infty} (\Delta_n(K))^{\frac{2}{n(n-1)}}$$

exists (see [446]), is known as the *transfinite diameter*, and is equal to the logarithmic capacity (see also [441], [487]).

Let K be a continuum (a closed connected set containing at least two points) in $\overline{\mathbb{C}}$ and $\Omega = \overline{\mathbb{C}} \setminus K$. Then, from the definition of the logarithmic capacity and the reduced modulus it is clear that $\text{cap } K = \text{cap } \partial K = \exp(-2\pi \, m(\Omega, \infty))$. It is well known that for $K = [0, 1]$ the capacity is given by $\text{cap } K = 1/4$.

If we have a condenser $C^{(h)} = \{K_1, K_2\}$ of the special type $K_1 \subset \mathbb{D}$, $K_2 = \overline{\mathbb{C}} \setminus \mathbb{D}$, then $\text{cap } C^{(h)}$ is said to be the hyperbolic capacity of K_1 and $\text{cap}^{(h)} K_1 = \text{cap } C^{(h)}$. One can define also $\text{cap}^{(h)} K$ by means of the *hyperbolic transfinite diameter*. Set

$$\Delta_n^{(h)}(K) = \max_{z_1,\ldots,z_n \in K} \prod_{1 \le k < j \le n}^{n} \left| \frac{z_k - z_j}{1 - z_k \overline{z_j}} \right|.$$

Then,

$$\text{cap}^{(h)} K = \lim_{n \to \infty} (\Delta_n^{(h)}(K))^{\frac{2}{n(n-1)}}.$$

Finally, if K is a continuum in $\mathbb{D}$ and $\Omega = \mathbb{D} \setminus K$, then

$$\text{cap}^{(h)} K = \text{cap}^{(h)} \partial K = \exp(-2\pi M(\Omega)),$$

where $M(\Omega)$ is the modulus of the doubly connected domain Ω with respect to the family of separating curves.

5.1.3   Integral means and the radius-area problem

We consider the zero surface tension Hele-Shaw model with injection through a source at the origin and with a bounded initial phase domain Ω_0.

Let f be a univalent function $f(\zeta) = a\zeta + a_2\zeta^2 + \cdots$, $a > 0$, defined in the unit disk $\mathbb{D}$ and let S be the area of $f(\mathbb{D})$. An obvious inequality, which we call the *radius-area estimate*, is $S \ge \pi a^2$. It follows from the formula for the area $S = \pi(a^2 + \sum_{k=2}^{\infty} k|a_k|^2)$. Equality is attained for a trivial map $f(\zeta) = a\zeta$. There is no upper estimate of S, namely, the area can be infinite. As for the perimeter, the classical result by Pólya and Schiffer [442] states that $L \ge 2\pi a$. The upper estimate is ∞ in general and $8R(f(\mathbb{D}), \infty)$ for convex domains.

Let $S(t)$ be the area of the domain $\Omega(t)$ in the Hele-Shaw dynamics under the conditions of Section 1.4.2. A simple application of Green's theorem implies that the rate of the area change is expressed as $\dot{S} = Q$ under injection ($Q > 0$). From (1.24) we deduce that

$$\dot{a} = a \frac{1}{4\pi^2} \int_0^{2\pi} \frac{Q}{|f'(e^{i\theta}, t)|^2} d\theta \ge a \frac{1}{4\pi^2} \int_0^{2\pi} \text{Re} \, \frac{Q}{[f'(e^{i\theta}, t)]^2} d\theta = \frac{Q}{2\pi a} = \frac{\dot{S}}{2\pi a},$$

where $a = f'(0, t)$. In other words the area rate is controlled by the rate of the conformal radius of the domain $\Omega(t)$ with respect to the origin: $\dot{S} \leq 2\pi a \dot{a}$. Equality in the above inequalities is attained for $\Omega(t) = \{z : |z| < a(t)\}$.

The lower bound for $\dot{S}$ in terms of a is much more difficult. One must estimate the integral mean

$$\int_0^{2\pi} \frac{1}{|f'(e^{i\theta}, t)|^2} d\theta \qquad (5.5)$$

from above. The fact that cusps may develop shows that there is no uniform estimate with respect to t. But one can estimate (5.5) under some geometric constraints on the domain $\Omega(t)$ at an instant t. For example, assume that the domain $\Omega(t)$ is convex. The function f is also convex and, thus, $\frac{1}{2}$-starlike, i.e., Re $\zeta f'(\zeta, t)/f(\zeta, t) > 1/2$. Moreover the Koebe covering theorem for the convex functions says that $|f(\zeta, t)| \geq a/2$. This implies the estimate

$$\int_0^{2\pi} \frac{1}{|f'(e^{i\theta}, t)|^2} d\theta < \frac{32\pi}{a^2}, \qquad \text{and so } \dot{S} > \frac{\pi a}{8} \dot{a}.$$

Let us give a precise estimate for this integral mean in the case of convex functions f. If f is convex, then the function $g(\zeta) \equiv \zeta f'(\zeta)$ is starlike in $\mathbb{D}$ and the function $h(\zeta) \equiv 1/g(1/\zeta) = \frac{1}{a}(\zeta + c_0 + c_1/\zeta + \cdots)$ is starlike in the complement $\mathbb{D}^*$ of the closure of $\mathbb{D}$. The function $h(\zeta)$ is univalent, bounded in $\mathbb{D}$, and has no zeros in the closure $\bar{\mathbb{D}}$. Therefore,

$$\frac{1}{2\pi} \int_0^{2\pi} \frac{d\theta}{|f'(e^{i\theta})|^2} = \frac{1}{2\pi} \int_0^{2\pi} |h(e^{i\theta})|^2 d\theta = \frac{1}{a^2}\left(1 + |c_0|^2 + \sum_{k=1}^{\infty} |c_k|^2\right)$$

$$\leq \frac{1}{a^2}\left(1 + |c_0|^2 + \sum_{k=1}^{\infty} k|c_k|^2\right). \qquad (5.6)$$

We have $|c_0| \leq 2$, and by the Area Theorem (see, e.g., [213, 216, 446]) the right-hand side of (5.6) is $\leq 6/a^2$. This estimate is sharp. Finally, for domains $\Omega(t)$ which are convex at an instant t we have

$$\dot{S} \geq \frac{\pi a}{3} \dot{a}.$$

Problem 5.1.1. If the domain $\Omega(t)$ is convex, then during the time of the existence of the solution of (1.22) convexity may be lost at the next instant. It is better to find a geometric condition that is preserved during some time interval of the existence of the solution of (1.22) and that permits us to estimate the integral mean (5.5) from above.

Let a univalent function f be defined in $\mathbb{D}$, have the non-vanishing finite angular derivatives almost everywhere at the unit circle, and the boundary of $f(\mathbb{D})$ be reachable by outer angles $> \pi/2$. Then $1/f'$ is from the Hardy class H^2. Generally, we operate with univalent functions with analytic boundaries of $f(\mathbb{D})$. Of course, we can consider domains with angles on the boundary and weak solutions. For example, the following theorem was proved in [454].

Theorem 5.1.1. *Let a univalent map $z = f(\zeta) = \zeta + a_2\zeta^2 + \cdots$ be α-convex in $\mathbb{D}$. Then the angular derivative of f exists almost everywhere on the unit circle and*

$$\frac{1}{2\pi} \int\limits_0^{2\pi} \frac{1}{|f'(e^{i\theta})|^2} d\theta \leq \frac{2^{8(1-\alpha)}}{\pi} \mathbf{B}\left(\frac{5}{2} - 2\alpha, \frac{5}{2} - 2\alpha\right),$$

where $\mathbf{B}(\cdot, \cdot)$ stands for the Euler Beta-function. The inequality is sharp. In particular,

$$\frac{1}{2\pi} \int\limits_0^{2\pi} \frac{1}{|f'(e^{i\theta})|^2} d\theta \leq \frac{4^{1-4\alpha}}{2\pi} \frac{(3 - 4\alpha)(1 - 4\alpha)}{(1 - \alpha)(1 - 2\alpha)} \mathbf{B}\left(\frac{1}{2} - 2\alpha, \frac{1}{2} - 2\alpha\right)$$

for $0 \leq \alpha < 1/4$.

Proof. If a function f is α-convex in $\mathbb{D}$, then the analytic function $g(z) \equiv zf'(z)$ is α-starlike (S_α^*, see Section 4.2). Functions from S_α^* admit the following known integral representation

$$g(z) \in S_\alpha^* \Leftrightarrow g(z) = z \, \exp\left\{-2(1 - \alpha) \int\limits_{-\pi}^{\pi} \log(1 - e^{i\theta}z)d\mu(\theta)\right\},$$

where $\mu(\theta)$ is a non-decreasing function of $\theta \in [-\pi, \pi]$ and $\int\limits_{-\pi}^{\pi} d\mu(\theta) = 1$.

If $\mu(\theta)$ is a piecewise constant function, then we have a set of complex-valued functions $g_n(z)$ that admit the following representation

$$g_n(z) = \frac{z}{\prod\limits_{k=1}^{n}(1 - e^{i\theta_k}z)^{2(1-\alpha)\beta_k}} \in S_\alpha^*, \quad \theta_k \in [-\pi, \pi], \quad \beta_k \geq 0, \quad \sum_{k=1}^{n} \beta_k = 1.$$

$$(5.7)$$

Using known properties of Stieltjes' integral and Vitali's theorem it is easy to show that the set of functions (5.7) is dense in S_α^*, i.e., for every function $g(z) \in S_\alpha^*$ there exists a sequence $\{g_n(z)\}$ satisfying (5.7) that locally uniformly converges to $g(z)$ in $\mathbb{D}$. Therefore, we need to prove our result only for $g(z) = g_n(z)$.

Let us present a chain of inequalities

$$\frac{1}{2\pi}\int\limits_0^{2\pi}\frac{1}{|g_n(e^{i\theta})|^2}\,d\theta = \frac{1}{2\pi}\int\limits_0^{2\pi}\prod_{k=1}^n |1-e^{i(\theta-\theta_k)}|^{4(1-\alpha)\beta_k}\,d\theta$$

$$\leq \frac{1}{2\pi}\int\limits_0^{2\pi}\sum_{k=1}^n \beta_k |1-e^{i(\theta-\theta_k)}|^{4(1-\alpha)}\,d\theta$$

$$= \frac{1}{2\pi}\sum_{k=1}^n \beta_k \int\limits_0^{2\pi} |1-e^{i(\theta-\theta_k)}|^{4(1-\alpha)}\,d\theta$$

$$= \frac{1}{2\pi}\int\limits_0^{2\pi} |1-e^{i\theta}|^{4(1-\alpha)}\,d\theta = \frac{4^{1-\alpha}}{2\pi}\int\limits_0^{2\pi}(1-\cos\theta)^{2(1-\alpha)}\,d\theta$$

$$= \frac{2^{8(1-\alpha)}}{\pi}\mathbf{B}\left(\frac{5}{2}-2\alpha, \frac{5}{2}-2\alpha\right).$$

The last assertion of the theorem follows from the formulae of reduction of the Beta-function.   □

We summarize the results of this section in the following theorem.

Theorem 5.1.2. *Let $\Omega(t)$ be a phase domain occupied by a fluid injected through the origin, let the area of $\Omega(t)$ be $S(t)$, and $a(t)$ be the conformal radius of $\Omega(t)$ with respect to the origin. Then $\dot{S} \leq 2\pi a\dot{a}$. If, moreover, $\Omega(t)$ is α-convex at an instant t, then*

$$\frac{2\pi^2 a\dot{a}}{2^{8(1-\alpha)}\mathbf{B}(\frac{5}{2}-2\alpha, \frac{5}{2}-2\alpha)} \leq \dot{S} \leq 2\pi a\dot{a}.$$

In the case of a contracting bubble we have a similar estimate $\dot{S} \geq 2\pi a\dot{a}$, where $S(t)$ means the area of the bubble and $a = \operatorname{cap}\Gamma(t)$. The good thing is that the outer Hele-Shaw problem preserves the convex dynamics. More about estimates for integral means can be found, e.g., in [447].

5.2 Hele-Shaw cells with obstacles

Recent studies of Robin's function and Robin's capacity [157]–[162] showed their connections with several problems of potential theoretic nature as well as extremal length and minimal energy considerations. Our goal is to give another physical interpretation that comes the Hele-Shaw problem with an obstacle inside. We shall connect the rate of area change of the phase domain with the rate of change of Robin's reduced modulus of the free boundary.

5.2.1 Robin's capacity and Robin's reduced modulus

Let Ω be a finitely connected domain in $\overline{\mathbb{C}}$, and A be an arbitrary closed set of the boundary $\partial\Omega$. Let us denote by B the complementary part of $\partial\Omega$, so that $\partial\Omega = A \cup B$. For a fixed finite point $z_0 \in \Omega$, the complex *Robin function* $R(z, z_0)$ is defined by the following requirements:

- $R(z, z_0)$ is analytic in Ω except at the point z_0 where R has a logarithmic singularity: $R(z, z_0) = \frac{1}{2\pi} \log(z-z_0) + w_0(z)$, where $w_0(z)$ is a regular function in Ω;
- Re $R(z, z_0) = 0$ for all $z \in A$, while $\frac{\partial \mathrm{Re}\, R}{\partial n}(z, z_0) = 0$ for all $z \in B$.

For $z_0 = \infty$ the definition is modified by requiring $R(z, \infty) - \frac{1}{2\pi} \log z$ to be regular in a neighbourhood of infinity. The real part of this function $R_{re}(z, z_0) =$ Re $R(z, z_0)$ is the classical Robin function that was studied deeply in [157]–[162]. The main property which we use here is its conformal invariance. For basic properties of Robin's function we refer the reader to [157].

Let us define *Robin's reduced modulus* $M_\Omega(A, z_0)$ of the set A with respect to the domain Ω and the point $z_0 \in \Omega$ as

$$M_\Omega(A, z_0) = \lim_{r \to 0} R_{re}(z, z_0) + \frac{1}{2\pi} \log r, \quad |z - z_0| = r$$

in the case of a finite z_0, and

$$M_\Omega(A, \infty) = \lim_{r \to \infty} R_{re}(z, \infty) - \frac{1}{2\pi} \log r, \quad |z| = r$$

otherwise. We note that $\delta_\Omega(A) := \exp(2\pi M_\Omega(A, \infty))$ is Robin's capacity of the set A with respect to Ω. In particular, if $B = \partial\Omega \setminus A = \emptyset$, then Robin's capacity coincides with the usual logarithmic capacity $d(A)$ and Robin's reduced modulus is exactly the reduced modulus of the domain Ω with respect to the finite point z_0.

Another description of Robin's capacity and Robin's reduced modulus is provided by means of the modulus of a family of curves.

Let $C_r(z_0) = \{z : |z - z_0| = r\}$ and $C_r = C_r(0)$. For r sufficiently small and a finite $z_0 \in \Omega$ let us consider the family Γ of curves that connect the set A with $C_r(z_0)$. Then the limit

$$\lim_{r \to 0} \frac{1}{m(\Omega, \Gamma)} + \frac{1}{2\pi} \log r$$

exists and is exactly Robin's reduced modulus $M_\Omega(A, z_0)$. Analogously, for $z_0 = \infty \in \Omega$ and r sufficiently large we define Γ to be the family of rectifiable arcs that connect A with C_r. Then the limit

$$\lim_{r \to \infty} \frac{1}{m(\Omega, \Gamma)} - \frac{1}{2\pi} \log r$$

exists and is Robin's reduced modulus $M_\Omega(A, \infty)$. From this definition it follows that Robin's modulus is changed under a conformal map $f : \Omega \to \Omega'$ by the following rule: for finite points $w_0 = f(z_0)$

$$w = f(z) = w_0 + a(z - z_0) + \cdots, \quad M_{\Omega'}(f(A), w_0) = M_\Omega(A, z_0) + \frac{1}{2\pi} \log |a|,$$

for infinite points $(z_0 = w_0 = \infty)$:

$$w = f(z) = az + a_0 + \frac{a_{-1}}{z} + \cdots, \quad M_{\Omega'}(f(A), \infty) = M_\Omega(A, \infty) - \frac{1}{2\pi} \log |a|,$$

and for the mixed case $(z_0 = \infty, w_0$ finite$)$:

$$w = f(z) = w_0 + \frac{a}{z} + \cdots, \quad M_{\Omega'}(f(A), w_0) = M_\Omega(A, \infty) + \frac{1}{2\pi} \log |a|.$$

Let us mention some results about distortion of Robin's capacity under an "admissible" conformal map $f(z) = z + a_0 + \frac{a_{-1}}{z} + \cdots$. Ch. Pommerenke [447], [444], [446] proved that for an arbitrary closed set A on the unit circle the sharp estimate $d(f(A)) \geq (d(A))^2$ holds. Later on, P. Duren and M.M. Schiffer [159] generalized this result to an arbitrary multiply connected domain giving Robin's interpretation to the inequality $d(f(A)) \geq \delta_\Omega(A)$, which is sharp. Let us give two elementary examples of Robin's capacity and Robin's reduced modulus.

Examples

- Let $\bar{\mathbb{D}}$ be the closed unit disk and A be an arc on the boundary which subtends an angle 2α at the center. Then $d(A) = \sin\frac{\alpha}{2}$, while $\delta_{\bar{\mathbb{D}}}(A) = \left(\sin\frac{\alpha}{2}\right)^2$. Besides, $\delta_{\bar{\mathbb{D}}}(A) + \delta_{\bar{\mathbb{D}}}(B) = 1 = d(A \cup B)$ (see [159]).

- Let $\mathbb{D}' = \mathbb{D} \setminus (-1, -r]$, $r \in (0, 1)$, and A be the unit circle. Then Robin's reduced modulus $M_{\mathbb{D}'}(A, 0) = 0$ of the set A with respect to $\mathbb{D}'$, whereas the usual reduced modulus of the domain $\mathbb{D}'$ with respect to the origin is $\frac{1}{2\pi} \log \frac{4r}{(1+r)^2} < 0$. To see this we use the standard Pick function $\zeta = \varphi(w)$,

$$\varphi(w) = \frac{\left(\beta(1 - w) - \sqrt{\beta^2(1 - w)^2 + 4w}\right)^2}{4w}$$

$$= \frac{1}{\beta^2} w + \cdots$$

$$= \frac{4w}{\left(\beta(1 - w) + \sqrt{\beta^2(1 - w)^2 + 4w}\right)^2},$$

$\beta \geq 1$, that maps the unit disk $\mathbb{D}$ onto $\mathbb{D}'$ with

$$r = \frac{1}{\beta + \sqrt{\beta^2 - 1}} = \beta - \sqrt{\beta^2 - 1}.$$

The arc $\gamma = \{e^{i\theta}, \theta \in [\pi - \alpha, \pi + \alpha]\}$, $\cos(\alpha/2) = 1/\beta$, is mapped onto the slit $[-1, -r]$. Robin's modulus of γ is $M_{\mathbb{D}}(\gamma, 0) = \frac{1}{2\pi} \log \beta^2$. Therefore, making use of the formula of the modulus transformation we have $M_{\mathbb{D}'}(A, 0) = \frac{1}{2\pi} \log 1/\beta^2 + \frac{1}{2\pi} \log \beta^2 = 0$.

Sometimes the notion of reduced modulus of the domain Ω with respect to the point z_0 is replaced by the notion of conformal radius. These concepts are linked by the formula

$$M_\Omega(z_0) = M(\Omega, z_0) = \frac{1}{2\pi} \log R(\Omega, z_0).$$

Similarly we define Robin's radius of the set A with respect to the domain Ω and the point z_0 as

$$M_\Omega(A, z_0) = \frac{1}{2\pi} \log R_\Omega(\Gamma, z_0).$$

5.2.2   A problem with an obstacle

Let a viscous fluid be injected into a Hele-Shaw cell through a point well producing a simply connected evolution until it meets a straight wall. Then it starts sliding along the wall. We denote by $\Omega(t)$ the bounded plane domain in the phase z-complex plane occupied by the moving fluid at instant t. The source is located at the origin and is of strength Q, $Q > 0$. The unique force in consideration is the dimensionless pressure p scaled so that $p = 0$ corresponds to the atmospheric pressure. The initial moment we choose to be when the fluid reaches the wall. This stationary infinite straight wall placed in the Hele-Shaw cell so that $\partial \Omega(t)$ splits into two parts: $\Gamma(t)$ is the free boundary and $\Pi(t)$ is the complementary arc on the wall. The potential function solves the mixed boundary value problem

$$-\Delta p = Q\delta_0 \qquad\qquad \text{in} \quad \Omega(t), \qquad\qquad (5.8)$$

$$p = 0, \quad V_n = -\frac{\partial p}{\partial n} \quad \text{on} \quad \Gamma(t), \qquad\qquad (5.9)$$

$$V_n = 0 \qquad\qquad \text{on} \quad \Pi(t). \qquad\qquad (5.10)$$

The complex potential is exactly given by Robin's function as $W = -QR(z, 0)$ (of course R depends on t).

Richardson [473], considered a similar problem in a wedge assuming circular initial evolution.

We consider the case when the boundary $\Pi(t)$ is an interval during the time of consideration. Observation of the velocities at the contact point between $\Gamma(t)$ and $\Pi(t)$ suggests the contact angle to be $\pi/2$. To derive the equation for the free boundary $\Gamma(t)$ we involve an auxiliary parametric complex ζ-plane, $\zeta = \xi + i\eta$. The Riemann Mapping Theorem yields that there exists a unique conformal univalent map $f(\zeta, t)$ from the unit disk minus a radial slit $\mathbb{D}' = \mathbb{D} \setminus (-1, -r(t)]$, $r(t) \in (0, 1]$, $\mathbb{D} = \{\zeta : |\zeta| < 1\}$, onto the phase domain $f : \mathbb{D}' \to \Omega(t)$,

$f(0,t) = 0$, $a(t) = f'(0,t) > 0$, so that $\Gamma(t) = \{f(e^{i\theta}), \theta \in (-\pi,\pi)\}$, and $\Pi(t) = \{f(\zeta), \zeta \in (-1-0i, -r(t)-0i) \cup (-r(t)+0i, -1+0i)\}$. The point $\zeta = r(t)$ corresponds to a stagnation point at $\Pi(t)$. The function $f(\zeta,0) = f_0(\zeta)$ produces a parametrization of $\partial\Omega(0) = \overline{\Gamma(0) \cup \Pi(0)}$. The moving boundary is parameterized by $\Gamma(t) = f(\partial\mathbb{D}, t)$. The normal outer vector is given by the formula

$$\mathbf{n} = \zeta\frac{f'}{|f'|}, \quad \zeta = e^{i\theta}, \quad \theta \in (-\pi,\pi),$$

and $\mathbf{n} = -1$ on $\Pi(t)$. Therefore, the normal velocity at the free boundary is obtained as

$$V_n = \mathbf{V} \cdot \mathbf{n} = \begin{cases} -\mathrm{Re}\left(W'\zeta\frac{f'}{|f'|}\right), & \text{for } z \in \Gamma,\, \zeta = e^{i\theta},\, \theta \in (-\pi,\pi), \\ \mathrm{Re}\, W', & \text{for } z \in \Pi(t). \end{cases}$$

The superposition $(W \circ f)(\zeta,t) \equiv -Q\, R \circ f(\zeta,t)$ is Q times Robin's function of the domain $\mathbb{D}'$ because of the conformal invariance. The set A for the function $R \circ f$ is the unit circle and the set B is the radial segment $[-1, -r(t)]$. Robin's function for $\mathbb{D}'$ with the chosen A and B is simply $\frac{1}{2\pi}\log\zeta$. Hence, $W'f' = -\frac{Q}{2\pi\zeta}$. On the other hand, we have $V_n = \mathrm{Re}\,\dot{f}\,\overline{\zeta f'}/|f'|$, for $\zeta = e^{i\theta}$, $\theta \in (-\pi,\pi)$ and

$$\mathrm{Re}\,\frac{Q}{2\pi\zeta f'} = \mathrm{Re}\, W' = \mathrm{Re}\,\dot{f} = 0$$

on the wall. This implies that $\dot{f}$ and $1/\zeta f'$ are imaginary or $\dot{f}/\zeta f'$ is real. Finally we deduce the Polubarinova–Galin type equations

$$\mathrm{Re}\,[\dot{f}(\zeta,t)\overline{\zeta f'(\zeta,t)}] = \frac{Q}{2\pi}, \quad |\zeta| = 1,\, \arg\zeta \in (-\pi,\pi), \tag{5.11}$$

$$\mathrm{Im}\,[\dot{f}(\zeta,t)\overline{\zeta f'(\zeta,t)}] = 0 \quad \text{on the radial slit } [-1, -r(t)]. \tag{5.12}$$

The length of the radial slit $1 - r(t)$ is such that the conformal radius of the domain $\Omega(t)$ with respect to the origin is equal to $4r(t)a(t)(1 + r(t))^{-2}$.

If $\Pi(t)$ is the union of intervals, then the function $f(\zeta,t)$ maps the unit disk minus several slits onto the phase domain. Each slit corresponds to a connected component of $\Pi(t)$.

Now we apply Robin's reduced modulus to estimate the area growth of the phase domain of the injecting fluid in the Hele-Shaw problem. First of all, let us recall that the boundary $\partial\Omega(t)$ of the domain occupied by viscous fluid contains a free part $\Gamma(t)$ and the solid part $\Pi(t)$. The fluid is injected through the origin $0 \in \Omega(t)$. The parametric function $f(\zeta,t)$ maps $\mathbb{D}' = \mathbb{D} \setminus (-1, -r(t)]$ onto $\Omega(t)$ and satisfies the equations (5.11)–(5.12).

A simple application of Green's theorem implies that the rate of the area change is expressed as $\dot{S} = Q$, where $S(t)$ is the Euclidean area of $\Omega(t)$. For injection we have $Q > 0$.

Let the Pick function $\zeta = \varphi(w,t)$ map $\mathbb{D}$ onto $\mathbb{D}'$,

$$\varphi(w,t) = \frac{4r(t)}{(1+r(t))^2}w + b_2w^2 + \cdots \equiv bw + b_2w^2 + \cdots, \cdots b = 1/\beta^2$$

so that the arc $\{e^{i\theta}, \theta \in (-\alpha(t), \alpha(t))\}$ is mapped onto $\partial\mathbb{D} \setminus \{-1\}$ and the arc $\{e^{i\theta}, \theta \in (\alpha(t), 2\pi - \alpha(t))\}$ is mapped onto the radial slit $(-1, -r(t))$.

We introduce the analytic function

$$\Phi(w,t) = \frac{\dot{f} \circ \varphi}{\varphi(f' \circ \varphi)}(w,t),$$

defined in $\mathbb{D}$. The mixed boundary value problem (5.11)–(5.12) can be reformulated as the Riemann–Hilbert problem for the analytic function Φ as

$$\mathrm{Re}\ \Phi(e^{i\theta},t) = \frac{Q}{2\pi|(f' \circ \varphi)(e^{i\theta},t)|^2}, \quad \theta \in (-\alpha, \alpha),$$

$$\mathrm{Im}\ \Phi(e^{i\theta},t) = 0, \quad \theta \in (\alpha, 2\pi - \alpha),$$

with bounded values of $|\lim_{\theta \to \pm\alpha^\pm} \Phi(e^{i\theta},t)|$. The solution to this problem is given by the integral representation

$$\Phi(w,t) = \frac{1}{2\pi} \frac{\sqrt{(w-e^{i\alpha})(w-e^{-i\alpha})}}{w+1}$$

$$\times \int_0^{2\pi} \frac{e^{i\theta}+1}{\sqrt{(e^{i\theta}-e^{i\alpha})(e^{i\theta}-e^{-i\alpha})}} h(e^{i\theta},t)\frac{e^{i\theta}+w}{e^{i\theta}-w}d\theta,$$

where the branch of the root is chosen so that $\sqrt{1} = 1$, and

$$h(e^{i\theta},t) = \frac{Q}{2\pi|(f' \circ \varphi)(e^{i\theta},t)|^2}, \quad \theta \in (-\alpha, \alpha),$$

and vanishes in the complementary arc of $\partial\mathbb{D}$, see, e.g., [198]. We deduce that

$$\frac{\dot{a}}{a} = \frac{1}{2\pi} \int_0^{2\pi} \frac{\sqrt{2}\cos\frac{\theta}{2}}{\sqrt{\cos\theta - \cos\alpha}} h(e^{i\theta},t)d\theta.$$

Obviously,

$$\frac{\sqrt{2}\cos\frac{\theta}{2}}{\sqrt{\cos\theta - \cos\alpha}} \geq \frac{1}{\sin\frac{\alpha}{2}} = \beta, \quad \text{for } \theta \in (-\alpha, \alpha).$$

Therefore,

$$\frac{\dot{a}}{a} \geq \frac{Q\beta}{4\pi^2} \int_0^{2\pi} \chi_{[-\alpha(t),\alpha(t)]} \frac{1}{|f'(\varphi(e^{i\theta},t),t)|^2}d\theta,$$

where $\chi_{[-\alpha(t),\alpha(t)]}$ is the characteristic function of the segment $[-\alpha(t),\alpha(t)]$, and $\varphi(e^{i\alpha(t)},t) = -1$. The Hölder inequality implies

$$\frac{\dot{a}}{a} \geq \frac{Q\beta}{8\pi^3} \left(\int_0^{2\pi} \chi_{[-\alpha(t),\alpha(t)]} \frac{1}{|f'(\varphi(e^{i\theta},t),t)|} d\theta \right)^2.$$

In its turn,

$$\frac{Q\beta}{8\pi^3} \left(\int_0^{2\pi} \chi_{[-\alpha(t),\alpha(t)]} \frac{1}{|f'(\varphi(e^{i\theta},t),t)|} d\theta \right)^2$$

$$\geq \frac{Q\beta}{8\pi^3} \left(\int_0^{2\pi} \chi_{[-\alpha(t),\alpha(t)]} \mathrm{Re}\left[\frac{1}{f'(\varphi(e^{i\theta},t),t)} \right] d\theta \right)^2.$$

But $\mathrm{Re}\, f'(\varphi(e^{i\theta},t),t) = 0$ for $\theta \in (\alpha(t), 2\pi - \alpha(t))$. Therefore,

$$\frac{Q\beta}{8\pi^3} \left(\int_0^{2\pi} \chi_{[-\alpha(t),\alpha(t)]} \mathrm{Re}\left[\frac{1}{f'(\varphi(e^{i\theta},t),t)} \right] d\theta \right)^2$$

$$= \frac{Q\beta}{8\pi^3} \left(\int_0^{2\pi} \mathrm{Re}\left[\frac{1}{f'(\varphi(e^{i\theta},t),t)} \right] d\theta \right)^2 = -\frac{Q\beta}{2\pi a^2}.$$

Finally, we have an estimate $2\pi \dot{a} a \sin \frac{\alpha}{2} \geq \dot{S}$, where $a = f'(0,t)$. The conformal radius $R(\Omega(t),0)$ of the domain $\Omega(t)$ is just $ab = a/\beta^2 = a\sin^2 \frac{\alpha}{2}$. Robin's radius of the arc $\Gamma(t)$ is $R_{\Omega(t)}(\Gamma(t),0) = a$. Therefore, we have our main isoperimetric inequality

$$\dot{S} \leq 2\pi \dot{R}_{\Omega(t)}(\Gamma(t),0) \sqrt{R_{\Omega(t)}(\Gamma(t),0)\ R(\Omega(t),0)}.$$

In other words this means that the rate of area change of the phase domain $\Omega(t)$ is controlled by Robin's radius of the free boundary as well as by its conformal radius.

Finally, let us remark that a general case of disconnected boundary component $\Pi(t)$ can be treated in the same way. The solution to the corresponding Riemann–Hilbert problem yields more complicated formulations, so we have considered only the simplest case.

5.3   Isoperimetric inequality for a corner flow

In this section we shall obtain an analogue of the right-hand side estimate given in
Theorem 5.1.2 for the corner flow. In Section 2.6 we already considered such flows
and derived the governing equations for the conformal map that parameterizes the
phase domain. Here we use slightly different parametrization that fits better for
our concrete purpose.

Similarly to the model analysed in Section 2.6 we consider a Hele-Shaw cell
where the viscous fluid occupies a simply connected domain $\Omega(t)$ in the phase z-
plane whose boundary $\Gamma(t)$ at an instant t consists of two walls $\Gamma_1(t)$ and $\Gamma_2(t)$ of
the corner and a free interface $\Gamma_3(t)$ between them. The inviscid fluid (or air) fills
the complement of $\Omega(t)$. The simplifying assumption of constant pressure at the
interface between the fluids means that the surface tension effect is neglected. We
let the positive real axis x contain one of the walls and fix the angle between walls
as $\alpha \in (0, 2\pi)$. The motion of the boundary $\Gamma_3(t)$ is due to injection of strength
$Q > 0$ through the vertex of the corner placed at the origin. The initial domain
$\Omega(0)$ fills the vertex. In our model we consider the local behavior of $\Gamma_3(t)$ and
agree that $\Gamma_3(t)$ is connected. At the wall-fluid contact points where Γ_1 or Γ_2 join
with Γ_3 the velocity vector is directed along the walls that implies that Γ_1 and Γ_2
are perpendicular to Γ_3 at these points.

As before, the pressure field p satisfies the Laplacian equation and the boundary conditions split into the free boundary condition (given on Γ_3) for pressure
and the wall conditions for pressure's normal derivative. The potential p behaves
near the origin as

$$p(z,t) \sim -\frac{Q}{\alpha} \log |z|, \quad \text{as } |z| \to 0.$$

Let us consider an auxiliary parametric complex ζ-plane, $\zeta = \xi + i\eta$. We set
$D = \{\zeta : |\zeta| < 1, 0 < \arg \zeta < \pi\}$, $D_3 = \{z : z = e^{i\theta}, \theta \in (0, \pi)\}$, $D_1 = \{z :
z = -r, r \in (0, 1)\}$, $D_2 = \{z : z = r, r \in (0, 1)\}$, $\partial D = D_1 \cup D_2 \cup D_3$. Construct
a conformal univalent time-dependent map $z = f(\zeta, t)$, $f : D \to \Omega(t)$, such that
being continued onto ∂D, $f(0, t) \equiv 0$, and the circular arc D_3 of ∂D is mapped
onto Γ_3. This map has the expansion $f(\zeta, t) = \zeta^{\alpha/\pi} \sum_{k=0}^{\infty} a_k(t) \zeta^k$ near the origin,
and $a_0(t) > 0$. The function f parameterizes the boundary of the domain $\Omega(t)$ by
$\Gamma_j = \{z : z = f(\zeta, t), \zeta \in D_j\}$, $j = 1, 2, 3$.

Using standard steps of Section 2.6 we arrive at the free boundary condition
expressed in terms of the function f as

$$\text{Re}\,(\dot{f}\,\overline{\zeta f'}) = \frac{Q}{\pi}, \quad \text{for } \zeta \in D_3. \tag{5.13}$$

The wall conditions imply that

$$\text{Im}\,(\dot{f}e^{-i\alpha}) = 0 \quad \text{for } \zeta \in D_1; \quad \text{Im}\,(\dot{f}) = 0 \quad \text{for } \zeta \in D_2. \tag{5.14}$$

We note that the derivative of the mapping function $f'(\zeta, t)$ satisfies the following conditions at D_1 and D_2:

$$\arg(\zeta f'(\zeta, t)) = \pi + \alpha \quad \text{for } \zeta \in D_1,$$

and

$$\operatorname{Im}(\zeta f'(\zeta, t)) = 0 \qquad \text{for } \zeta \in D_2.$$

Hence, we can rewrite the conditions (5.13)–(5.14) as a mixed boundary value problem for the analytic function

$$\Phi(\zeta, t) := \frac{\dot{f}(\zeta, t)}{\zeta f'(\zeta, t)},$$

given by

$$\operatorname{Re}(\Phi(\zeta, t)) = \frac{Q}{\pi |f'(\zeta, t)|^2}, \quad \text{for } \zeta \in D_3, \tag{5.15}$$

$$\operatorname{Im}(\Phi(\zeta, t)) = 0, \qquad \text{for } \zeta \in D_1 \cup D_2. \tag{5.16}$$

Firstly, we solve the mixed boundary value problem (5.15)–(5.16). Making use of an auxiliary Joukowski transform

$$\omega(\zeta) := \frac{1}{2}\left(\zeta + \frac{1}{\zeta}\right), \quad \text{or } \zeta(\omega) := \omega - \sqrt{\omega^2 - 1},$$

we reduce this problem to a Riemann–Hilbert problem in the upper ω-half-plane. Applying the Keldysh–Sedov formula (see, e.g., [198]) for the analytic function $\Phi(\zeta(\omega), t)$ which is bounded at ± 1, we get

$$\Phi(\zeta(\omega), t) = \frac{\sqrt{\omega^2 - 1}}{\pi i} \int_{-1}^{1} \frac{Q}{\pi |f'(\zeta(\tau), t)|^2 \sqrt{\tau^2 - 1}} \frac{d\tau}{\tau - \omega}.$$

The analytic function in the right-hand side is defined in $\mathbb{C} \setminus [-1, 1]$, therefore, choosing a suitable branch of the root we can calculate

$$\lim_{\omega \to \infty} \Phi(\zeta(\omega), t) = \frac{1}{\pi} \int_{-1}^{1} \frac{Q}{\pi |f'(\zeta(\tau), t)|^2 \sqrt{1 - \tau^2}} d\tau.$$

Secondly, we return back to the variable ζ and obtain

$$\frac{\dot{a}_0(t)}{a_0(t)} = \frac{\alpha Q}{\pi^3} \int_0^{\pi} \frac{1}{|f'(e^{i\theta}, t)|^2} d\theta.$$

Certainly,

$$\frac{\dot{a}_0(t)}{a_0(t)} \geq \frac{\alpha Q}{\pi^3} \operatorname{Im} \int\limits_0^\pi \frac{e^{2i\alpha\theta/\pi}}{(e^{i\theta} f'(e^{i\theta}, t))^2} \cdot ie^{i\theta} \frac{d\theta}{e^{i\theta}},$$

or

$$\frac{\dot{a}_0(t)}{a_0(t)} \geq \frac{\alpha Q}{\pi^3} \operatorname{Im} \int\limits_{D_1 \cup D_2 \cup D_3} \frac{\zeta^{2\alpha/\pi}}{(\zeta f'(\zeta, t))^2} \frac{d\zeta}{\zeta}.$$

The function

$$\frac{\zeta^{2\alpha/\pi}}{(\zeta f'(\zeta, t))^2}$$

is analytic about the origin and symmetric with respect to the real axis. Hence, we can take a small circle $S_\varepsilon = \{\zeta : |\zeta| = \varepsilon\}$ and write

$$\operatorname{Im} \int\limits_{D_1 \cup D_2 \cup D_3} \frac{\zeta^{2\alpha/\pi}}{(\zeta f'(\zeta, t))^2} \frac{d\zeta}{\zeta} = \operatorname{Im} \int\limits_{D_3} \frac{\zeta^{2\alpha/\pi}}{(\zeta f'(\zeta, t))^2} \frac{d\zeta}{\zeta}$$

$$= \frac{1}{2} \operatorname{Im} \int\limits_{S_\varepsilon} \frac{\zeta^{2\alpha/\pi}}{(\zeta f'(\zeta, t))^2} \frac{d\zeta}{\zeta}$$

$$= \frac{\pi^3}{\alpha^2 a_0^2}.$$

So we have the inequality $Q \leq \alpha a_0 \dot{a}_0$. The constant Q corresponds to the rate of the area growth. However, one can obtain this directly using Green's theorem. In fact, if $S(t)$ means the area of $\Omega(t)$ and $\Gamma = \Gamma_1 \cup \Gamma_2 \cup \Gamma_3$, then

$$S(t) = \frac{1}{2} \operatorname{Im} \int\limits_\Gamma \bar{z} dz = \frac{1}{2} \operatorname{Im} \int\limits_0^\pi \bar{f} f' ie^{i\theta} d\theta.$$

Therefore,

$$\dot{S} = \frac{1}{2} \operatorname{Im} \int\limits_0^\pi \dot{\bar{f}} f' ie^{i\theta} d\theta + \frac{1}{2} \operatorname{Im} \int\limits_0^\pi \bar{f} \dot{f}' ie^{i\theta} d\theta.$$

Integrating by parts the second term and using (5.13) we come to the equality $\dot{S} = Q$. Finally, we obtain the desirable inequality

$$\dot{S} \leq \alpha a_0 \dot{a}_0, \tag{5.17}$$

which is known for $\alpha = 2\pi$ (see Theorem 5.1.2).

To make inequality (5.17) isoperimetric we interpret a_0 as a certain entity related to the free boundary Γ_3. Let D be a hyperbolic simply connected domain in $\mathbb{C}$ with three finite fixed boundary points z_1, z_2, and a on its piecewise smooth boundary. Denote by D_ε the domain $D \setminus \mathbb{D}(a, \varepsilon)$ for a sufficiently small ε, where $\mathbb{D}(a, \varepsilon) = \{z : |z - a| < \varepsilon\}$. Denote by $M(D_\varepsilon)$ the modulus of the family of arcs in D_ε joining the boundary arc of $\mathbb{D}(a, \varepsilon)$ that lies in the circumference $|z - a| = \varepsilon$ with the leg of the triangle D which is opposite to a (we choose a unique arc of the circle so that it can be connected in D_ε with the leg (z_1, z_2) for any $\varepsilon \to 0$). If the limit

$$M_\Delta(D, a) = \lim_{\varepsilon \to 0} \left(\frac{1}{M(D_\varepsilon)} + \frac{1}{\varphi_a} \log \varepsilon \right)$$

exists, where $\varphi_a = \sup \Delta_a$ is the inner angle and Δ_a are the Stolz angles inscribed in D at a, then it is called the *reduced modulus of the triangle* D. The conditions for the reduced modulus to exist are found in [448], [528], [562]. It turns out that the reduced modulus exists if D is conformal at a. Let there exist a conformal map $f(z)$ of the triangle D onto a triangle D' such that there is an angular limit $f(a)$ (see definitions in [447]) with the inner angle ψ_a at the vertex $f(a)$. If the function f has the angular finite non-zero derivative $f'(a)$, then $\varphi_a = \psi_{f(a)}$ and the reduced modulus of D exists and changes [528, 562] according to the rule

$$M_\Delta(f(D), f(a)) = M_\Delta(D, a) + \frac{1}{\psi_a} \log |f'(a)|.$$

If we suppose, moreover, that f has the expansion

$$f(z) = w_1 + (z - a)^{\psi_a / \varphi_a} (c_1 + c_2(z - a) + \cdots)$$

in a neighborhood of the point a, then the reduced modulus of D changes according to the rule

$$M_\Delta(f(D), f(a)) = M_\Delta(D, a) + \frac{1}{\psi_a} \log |c_1|.$$

Similarly to the connection between the conformal radius and the usual reduced modulus of a simply connected domain with respect to an inner point, we introduce the conformal triangle radius $R_\Delta(D, a)$ as

$$R_\Delta(D, a) = \exp[\varphi_a M_\Delta(D, a)].$$

The conformal triangle radius of the half-disk $\{|z| < 1\} \cap \{\operatorname{Im} z > 0\}$ with marked vertices $0, \pm 1$ with respect to the origin is 1. The phase domain Ω is conformal at the origin. Using this interpretation we rewrite inequality (5.17) as

$$\dot{S} \le \alpha \, R_\Delta(\Omega(t), 0) \dot{R}_\Delta(\Omega(t), 0).$$

This is the isoperimetric inequality we were looking for.

5.4 Melting of a bounded crystal

In Section 4.6 we have already discussed the governing equation for a melting crystal. In this section we consider a bounded initial crystal that is melting in forced flow. The fluid moves to the right and there are two stagnating points on the interface of the crystal. The governing equations are the same (4.59) with the initial conditions

$$\lim_{x \to \pm\infty} \theta = 1, \quad \lim_{y \to \pm\infty} \frac{\partial \theta}{\partial y} = 0.$$

The Boussinesq transformation applied to the convective heat transfer equation (4.59) leads to uncoupling of the problem. There is a conformal univalent map from the phase domain $\Omega(t)$ onto the plane of the complex potential $W = \varphi + i\psi$. Under this transformation the boundary of the crystal cross-section is mapped into the slit along the positive real axis $\psi = 0$, $\varphi \in [-2a, 2, a]$ in the W-plane. Thus, the problem admits the form:

$$Pe \frac{\partial \theta}{\partial \varphi} = \Delta\theta, \quad W \in D, \tag{5.18}$$

where $D = \mathbb{C} \setminus [-2a, 2a]$. The boundary conditions are

$$\lim_{\varphi \to \pm\infty} \theta = 1, \quad \lim_{\psi \to \pm\infty} \frac{\partial \theta}{\partial \psi} = 0, \quad \theta = 0, \quad W \in \partial D. \tag{5.19}$$

We introduce the auxiliary parametric complex ζ-plane, $\zeta = \xi + i\eta$. The Riemann Mapping Theorem yields that there exists a conformal univalent map $f(\zeta, t)$ of the exterior part $\mathbb{D}^*$ of the unit disk $\mathbb{D}$ onto the phase domain $f : \mathbb{D}^* \to \Omega(t)$, normalized by $f(\zeta, \cdot) = a\zeta + a_0 + a_{-1}/\zeta + \cdots$. This problem does not admit separation of variables as in the previous case. In [318] it was shown that the heat flux density at the slit on the plane of the complex potential $W = \varphi + i\psi$ can be expressed as $|\partial\theta/\partial\psi| = (4a^2 - \varphi)^{-1/2}\mu(\varphi/2a)$. The Joukowski function $W = a(\zeta + 1/\zeta)$ permits us finally to come to the Polubarinova–Galin type equation for the free boundary:

$$\text{Re} \left[\dot{f}(\zeta, t)\overline{\zeta f'(\zeta, t)} \right] = -\mu(\cos\theta), \quad \zeta = e^{i\theta}. \tag{5.20}$$

The function μ in (5.20) satisfies the integral equation [318] for a crystal that admits reflection with respect to the real axis and for small Péclet numbers:

$$\int_{-1}^{1} \frac{\mu(\xi)}{\sqrt{1 - \xi^2}} \ln \left| \frac{\varphi}{2a} - \xi \right| d\xi = \pi - \frac{Q}{2} \ln \frac{ae^\gamma Pe}{2},$$

where $\varphi \in [-2a, 2a]$, γ is Euler's constant, and

$$Q = 2 \int_{-1}^{1} \frac{\mu(\xi)}{\sqrt{1 - \xi^2}} d\xi = \int_0^{2\pi} \mu(\cos\theta) d\theta$$

is the total heat flux. Obviously, the sign of the function μ is connected with the sign of the normal velocity. Therefore, for a melting crystal we have $\mu(\cos\theta) \geq 0$ for all $\theta \in [0, 2\pi)$. A simple application of Green's Theorem yields that the rate of the area change of the nucleus $\dot{S}$ is exactly equal to the total heat flux taken with a minus sign: $\dot{S} = -Q$. In fact, we have

$$
2S(t) = - \int_{\Gamma(t)} \mathrm{Im}\,(f d\bar{f}) = \mathrm{Re} \int_0^{2\pi} f(e^{i\theta}, t) \overline{\frac{\partial f(\zeta, t)}{\partial \zeta}} \bigg|_{\zeta = e^{i\theta}} e^{-i\theta} d\theta.
$$

Then,

$$
2\frac{dS}{dt} = - \int_0^{2\pi} \mu(\cos\theta) d\theta - \mathrm{Im} \int_0^{2\pi} f \frac{\partial}{\partial t} \left(\frac{\partial f}{\partial \theta} \right) d\theta.
$$

Integrating the last term by parts we obtain that $\dot{S} = - \int_0^{2\pi} \mu(\cos\theta) d\theta = -Q$.

The equation (5.20) implies

$$
\dot{a} = -\frac{a}{2\pi} \int_0^{2\pi} \frac{\mu(\cos\theta)}{|f'(e^{i\theta}, t)|^2} d\theta.
$$

Since we have the inequality $|f'(e^{i\theta}, t)| \leq 2a$ for functions that map $\mathbb{D}^*$ onto a convex domain, the radius-area estimate $\dot{S} \geq 8\pi a \dot{a}$ can be given.

Theorem 5.4.1. *If the initial nucleus is convex, then locally in time we have the estimate $\dot{S} \geq 8\pi a \dot{a}$ where $a = \mathrm{cap}\,\Gamma(t)$.*

Problem 5.4.1. Give an analogous estimate without the assumption about convexity.

Problem 5.4.2. Give an estimate of $\dot{a}$ or of the above integral mean depending on the total heat flux Q.

5.5   Asymptotics of roots and coefficients in the polynomial case

In this section we shall say something about the asymptotic behaviour of the roots and coefficients of $g = f'$ for polynomial solutions f of the Polubarinova–Galin equation (1.22) with $Q > 0$. We write

$$
f(\zeta, t) = \sum_{j=1}^{m+1} a_j(t) \zeta^j \tag{5.21}
$$

and assume $a_{m+1}(0) \neq 0$, by which $a_{m+1}(t) \neq 0$ for all t (see after (5.23) below). Thus $g(\zeta, t)$ will have m complex zeros (roots) for each t. Since we are going to study asymptotics we assume that the solution (5.21) exists, as a locally univalent solution of (1.22), for all $0 < t < \infty$. By Theorem 4.2.1, it is then actually univalent all the time.

The principal result, Theorem 5.5.1, says that the roots of g asymptotically, as t is large, are equidistributed on an expanding circle, even though the initial behaviour may be much more complicated. The proof uses estimates of the coefficients in (5.21), which are of independent interest (Lemma 5.5.1). The results in this section are due to Y.-L. Lin [360, 361, 239]. For some further results on asymptotics, and stability, see [575, 417, 418].

We first recall Richardson's formula [470] for the harmonic moments M_k (see (1.33), 1.37):

$$M_k = \sum_{(i_1,\ldots,i_{k+1})} i_1 a_{i_1} \cdots a_{i_{k+1}} \overline{a_{i_1+\cdots+i_{k+1}}}. \tag{5.22}$$

Here $k = 0, 1, 2, \ldots$ and the summation runs over all $(k + 1)$-tuples $(i_1, \ldots, i_{k+1})$ of integers with $i_j \geq 1$, with the convention that $a_j = 0$ for $j > m + 1$ in the present polynomial case (5.21). See [470] and [336] for more details. Note that the final non-zero moment has a quite simple expression:

$$M_m = a_1^{m+1} \overline{a_{m+1}}. \tag{5.23}$$

Since M_m is conserved in time this shows that $a_{m+1}(t) \neq 0$ if $a_{m+1}(0) \neq 0$. Lemma 5.5.1 below will show that asymptotically, as $t \to \infty$, the formula (5.23) will be almost true for all moments.

Let $\omega_1(t), \ldots, \omega_m(t)$ denote the zeros of

$$g(\zeta, t) = \sum_{j=1}^{m+1} j a_j(t) \zeta^{j-1} = \sum_{j=0}^{m} (j+1) a_{j+1}(t) \zeta^j.$$

Besides these we introduce two other sets of zeros, for comparison:

- Let

$$\tilde{\omega}_k(t) = \omega_k(t) a_1(t)^{-\frac{m+2}{m}} \tag{5.24}$$

  be corresponding **rescaled zeros**, namely the zeros of $\tilde{g}(\zeta, t) = g(a_1^{\frac{m+2}{m}} \zeta, t)$.
- Let

$$\hat{\omega}_k = \frac{1}{\sqrt[m]{-(m+1)\overline{M}_m}},$$

  be the zeros of the polynomial $a_1 + (m+1)a_1^{m+2} a_{m+1} \zeta^m$ consisting of the first and last term in $\tilde{g}(\zeta, t)$.

The $\hat{\omega}_k$ do not depend on time. On the other hand, the rescaled zeros $\tilde{\omega}_k(t)$ are normalized in such a way that they asymptotically stabilize, and approach the $\hat{\omega}_k$ (Theorem 5.5.1). Note also that the product $\tilde{\omega}_1(t) \cdots \tilde{\omega}_m(t) = -\frac{1}{(m+1)\overline{M}_m}$ is a conserved quantity.

Lemma 5.5.1. *For any locally univalent polynomial solution* (5.21) *of* (1.22) *we have*

$$\left|a_j(t)a_1(t)^j - \overline{M_{j-1}}\right| \le \frac{F_j}{a_1(t)^4}, \quad 0 < t < \infty, \ 1 \le j \le m+1 \tag{5.25}$$

where the constants F_j, $1 \le j \le m+1$, *depend only on the initial data* $a_1(0)$, ..., $a_{m+1}(0)$. *In particular we have estimates of the kind*

$$|N_0(t)| \le \frac{C_1}{a_1(t)^4}, \tag{5.26}$$

$$|a_j(t)| \le \frac{C_j}{a_1(t)^j}, \quad 2 \le j \le m+1, \tag{5.27}$$

for suitable constants C_j *depending only on the initial data* (N_0 *is defined in* (4.3)).

Proof. Using (5.22) we first prove by induction on decreasing j that there exist constants G_j, $1 \le j \le m+1$, such that

$$\left|a_j(t)a_1(t)^j - \overline{M_{j-1}}\right| \le \frac{G_j}{a_1(t)^2}. \tag{5.28}$$

We start by setting $G_{m+1} = 0$, which makes (5.28) hold for $j = m+1$. Now we show (5.28) for $j = k$ assuming it holds for all $j \ge k+1$. The arguments will work for any $1 \le k \le m$. Using (5.22) and suppressing dependence on t from notation, we have

$$\begin{aligned}
M_{k-1} - \bar{a}_k a_1^k &= \sum_{(i_1,\ldots,i_k) \ne (1,\ldots,1)} i_1 a_{i_1} \cdots a_{i_k} \cdot \overline{a_{i_1+\cdots+i_k}} \\
&= \sum_{(i_1,\ldots,i_k) \ne (1,\ldots,1)} \frac{i_1 a_{i_1} \cdots a_{i_k}}{a_1^{i_1+\cdots+i_k}} \cdot \overline{a_{i_1+\cdots+i_k} a_1^{i_1+\cdots+i_k}} \\
&= \frac{1}{a_1^2} \sum_{(i_1,\ldots,i_k) \ne (1,\ldots,1)} \frac{i_1 a_{i_1} \cdots a_{i_k}}{a_1^{i_1+\cdots+i_k-2}} \cdot \overline{a_{i_1+\cdots+i_k} a_1^{i_1+\cdots+i_k}}.
\end{aligned}$$

We shall estimate the terms in the above sum. Since $(i_1,\ldots,i_k) \ne (1,\ldots,1)$, at least one of the i_j is ≥ 2, hence it follows from (4.6), (4.2) that the first factors can be estimated as

$$\left|\frac{i_1 a_{i_1} \cdots a_{i_k}}{a_1^{i_1+\cdots+i_k-2}}\right| \le (m+1) \sum_{0 \le \alpha,\beta \le m+1} \frac{(\sqrt{N_0(0)})^\alpha}{a_1(0)^\beta}. \tag{5.29}$$

Moreover $i_1 + \cdots + i_k \ge k+1$, so by the induction hypothesis we have

$$\left|a_{i_1+\cdots+i_k} a_1^{i_1+\cdots+i_k} - \overline{M_{i_1+\cdots+i_k-1}}\right| \le \frac{G_{i_1+\cdots+i_k}}{a_1(0)},$$

in particular

$$|a_{i_1+\cdots+i_k} a_1^{i_1+\cdots+i_k}| \leq \frac{G_{i_1+\cdots+i_k}}{a_1(0)^2} + \sum_{j=1}^{m} |M_j|.$$

Therefore we can estimate also the second factors in the above expression for $a_k a_1^k - \overline{M_{k-1}}$. From this we easily deduce (5.28) for $j = k$ knowing that it is true for $j \geq k+1$. This completes the induction step and hence proves (5.28) for $1 \leq j \leq m$.

Now, already (5.28) shows that the estimate (5.27) holds. Since this estimate improves (4.6) by a factor at least $a_1(t)^2$ in the denominator, we can 'bootstrap' the previous argument: using the new estimate in (5.29) makes the induction process work with the factor $1/a_1(t)^2$ in (5.28) replaced by $1/a_1(t)^4$. Thus (5.25) follows, and since (5.26) is just the special case $j = 1$ the lemma is proved.   □

Theorem 5.5.1. *With notations as above, assume that $f(\zeta, t)$ is a global polynomial solution of (1.22). Then*

$$|\tilde{\omega}_k(t) - \hat{\omega}_k| \to 0 \quad as \ t \to \infty$$

if the roots $\tilde{\omega}_k(t)$ and $\hat{\omega}_k$ are ordered appropriately. Furthermore, all roots eventually move away from the origin as time is large enough.

Proof. The monic polynomial vanishing at the rescaled roots (5.24) is

$$\prod_{j=1}^{m}(\zeta - \tilde{\omega}_j) = \sum_{j=1}^{m+1} \left(a_1^{-\frac{m+2}{m}}\right)^{m+1-j} \frac{ja_j}{(m+1)a_{m+1}} \zeta^{j-1}$$

$$= \sum_{j=1}^{m+1} \frac{j}{m+1} \frac{a_j a_1^j}{a_{m+1} a_1^{m+1}} a_1^{-\frac{2}{m}(m+1-j)} \zeta^{j-1}$$

$$= \frac{1}{(m+1)\overline{M_m}} + \sum_{j=2}^{m} \frac{j}{m+1} \frac{a_j a_1^j}{\overline{M_m}} a_1^{-\frac{2}{m}(m+1-j)} \zeta^{j-1} + \zeta^m.$$

Due to Lemma 5.5.1, and because the exponents $-\frac{2}{m}(m+1-j)$ are strictly negative, the coefficients of the middle terms tend to zero:

$$\left| \frac{j}{m+1} \frac{a_j a_1^j}{\overline{M_m}} a_1^{-\frac{2}{m}(m+1-j)} \right| \leq \left| \frac{j}{m+1} \frac{F_j}{a_1^4 \overline{M_m}} a_1^{-\frac{2}{m}(m+1-j)} \right|$$

$$+ \left| \frac{j}{m+1} \frac{M_{j-1}}{\overline{M_m}} a_1^{-\frac{2}{m}(m+1-j)} \right| \to 0 \quad as \ t \to \infty \quad (2 \leq j \leq m). \tag{5.30}$$

From this it follows that $|\tilde{\omega}_k(t) - \hat{\omega}_k| \to 0$ as $t \to \infty$.

It also follows that the roots move away from the origin as time is large, because the speed of the roots only depends on the position of the roots, and it is clear that for a symmetric configuration of roots, like $\hat{\omega}_k$, the speed points radially away from the origin.   □

Chapter 6

Laplacian Growth and Random Matrix Theory

The link between Laplacian growth and stochastic processes in the complex plane was discovered rather unexpectedly [581, 551], through their common relation to the multi-particle wavefunction description of the Quantum Hall Effect, in the single-Landau level approximation. As pointed out in [551], the classical Laplacian growth and its stochastic variant based on the normal random matrix theory (NRMT) can be identified to the dispersionless limit of a certain integrable hierarchy and its dispersionful versions, respectively. We discuss this formulation of the relation between the two models in the last chapters of this book; in the present chapter, we only mention the works where this relationship was already implicit, although not recognized as such.

Looking beyond the interest presented by the connections between these seemingly different mathematical objects, expressing classical Laplacian growth as a continuum limit of a stochastic process with a finite length scale (whose zero limit produces the classical problem), has obvious and important applications, in particular as a regularization approach for the critical cases of Laplacian growth, see Chapter 3. As discussed, since the critical cases describe the approach towards viscous fingering, a finite-time singularity of the singular perturbation class, having a regularization method which preserves the integrable structure of classical Laplacian growth (by embedding it into a larger integrable hierarchy) is the ideal candidate for the proper way of continuing the growth process beyond the classical critical case, and to further compute the geometric characteristics of the viscous finger-splitting phenomenon.

Before formulating the stochastic model in the complex plane known as normal random matrix theory, we give a very brief outline of the main thrust of random matrix theory, which for the most part (during the last four decades of the last century) was developed for classes of matrices whose equilibrium spectrum distribution lies on analytical arcs, rather than domains in $\mathbb{C}$, such as (but not reduced to) the Hermitian and unitary cases.

During the second half of the last century and continuing through the present, random matrix theory has grown from a special method of theoretical physics, meant to approximate energy levels of complex nuclei [585, 586, 587, 165, 166, 167, 168], into a vast mathematical theory with many different applications in physics, computer and electrical engineering. Simply describing all the developments and methods currently employed in this context would result in a monography much more extensive than this one. Therefore, we will only briefly mention topics which are themselves very interesting, but lie beyond the scope of this work.

The applications of random matrix theory (RMT) have been extended from the original subject, spectra of heavy nuclei, to descriptions of large N $SU(N)$ gauge theory [273, 74], critical statistical models in two dimensions [299, 321, 63] disordered electronic systems [425, 170, 264, 197, 21], quantum chromo-dynamics (QCD) [567, 16], to name only a few. Non-physics applications range from communication theory [557] to stochastic processes out of equilibrium [461, 462] and even more exotic topics [135].

A number of important results, both at theoretical and applied levels, were obtained from the connection between random matrices and orthogonal polynomials, especially in their weighted limit [534, 487, 74, 54, 55]. These works explored the relationship between the branch cuts of spectral (Riemann) curves of systems of differential equations and the support of limit measures for weighted orthogonal polynomials. Yet another interesting connection stemming from this approach is with the general (matrix) version of the Riemann–Hilbert problem with finite support [57, 134].

In [581, 551], it was shown that such relationships also hold for the class of normal random matrices. Unlike in previous works, for this ensemble, the support of the equilibrium distribution for the eigenvalues of matrices in the infinite-size limit, is two-dimensional, which allows us to interpret it as a growing cluster in the plane. Thus, a direct relation to the class of models known as Laplacian growth (both in the deterministic and stochastic formulations), was derived, with important consequences. In particular, this approach allowed us to study formation of singularities in models of two-dimensional growth. Moreover, these results allowed us to define a proper way of continuing the solution for singular Laplacian growth, beyond the critical point.

From the point of view of the dimensionality of the support for random matrix eigenvalues, it is possible to distinguish between one-dimensional situations (which characterize 1- and 2-matrix models), and two-dimensional situations, like in the case of normal random matrix theory. In fact, very recent results point to intermediate cases, where the support is a set of dimensions between 1 and 2. This situation is very similar to the description of disordered, interacting electrons in the plane, in the vicinity of the critical point which separates localized from de-localized behavior [170].

6.1 Random matrix theory in 1D

6.1.1 The symmetry group ensembles and their physical realisations

Following [385], we reproduce the standard introduction of the symmetry-groups ensemble of random matrices. The traditional ensembles (orthogonal, unitary and symplectic) were introduced mainly because of their significance with respect to symmetries of Hamiltonian operators in physical theories: time-reversal and rotational invariance corresponds to the orthogonal ensemble (which, for Gaussian measures, is naturally abbreviated GOE), while time-reversal alone and rotational invariance alone correspond to the symplectic and unitary ensembles, respectively (GSE and GUE for Gaussian measures).

An invariant measure is defined for each of these ensembles, in the form

$$d\widetilde{\mu}(M) \equiv P(M)d\mu(M) \equiv Z^{-1}e^{\mathrm{Tr}[W(M)]}d\mu(M), \qquad (6.1)$$

where M is a matrix from the ensemble, Z is a normalization factor (partition function), $\mathrm{Tr}[W(M)]$ is invariant under the symmetries on the ensemble, and $d\mu(M)$ is the appropriate flat (Haar) measure for that ensemble: $\prod_{i\leq j} dM_{ij}$ for orthogonal, $\prod_{i\leq j} d\,\mathrm{Re}\,M_{ij}\prod_{i<j} d\,\mathrm{Im}\,M_{ij}$ for unitary, and $\prod_{i\leq j} dM_{ij}^{(0)}\prod_{k=1}^{3}\prod_{i<j} dM_{ij}^{(k)}$ for symplectic (where each matrix element is an element of the real Klein group, $M_{ij} = M_{ij}^{(0)}\cdot 1 + \sum_{k=1}^{3} M_{ij}^{(k)}\cdot\sigma_k$). Correspondingly, to each of these ensembles, a parameter β, indicating the number of independent real parameters necessary to describe the pair of values M_{ij}, M_{ji}, is introduced, with values $\beta = 1, 2, 4$ for orthogonal, unitary and symplectic ensembles, respectively.

The invariance under transformations from the appropriate symmetry group leads to the following simplification of the measure: for any of these ensembles, the generic matrix M can be diagonalized by a transformation $M = U^{-1}\Lambda U$, with U from the same group, and $\Lambda = \mathrm{diag}(\lambda_1, \ldots, \lambda_N)$. The Jacobian of the transformation $M \to \Lambda, U$ (where U is said to carry the "angular" degrees of freedom of M) is $J = \prod_{i<j}|\lambda_i - \lambda_j|^{\beta} = |\Delta(\Lambda)|^{\beta}$, with Δ the Vandermonde determinant. The angular degrees of freedom can be integrated out (a trivial redefinition of the normalization factor), giving the simplified measure

$$\rho(\lambda_1, \ldots, \lambda_N)\prod_{i=1}^{N} d\lambda_i = Z^{-1}e^{\mathrm{Tr}[W(\Lambda)]}|\Delta(\Lambda)|^{\beta}\prod_{i=1}^{N} d\lambda_i. \qquad (6.2)$$

For example, in the case of Gaussian measure $W(M) = -M^2$, the joint probability distribution function of eigenvalues, ρ, becomes (up to normalization)

$$\rho(\lambda_1, \ldots, \lambda_N) = \exp\left[-\sum_{i=1}^{N}\lambda_i^2 + \beta\sum_{i<j}\log|\lambda_i - \lambda_j|\right]. \qquad (6.3)$$

Clearly, this procedure is useful only if we are interested in computing expectation values of quantities which depend only on the distribution of eigenvalues, and not on the angular degrees of freedom. This is indeed the case for all situations of interest.

The next standard transformation (which we discuss for the case of unitary ensemble, $\beta = 2$) that is performed on the measure uses the well-known property of Vandermonde determinant $\Delta(\Lambda) = \det[\lambda_i^{j-1}]_{1 \leq i,j \leq N}$. Because of standard determinantal identities, this is equivalent with replacing each monomial λ_i^{j-1} by a *monic* polynomial of the same order, $P_{j-1}(\lambda_i) = \lambda_i^{j-1} + \cdots$. Finally, these polynomials may be chosen to be orthogonal with respect to the measure $e^{W(\lambda)}$, giving for the p.d.f. of eigenvalues the expression

$$\rho(\lambda_1, \ldots, \lambda_N) = |\det[P_{j-1}(\lambda_i)e^{W(\lambda_i)/2}]|^2. \tag{6.4}$$

This type of formula can be generalized to the case of matrix ensembles with two-dimensional support of eigenvalues.

Generalizations of group ensembles. Recently, various generalizations were proposed in order to extend the theory for ensembles of matrices which are not associated with symmetry groups. In particular, ensembles of matrices which may be reduced to a tridiagonal form (instead of standard diagonal) by a transformation which eliminates "angular" degrees of freedom, were introduced in [155]. As an interesting consequence, many results carry over to this case, while the parameter β is allowed to take any positive real value.

6.1.2   Critical ensembles

In this section we explain how, using properly chosen non-Gaussian measures, it is possible to construct ensembles of Hermitian matrices (corresponding again to the unitary symmetry) which are in a sense critical, i.e., for which a continuum limit ($N \to \infty$) may be defined. The discussion relies on the formulation based on orthogonal polynomials indicated above, and it follows (at a more elementary level) the general theory of Saff and Totik [487].

General formalism

Let $d\mu(x) = e^{W(x)}dx$ be a well-defined measure on the real axis, $W(x) \to -\infty$ as $|x| \to \infty$, and $P_n^{(1)}(x)$ the corresponding family of orthogonal polynomials

$$\int_{-\infty}^{\infty} P_n^{(1)}(x)P_m^{(1)}(x)d\mu(x) = \delta_{nm}. \tag{6.5}$$

Orthonormal functions are obtained through $\psi_n(x) = P_n(x)e^{W(x)/2}$, which are orthogonal with respect to the flat measure on $\mathbb{R}$. We consider a deformation of this ensemble through a positive real parameter $\lambda \geq 1$, so that $d\mu_\lambda(x) = e^{\lambda W(x)} dx$ and

$$\int_{-\infty}^{\infty} P_n^{(\lambda)}(x) P_m^{(\lambda)}(x) d\mu_\lambda(x) = \delta_{nm}. \tag{6.6}$$

Clearly, if $W(x)$ is a monomial of degree k, the deformation amounts to a simple rescaling

$$P_n^\lambda(x) = \lambda^{1/2k} P_n^{(1)}(\lambda^{1/k} x). \tag{6.7}$$

The first non-trivial example is a quartic polynomial of the type

$$W(x) = -(x^2 + gx^4), \quad g > 0, \tag{6.8}$$

for which the deformation is not a simple rescaling. In this case, it is possible to consider a special limit $n \to \infty, \lambda \to \infty, \lambda \to nr_c$, where r_c is a constant. As we will see, for a specific value of r_c, this limit yields a special asymptotic behavior of the orthonormal functions $\psi_n(x)$. However, even for the simplest, trivial monomial (a Gaussian), which yields the Hermite polynomials, the asymptotic behavior of the orthogonal functions is non-trivial, in the sense that there are no known good approximations for the case $r_c = O(1)$.

Generically, in this large n, λ limit, we can ask where the wave function $\psi_n(x)$ will reach its maximum value, in the saddle point approximation:

$$\max_{|x|} \partial_x |\psi_n(x)| = 0, \tag{6.9}$$

giving

$$\partial_x \left[\sum_{i=1}^{n} \log(x - \xi_i) - nr_c + W(x) \right] = 0, \tag{6.10}$$

so that

$$-r_c W'(x) = \frac{2}{n} \sum_{i=1}^{n} \frac{1}{x - \xi_i}, \tag{6.11}$$

where ξ_i, $i = 1, \ldots, n$, are the roots of the nth polynomial.

Let

$$\omega(z) = \frac{1}{n} \sum_{i=1}^{n} \frac{1}{\xi_i - z}, \tag{6.12}$$

multiply (6.11) by $(\xi_i - z)^{-1}$ and sum over i, and obtain

$$\omega^2(z) - r_c W'(z)\omega(z) = -\frac{r_c}{n} \sum_{i=1}^{n} \frac{W'(z) - W'(\xi_i)}{z - \xi_i}. \tag{6.13}$$

Equation (6.13) can be solved in the large n limit by assuming that the roots will be distributed with density $\rho(\xi)$ on some compact (possibly disconnected) set $I \subset \mathbb{R}$. Defining

$$R(z) = -\frac{4}{r_c} \int_I \frac{W'(z) - W'(\xi)}{z - \xi} \rho(\xi) d\xi, \tag{6.14}$$

we obtain

$$w^2(z) - r_c W'(z)w(z) + \left(\frac{r_c}{2}\right)^2 R(z) = 0. \tag{6.15}$$

The proper solution of (6.13) (considering the behavior at ∞ of the function $w(z)$), is

$$w(z) = \frac{r_c}{2}\left[W'(z) + \sqrt{(W'(z))^2 - R(z)}\right], \tag{6.16}$$

and (since the function $w(z)$ is the Cauchy transform of the density $\rho(x)$), it gives us the asymptotic distribution of zeros as

$$\rho(x) = \frac{1}{2\pi i}\lim_{\epsilon \to 0}[w(x + i\epsilon) - w(x - i\epsilon)]. \tag{6.17}$$

Finally, to obtain the asymptotic form of wave functions $\psi_n(x)$, we can write

$$n^{-1}\log\psi_n(x) \to \int \rho(\xi)\log(x - \xi)d\xi - \frac{r_c}{2} + W(x). \tag{6.18}$$

Continuum limit and integrable equations

There are two related problems for the large n limit of deformed ensembles described in the previous section. The first is the determination of the support of zeros I; the second is the scaling behavior of the orthogonal functions $\psi_n(x)$. In general, the limiting support I may consist of several disconnected segments I_k, $I = \cup_{k=1}^{k=d}I_k$. In the simplest case, it is just one interval $I = [a, b] \subset \mathbb{R}$. In this section we indicate how to determine this support as well as the density $\rho(x)$, and what this yields for the orthogonal functions.

Let the function $W(x)$ be a polynomial of even degree d. From (6.14) we see that $R(z)$ is a polynomial of degree $d - 2$, and therefore solution (6.16) has generically $2(d - 1)$ branch points. Thus, the function $w(z)$ typically has $d - 1$ branch cuts, which constitute the disconnected support of distribution $\rho(z)$.

We are interested in a special case, when $d - 2$ of these cuts degenerate into double points, and there is a single interval $[a, b]$ which is the support of $\rho(z)$. This special case is called *critical* and it provides new asymptotic limits for the orthogonal functions. We will also refer to this solution as the "single-cut" solution.

From the equation

$$w(x + i0) + w(x - i0) = r_c W'(x), \tag{6.19}$$

we obtain for the single-cut solution

$$w(z) = -\frac{r_c\sqrt{(z - a)(z - b)}}{2\pi}\int_a^b \frac{W'(\xi)}{\sqrt{(b - \xi)(\xi - a)}}\frac{d\xi}{\xi - z}.$$

The large $|z|$ behavior of this function is known from the continuum limit of (6.12), and it implies the absence of regular terms in the Laurent expansion:

$$\omega(z) = -\frac{1}{z} + O(z^{-2}),\tag{6.20}$$

so that we impose the conditions

$$0 = \int_a^b \frac{W'(\xi)}{\sqrt{(b-\xi)(\xi-a)}}d\xi,\tag{6.21}$$

$$2\pi = -r_c \int_a^b \frac{\xi W'(\xi)}{\sqrt{(b-\xi)(\xi-a)}}d\xi.\tag{6.22}$$

Gaussian measure and the Hermite polynomials. Let $d = 2$ and $-W(x) = ax^2$, $a > 0$. Then conditions (6.21), (6.22) give a symmetric support $[-b, b]$ where $b^2 = 2/(ar_c)$.

More generally, using the saddle point equation for $|\psi_n(x)|$ at $x = a, b$ and (6.18), we conclude that

$$\frac{\log \psi_n(b+\zeta)}{n} = C_b - \frac{r_c}{2}\int_0^\zeta \sqrt{[W'(b+\eta)]^2 - R}\,d\eta.\tag{6.23}$$

Since the integrand behaves like $\eta^{d-3/2}$, we obtain

$$\psi_n(b+\zeta) = \psi_n(b)\exp\left[-\frac{nr_c}{2d-1}\zeta^{d-1/2}\right].\tag{6.24}$$

We immediately conclude that for $d = 2$ (Hermite polynomials), the asymptotic behavior is given by the Airy function, $\exp z^{3/2}$. The full scaling is achieved by considering the region around the end-point b, of order $\zeta = O(n^{-2/(2d-1)})$. Then we obtain

$$\psi_n(b+\tilde\zeta n^{-\frac{2}{2d-1}}) \sim \exp\left[-\frac{r_c}{2d-1}\tilde\zeta^{d-1/2}\right].\tag{6.25}$$

Scaled limits of orthogonal polynomials and equilibrium measures

The distribution of eigenvalues investigated in the previous sections illustrates the general approach developed by Saff and Totik [487] for holomorphic polynomials orthogonal on curves in the complex plane. We sketch here the more general result because of its relevance to the main topic of this review.

Given a closed set $\Sigma \subset \mathbb{C}$ and a properly-defined measure on it, $w(z) = e^{-Q(z)}$, we construct the holomorphic orthogonal polynomials $P_n(z)$, with respect to w. We then pose the question of finding the "extremal" measure (its support S_w and density μ_w), such that the F-functional $F(K) \equiv \log \text{cap}(K) - \int Q d\omega_K$, with $\text{cap}(K)$ and ω_K the capacity, respectively the equilibrium measure of the set K, is maximized by S_w. Furthermore, μ_w satisfies energy and capacity constraints on S_w.

The remarkable fact noticed in [487] is that if the extremal value $F(S_w)$ is approximated by the weighted *monic* polynomials $\tilde{P}_n(z)$ as $(w^n \tilde{P}_n)^{1/n} \rightharpoonup \exp(-F_w)$ in the weak star topology, then the asymptotic zero distribution of $\tilde{P}_n$ gives the support S_w. Hence, (6.17) may be interpreted as giving both the support of the extremal measure (labeled ρ in this formula), as well as its actual density.

6.2   Random matrix theory in higher dimensions

In this chapter, we show how to generalize the concepts of equilibrium measure, extremal measure, and their relations to orthogonal polynomials and ensembles of random matrices, in the case of two-dimensional support. The applications of this theory to planar growth processes will be discussed in the following two chapters.

6.2.1   The Ginibre–Girko ensemble

We begin with a brief discussion on the oldest and simplest ensemble of random matrices with planar support. The ensemble of complex, $N \times N$ random matrices with identical, independent, zero-mean Gaussian-distributed entries, was first studied by J. Ginibre in 1965 [209], and then it was generalized for non-zero mean Gaussian by Girko in 1985 [210]. Consider $N \times N$ random matrices with eigenvalues $z_k \in \mathbb{C}$, and joint p.d.f.

$$dP_N \sim \prod_{1 \le i < j \le N} |z_i - z_j|^2 \prod_{1 \le k \le N} d\mu_N(z_k), \qquad (6.26)$$

where $d\mu_N(z_k) = e^{-N|z_k|^2} d\operatorname{Re} z_k d\operatorname{Im} z_k$. Then, in the large N limit, the measure $\frac{1}{N} \sum_k \delta(z - z_k)$ converges weakly to the uniform measure on the unit disk. This is known as the Circular Law. If the exponent of the pure Gaussian is perturbed by a quadratic term, the result holds for a corresponding elliptical domain, giving the Elliptical Law. The same limiting curves (circular and elliptical) describe the graph of the distribution of *real* eigenvalues for Hermitian ensembles, with pure and perturbed Gaussian measures. In that case, the laws are known as Wigner–Dyson [585, 165] and Marchenko–Pastur, respectively (although the last one was originally derived for covariance matrices built from sparse regression matrices [424]).

Extensions and exceptions from the circular and elliptical laws were found by relaxing the conditions of the theorems. In particular, deviations from uniformity for angular statistics in the case of Gaussian measure were derived in [475], while the case of heavy-tails distributions was investigated in [529, 593] and subsequent publications.

6.2.2   Normal matrix ensembles

A special case of matrices with complex eigenvalues is given by *normal* matrices. A matrix M is called normal if it commutes with its Hermitian conjugate:

$[M, M^\dagger] = 0$, so that both M and $M^\dagger$ can be diagonalized simultaneously. The statistical weight of the normal matrix ensemble is given through a general potential $W(M, M^\dagger)$ [97]:

$$e^{\frac{1}{\hbar}\text{tr}\, W(M,M^\dagger)}d\mu(M). \tag{6.27}$$

Here $\hbar$ is a real, positive parameter, and the measure of integration over normal matrices is induced by the flat metric on the space of all complex matrices $d_\mathbb{C}M$, where $d_\mathbb{C}M = \prod_{ij} d\,\text{Re}\, M_{ij} d\,\text{Im}\, M_{ij}$. Using a standard procedure, one passes to the joint probability distribution of eigenvalues of normal matrices $z_1, \ldots, z_N$, where N is the size of the matrix.

The fundamental problem of NRMT is the full determination of the reduced distribution functions $\rho_{N,W}^{(k)}(z_1, z_2, \ldots, z_k), 1 \le k < N$, for the ensemble of $N \times N$ normal random matrices, described by the joint distribution function of eigenvalues

$$\rho_{N,W}^{(N)}(z_1, z_2, \ldots, z_N) \equiv \frac{1}{\tau_N(W)} \left| \det \left(z_j^{i-1}\right)_{1\le i,j\le n} \right|^2 \prod_{k=1}^{N} e^{\frac{1}{\hbar}W(z_k)}, \tag{6.28}$$

where $W(z) : \mathbb{C} \to \mathbb{R}$ (referred to as "confining potential" in the literature) is a C^2-function chosen to ensure convergence of the integrals

$$\tau_N(W) \equiv \int_{\mathbb{C}^N} \left|\Delta_N(z_k)\right|^2 \prod_{k=1}^{N} e^{\frac{1}{\hbar}W(z_k)}d^2 z_k < \infty, \tag{6.29}$$

where $\Delta_N(z_k) \equiv \det \left(z_j^{i-1}\right)_{1\le i,j\le N}$ is the Vandermonde determinant, and the reduced distribution functions are (the symmetric, positive, and integrable functions) given by

$$\rho_{N,W}^{(k)}(z_1, \ldots, z_k) \equiv \int_{\mathbb{C}^{N-k}} \rho_{N,W}^{(N)}(z_1, \ldots, z_k; z_{k+1}, \ldots, z_n) \prod_{s=k+1}^{n} d^2 z_s. \tag{6.30}$$

For a generic confining potential W and finite $N \in \mathbb{N}$, this problem is still open. A simpler version of the problem concerns the "density" function $\rho_{N,W}^{(1)}(z_1)$, in the limit $N \to \infty, \hbar \to 0$, asymptotically in $1/N(N \to \infty)$. For a large class of confining potentials, the limit $\rho_W(z) \equiv \lim_{N\to\infty} \rho_{N,W}^{(1)}(z)$ exists, and it determines the so-called "classical limit" of the problem.

A particularly important special case arises if the potential W has the form

$$W = -|z|^2 + V(z) + \overline{V(z)}, \tag{6.31}$$

where $V(z)$ is a holomorphic function in a domain which includes the support of eigenvalues. In this case, a normal matrix ensemble gives the same distribution as a general complex matrix ensemble. A general complex matrix can be decomposed as $M = U(Z+R)U^\dagger$, where U and Z are unitary and diagonal matrices, respectively, and R is an upper triangular matrix. The distribution (5.5) holds for the elements of the diagonal matrix Z which are eigenvalues of M. Here we mostly focus on the special potential (6.31), and also assume that the field

$$A(z) = \partial_z V(z) \qquad (6.32)$$

is a rational function.

6.2.3   Droplets of eigenvalues

In the large N limit ($\hbar \to 0$, $N\hbar = t_0$ fixed), the eigenvalues of matrices from the ensemble densely occupy a closed connected set D in the complex plane, or, in general, several such sets. This set (called the support of eigenvalues) has sharp edges (Figure 6.1). We refer to the connected components D_α of the domain D as *droplets*.

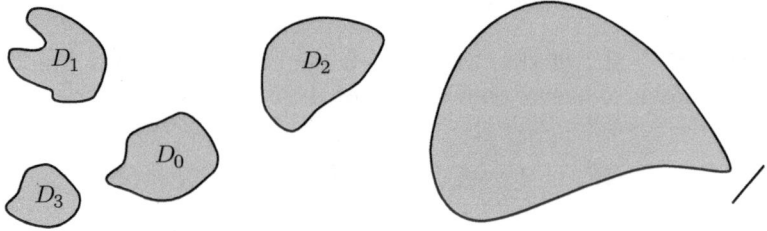

Fig. 6.1: A support of eigenvalues consisting of four disconnected
components (left). The distribution of eigenvalues for potential $V(z) = -\alpha \log(1 - z/\beta) - \gamma z$ (right)

For algebraic domains (see Section 3.3.1), the eigenvalues are distributed with the density $\rho = -\frac{1}{4\pi}\Delta W$, where $\Delta = 4\partial_z \bar{\partial}_z$ is the 2D Laplace operator [581]. For the potential (6.31) the density is uniform. The shape of the support of eigenvalues is the main subject of this chapter. For example, if the potential is Gaussian [209],

$$A(z) = 2t_2 z, \qquad (6.33)$$

the domain is an ellipse. If A has one simple pole,

$$A(z) = \frac{\alpha}{z - \beta} + \gamma \qquad (6.34)$$

the droplet (under certain conditions discussed below) has the profile of an aircraft wing given by the Joukowsky map (Figure 6.1, right). If A has two or more simple poles, there may be more than one droplet. This support and density represent the equilibrium solution to an electrostatic problem, as we will indicate in a later section.

6.2.4 Orthogonal polynomials and distribution of eigenvalues

We define the exact N-particle wave function (up to a phase), by

$$\Psi_N(z_1,\ldots,z_N) = \frac{1}{\sqrt{\tau_N}}\Delta_N(z)\,e^{\sum_{j=1}^{N}\frac{1}{2\hbar}W(z_j,\bar{z}_j)}. \qquad (6.35)$$

The joint probability distribution (6.28) is then equal to $|\Psi_N(z_1,\ldots,z_N)|^2$. Let the number of eigenvalues (particles) increase while the potential stays fixed. If the support of eigenvalues is simply-connected, its area grows as $\hbar N$.

We introduce a set of orthonormal one-particle functions on the complex plane as matrix elements of transitions between N and $(N+1)$-particle states:

$$\psi_N(z) = \int \Psi_{N+1}(z,z_1,z_2,\ldots,z_N)\overline{\Psi_N(z_1,z_2,\ldots,z_N)}d^2z_1\cdots d^2z_N. \qquad (6.36)$$

Then the rate of the density change is

$$\rho_{N+1}(z) - \rho_N(z) = |\psi_N(z)|^2. \qquad (6.37)$$

The proof of this formula is based on the representation of the ψ_n through holomorphic biorthogonal polynomials with the weight function $e^{W/\hbar}$.

For simplicity, we set in the following the parameter $N\hbar$ to 1. Under the convergence conditions

$$\int_{\mathbb{C}}|z|^n e^{NW(z)}d^2z < \infty \qquad n = 0,1,2,\ldots$$

for all for values of the scaling parameter $N > 0$, for fixed N, the orthogonal polynomials $\{P_n^{(N)}(z)\}$ of the weight function $e^{NW(z)}$ are defined by

$$\int_{\mathbb{C}} P_n^{(N)}(z)\overline{P_m^{(N)}(z)}e^{NW(z)}d^2z = \delta_{nm}. \qquad (6.38)$$

Then, up to a phase,

$$\psi_n(z) = e^{\frac{N}{2}W(z,\bar{z})}P_n^{(N)}(z). \qquad (6.39)$$

The approximation method presented in this work is based on the following statement: as

$$n \to \infty, \quad N \to \infty, \quad \frac{n}{N} \to t_0 \leq 1, \qquad (6.40)$$

the weighted polynomials $|P_n^{(N)}(z)|^2 e^{NW(z)}$ (which we denote by $\delta_n^{(N)}(z)$ in the following) converge to the equilibrium measure of the set $D(t_0)$, with support $\Gamma(t_0)$:

$$\delta_n^{(N)}(z) = |\partial_z f^{-1}(z)|\left[1 + O\left(\frac{1}{N}\right)\right], \quad n, N \to \infty, \qquad (6.41)$$

where $f : \mathbb{D}^* \to \Omega$ is the conformal map from the exterior of the unit disk to the (exterior) domain $\Omega = \overline{\mathbb{C}} \setminus D(t_0)$. The proof of this result appeared first in [551]. We do not repeat the entire argument, as it would require too much space, but recollect the main ideas: starting from the differential equations satisfied by the weighted functions $\psi_n(z) = P_n^{(N)}(z)e^{NW(z)/2}$, with respect to variables n/N and z, we integrate perturbatively in powers of N^{-1}, and obtain the expression ([551], equation (76)):

$$\psi(z) \sim \sqrt{[f^{-1}(z)]'} \exp\left[N\left(-\frac{|z|^2}{2} + \mathrm{Re}\int^z S(\zeta)d\zeta\right)\right],$$

where $S(z)$ is the *Schwarz function* of the domain Ω (or of $\Gamma(t_0)$).

Since the exponent in the asymptotic expression of $\psi(z)$ vanishes on the boundary Γ and gives a Gaussian decay away from it, the weighted polynomials $\delta_n^{(N)}(z)$ are described, in the $n, N \to \infty$ limit, by the harmonic measure of the domain. However, this asymptotic result says very little about the behavior of $\delta_n^{(N)}(z)$ for *finite* values of n, N.

In the remainder of this section we present numerical evidence for the convergence properties of $\delta_n^{(N)}(z)$ at finite values of their order. We show that the agreement between $\delta_n^{(N)}(z)$ and $|[f^{-1}(z)]'|$ is excellent for values of n as little as $n = 20$.

We consider the case given by

$$A(z) = \gamma + \frac{\alpha}{z - \beta},$$

for which the conformal map f is given by the Joukowski map (see (6.62) below). To fix the scaling limit, let $N(n) = n/t_0$. For fixed n, we have to calculate the entries of the Gram matrix

$$g_{ij}^{(n)} = \int_{\mathbb{C}} z^i \bar{z}^j e^{n/t_0 W(z)} d^2 z \quad i, j = 0, \dots, n. \tag{6.42}$$

For potentials $W(z)$ that are converging rapidly enough to infinity as $|z| \to \infty$, the exponentially decaying weight makes the planar numerical integration a feasible task. The stabilized Gram–Schmidt algorithm provides the orthogonal polynomials $P_n^{(N)}(z)$, which is known to be very sensitive to the accuracy of the Gram matrix and thus requires very precise computation of $\{g_{ij}^{(n)}\}$. Then the density $\delta_n^{(N)}(z)$ is obtained from the polynomial $P_n^{(N)}(z)$.

Of course, the usefulness of this approximation scheme relies on the rapidity of the convergence of the numerical procedure, which may not seem to be very promising. However, our numerical experiment (Figure 6.2) shows that in the example above the "shape" of the conformal measure (the blue curve) is recovered very accurately by the weighted polynomial density of a degree as low as $n = 20$.

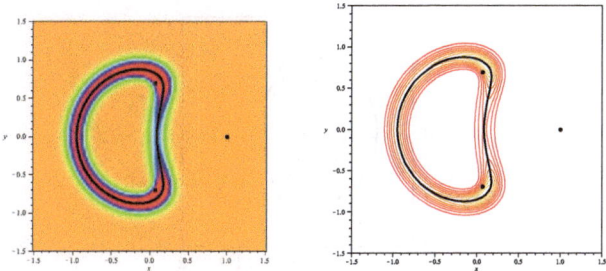

Fig. 6.2: Density, contour plots of functions $\delta_n^{(N(n))}$, $n = 20$

The asymptotic behaviour of the zeros of orthogonal polynomials in the scaling limit (6.40) was also investigated in this particular case. Since $f(\zeta)$ is a rational function of order two, the Cauchy transform $C_D(z)$ satisfies, in the exterior domain Ω, a quadratic equation (in y)

$$A_1(z)y^2 + B_1(z)y + C_1(z) = 0,$$

with rational coefficients in z depending on the parameters of $f(\zeta)$. As an algebraic function, $C_D(z)$ can be analytically continued on a plane with a branch cut connecting up the branchpoints

$$z_{1,2} = \frac{v}{A_1} + A_1 r \pm 2\sqrt{rv} \tag{6.43}$$

of the inverse mapping $f^{-1}(z)$. This "conjugate electric field" created by the uniformly charged domain D is mimicked by the field generated by the normalized counting measure of the zeros. However, these points seem to accumulate along some curve (as opposed to the equilibrium configuration in the presence of the background potential $W(z)$ – the so-called *Fekete points* – which are distributed asymptotically uniformly). Since the asymptotic zero distribution must be real and positive, the natural choice is dictated by the *Sokhotskiĭ–Plemelj* formula: the critical trajectory γ is selected by the condition that the jump between the two solutions $y_\pm = (-B_1 \pm \sqrt{B_1^2 - 4A_1C_1})/2A_1$ satisfies

$$\mathrm{Re}\,((y_+(z) - y_-(z))dz) = 0. \tag{6.44}$$

The critical trajectory can be found by calculating

$$\Phi(z) := \mathrm{Re}\left[\int_{z_1}^{z}(y_+(w) - y_-(w))dw\right] \tag{6.45}$$

and then plotting the contour $\Phi(z) = 0$. Three trajectories are emanating from each branchpoint: there are two trajectories that connect z_1 and z_2, and the one contained by the domain attracts the roots. As can be seen in Figure 6.3, the distribution of zeros (for $n = 50$) and the trajectory γ are almost indistinguishable.

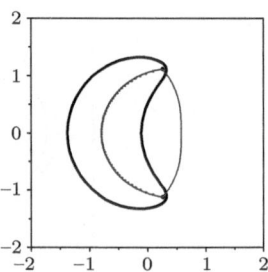

Fig. 6.3: The critical trajectory and
the zeros for $n = 50$

6.2.5   Equations for the wave functions

In this section we specify the potential to be of the form (6.31), and drop the
explicit dependence on N (or $\hbar$) wherever it is obvious. It is then convenient to
modify the exponential factor of the wave function. Namely, we define

$$\psi_n(z) = e^{-\frac{|z|^2}{2\hbar} + \frac{1}{\hbar}V(z)} P_n(z), \quad \text{and} \quad \chi_n(z) = e^{\frac{1}{\hbar}V(z)} P_n(z), \qquad (6.46)$$

where the holomorphic functions $\chi_n(z)$ are orthonormal in the complex plane
with the weight $e^{-|z|^2/\hbar}$. Like traditional orthogonal polynomials, the biorthogonal
polynomials P_n (and the corresponding wave functions) obey a set of differential
equations with respect to the argument z, and recurrence relations with respect to
the degree n. Similar equations for two-matrix models are discussed in numerous
papers (see, e.g., [23]).

We introduce the L-operator (the Lax operator) as multiplication by z in the
basis χ_n:

$$L_{nm}\chi_m(z) = z\chi_n(z) \qquad (6.47)$$

(summation over repeated indices is implied). Obviously, L is a lower triangular
matrix with one adjacent upper diagonal, $L_{nm} = 0$ as $m > n + 1$. Similarly, the
differentiation ∂_z is represented by an upper triangular matrix with one adjacent
lower diagonal. Integrating by parts the matrix elements of the ∂_z, one finds:

$$(L^\dagger)_{nm}\chi_m = \hbar\partial_z\chi_n, \qquad (6.48)$$

where $L^\dagger$ is the Hermitian conjugate operator.

The matrix elements of $L^\dagger$ are

$$(L^\dagger)_{nm} = \bar{L}_{mn} = A(L_{nm}) + \int e^{\frac{1}{\hbar}W} \bar{P}_m(\bar{z})\partial_z P_n(z) d^2 z,$$

where the last term is a lower triangular matrix. The latter can be written through
negative powers of the Lax operator. Writing $\partial_z \log P_n(z) = \frac{n}{z} + \sum_{k>1} v_k(n)z^{-k}$,

one represents $L^\dagger$ in the form

$$L^\dagger = A(L) + (\hbar n)L^{-1} + \sum_{k>1} v^{(k)}L^{-k}, \tag{6.49}$$

where $v^{(k)}$ and $(\hbar n)$ are diagonal matrices with elements $v_n^{(k)}$ and $(\hbar n)$. The coefficients $v_n^{(k)}$ are determined by the condition that lower triangular matrix elements of $A(L_{nm})$ are cancelled.

In order to emphasize the structure of the operator L, we write it in the basis of the shift operator[3] $\hat{w}$ such that $\hat{w}f_n = f_{n+1}\hat{w}$ for any sequence f_n. Acting on the wave function, we have:

$$\hat{w}\chi_n = \chi_{n+1}.$$

In the n-representation, the operators L, $L^\dagger$ acquire the form

$$L = r_n\hat{w} + \sum_{k\geq0} u_n^{(k)}\hat{w}^{-k}, \quad L^\dagger = \hat{w}^{-1}r_n + \sum_{k\geq0} \hat{w}^k\bar{u}_n^{(k)}. \tag{6.50}$$

Clearly, acting on χ_n, we have the commutation relation ("the string equation")

$$[L, L^\dagger] = \hbar. \tag{6.51}$$

This is the compatibility condition of equations (6.47) and (6.48).

Equations (6.50) and (6.51) completely determine the coefficients $v_n^{(k)}$, r_n and $u_n^{(k)}$. The first one connects the coefficients to the parameters of the potential. The second equation is used to determine how the coefficients $v_n^{(k)}$, r_n and $u_n^{(k)}$ evolve with n. In particular, the diagonal part of it reads

$$n\hbar = r_n^2 - \sum_{k\geq1}\sum_{p=1}^{k} |u_{n+p}^{(k)}|^2. \tag{6.52}$$

Moreover, we note that all the coefficients can be expressed through the τ-function τ_n and its derivatives with respect to parameters of the potential. This representation is particularly simple for r_n: $r_n^2 = \tau_n\tau_{n+1}^{-2}\tau_{n+2}$.

Finite-dimensional reductions

If the function $A(z)$ is rational, the coefficients $u_n^{(k)}$ are not all independent. The number of independent coefficients equals the number of independent parameters of the potential. For example, if the holomorphic part of the potential, $V(z)$, is a polynomial of degree d, the series (6.50) are truncated at $k = d - 1$.

---

[3]The shift operator $\hat{w}$ has no inverse. Below $\hat{w}^{-1}$ is understood as a shift to the left defined as $\hat{w}^{-1}\hat{w} = 1$. Same is applied to the operator L^{-1}. To avoid a possible confusion, we emphasize that although χ_n is a right-hand eigenvector of L, it is not a right-hand eigenvector of L^{-1}.

In this case the semi-infinite system of linear equations (6.48) and the recurrence relations (6.47) can be cast in the form of a set of finite-dimensional equations whose coefficients are rational functions of z, one system for every $n > 0$. The system of differential equations generalizes the Christoffel–Daurboux second-order differential equation valid for orthogonal polynomials. This fact has been observed in recent papers [56, 55] for biorthogonal polynomials emerging in the Hermitian two-matrix model with a polynomial potential. It is applicable to our case (holomorphic biorthogonal polynomials) as well.

In a more general case, when $A(z)$ is a general rational function with $d-1$ poles (counting multiplicities), the series (6.50) is not truncated. However, L can be represented as a "ratio",

$$L = K_1^{-1} K_2 = M_2 M_1^{-1}, \tag{6.53}$$

where the operators $K_{1,2}$, $M_{1,2}$ are polynomials in $\hat{w}$:

$$K_1 = \hat{w}^{d-1} + \sum_{j=0}^{d-2} A_n^{(j)} \hat{w}^j, \quad K_2 = r_{n+d-1} \hat{w}^d + \sum_{j=0}^{d-1} B_n^{(j)} \hat{w}^j, \tag{6.54}$$

$$M_1 = \hat{w}^{d-1} + \sum_{j=0}^{d-2} C_n^{(j)} \hat{w}^j, \quad M_2 = r_n \hat{w}^d + \sum_{j=0}^{d-1} D_n^{(j)} \hat{w}^j. \tag{6.55}$$

These operators obey the relation

$$K_1 M_2 = K_2 M_1. \tag{6.56}$$

It can be proven that the pair of operators $M_{1,2}$ is uniquely determined by $K_{1,2}$ and vice versa. We note that the reduction (6.53) is a difference analog of the "rational" reductions of the Kadomtsev–Petviashvili integrable hierarchy considered in [325].

The linear problems (6.47), (6.48) acquire the form

$$(K_2 \chi)_n = z \, (K_1 \chi)_n, \quad (M_2^\dagger \chi)_n = \hbar \partial_z (M_1^\dagger \chi)_n. \tag{6.57}$$

These equations are of *finite order* (namely, of order d), i.e., they connect values of χ_n on $d+1$ subsequent sites of the lattice.

The semi-infinite set $\{\chi_0, \chi_1, \dots\}$ is then a "bundle" of d-dimensional vectors

$$\underline{\chi}(n) = (\chi_n, \chi_{n+1}, \dots, \chi_{n+d-1})^{\text{t}}$$

(the index t means transposition, so $\underline{\chi}$ is a column vector). The dimension of the vector is the number of poles of $\tilde{A}(z)$ plus one. Each vector obeys a closed d-dimensional linear differential equation

$$\hbar \partial_z \underline{\chi}(n) = \mathcal{L}_n(z) \underline{\chi}(n), \tag{6.58}$$

where the $d \times d$ matrix $\mathcal{L}_n$ is a "projection" of the operator $L^\dagger$ onto the nth d-dimensional space. Matrix elements of the $\mathcal{L}_n$ are rational functions of z having the same poles as $A(z)$ and also a pole at the point $\overline{A(\infty)}$. (If $A(z)$ is a polynomial, all these poles accumulate to a multiple pole at infinity).

We briefly describe the procedure of constructing the finite-dimensional matrix differential equation. We use the first linear problem in (6.57) to represent the shift operator as a $d \times d$ matrix $\mathcal{W}_n(z)$ with z-dependent coefficients:

$$\mathcal{W}_n(z)\underline{\chi}(n) = \underline{\chi}(n+1). \tag{6.59}$$

This is nothing else than rewriting the scalar linear problem in the matrix form. Then the matrix $\mathcal{W}_n(z)$ is to be substituted into the second equation of (6.57) to determine $\mathcal{L}_n(z)$ (examples follow). The entries of $\mathcal{W}_n(z)$ and $\mathcal{L}_n(z)$ obey the Schlesinger equation, which follows from compatibility of (6.58) and (6.59):

$$\hbar\partial_z\mathcal{W}_n = \mathcal{L}_{n+1}\mathcal{W}_n - \mathcal{W}_n\mathcal{L}_n. \tag{6.60}$$

This procedure has been realized explicitly for polynomial potentials in recent papers [56, 55]. We will work it out in detail for the example considered above, corresponding to the Joukowski map.

6.2.6 Spectral curve

The Schottky double

We recall the relation between domains which admit a Schwarz function and the Schottky double, as introduced in Chapter 3.

There are two complementary ways to describe this Riemann surface. One is through the algebraic covering (of degree $d \geq 2$), in terms of local coordinates at each branching point. In this case we are working with d copies of the Riemann sphere, glued at the branch cuts according to the prescribed singularities of $S(z)$. Among d sheets we distinguish a *physical* sheet, selected by the condition that the differential $S(z)dz$ has the same poles and residues as the differential of the potential $A(z)dz$. It may happen that the condition $\bar{z} = S^{(i)}(z)$ defines a planar curve (or several curves, or a set of isolated points) for branches other than the physical one. We refer to the interior of these planar curves as *virtual* (or unphysical) droplets situated on sheets other than physical.

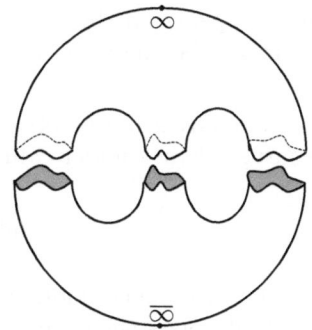

Fig. 6.4: The Schottky double. A Riemann surface with boundaries along the droplets (a front side) is glued to its mirror image (a back side)

Another way emphasizes the antiholomorphic involution. We say that the Schwarz function on the double is $S(z)$ if the point is on the front side, and $\bar{z}$ if the point belongs to the back side (here we understand $S(z)$ as a function defined on the complex curve, not just on the physical sheet).

The number of sheets of the curve is the number of poles (counted with their multiplicity) of the function $A(z)$, plus one. Indeed, poles of A are poles of the Schwarz function on the front side of the double. On the back side, there is also a pole at infinity. Since $S(z = \infty) = A(\infty)$, we have $\bar{S}(\bar{z} = A(\infty)) = \infty$. Therefore, the factor $a(z)$ is a polynomial with zeros at the poles of $A(z)$ and at $\overline{A(\infty)}$, and

$$d \equiv \text{number of sheets} = \text{number of poles of } A + 1.$$

The front and back sides meet at planar curves $\bar{z} = S(z)$. These curves are boundaries of the droplets. We repeat that not all droplets are physical. Some of them may belong to unphysical sheets, Figure 6.5.

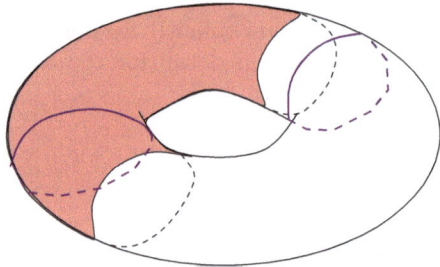

Fig. 6.5: Physical and unphysical droplets on a torus. The physical sheet (shaded) meets the unphysical sheet along the cuts. The cut situated inside the unphysical droplet appears on the physical sheet. The boundaries of the droplets (physical and virtual) belong to different sheets

Boundaries of droplets, physical and virtual, form a subset of the **a**-cycles on the curve. Their number cannot exceed the genus of the curve plus one:

$$\text{number of droplets} \leq g + 1.$$

The sheets meet along cuts located inside droplets. The cuts that belong to physical droplets show up on unphysical sheets. On the other hand, some cuts show up on the physical sheet (Figure 6.5). They correspond to droplets situated on unphysical sheets.

The Riemann–Hurwitz theorem computes the genus of the curve as

$$g = \text{half the number of branching points} - d + 1.$$

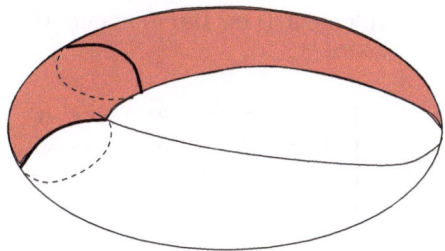

Fig. 6.6: Degenerate torus corresponds to the algebraic domain for the Joukowsky map

Degeneration of the spectral curve

Degeneration of the complex curve gives the most interesting physical aspects of
growth. There are several levels of degeneration. We briefly discuss them below.

Algebraic domains and double points. A special case occurs when the Schwarz
function on the physical sheet is meromorphic. It has no other singularities than
poles of $A(z)$. This is the case of algebraic domains. They appear in the semiclassical case. This situation occurs if cuts on the physical sheet, situated outside physical droplets, shrink to points, i.e., two or more branching points merge. Then the
physical sheet meets other sheets along cuts situated inside physical droplets only
and also at some points on their exterior (*double points*). In this case the Riemann
surface degenerates. The genus is given by the number of physical droplets only.

Critical degenerate curves. Algebraic domains appear as a result of merging of
simple branching points on the physical sheet. The double points are located
outside physical droplets. Remaining branching points belong to the interior of
physical droplets. Initially, they survive in the degeneration process. However, as
known in the theory of Laplacian growth, the process necessarily leads to a further
degeneration. Sooner or later, at least one of the interior branching points merges
with one of the double points in the exterior. Curves degenerated in this manner
are called *critical*. For the genus one curve, this degeneration is discussed below.

Since interior branching points can only merge with exterior branching points
on the boundary of the droplet, the boundary develops a cusp, characterized by a
pair p, q of mutually prime integers. In local coordinates around such a cusp, the
curve looks like $x^p \sim y^q$. The fact that the growth of algebraic domains always
leads to critical curves is known in the theory of Laplacian growth as finite time
singularities.

Example: genus one curve

We consider the case where $A(z)$ has one simple pole at $\beta \in \Omega$ of residue α, and
$S(z)$ also has a branch cut in the complement. Introducing the parameter t for the

normalized area of the complement (i.e., the moment M_0), the pole singularities of $S(z)$ are given by

$$S(z) = \begin{cases} -\frac{\alpha}{z-\beta} & \text{as} \quad z \to \beta_1, \\ \left(-\gamma + \frac{t-\alpha}{z-\alpha}\right) & \text{as} \quad z \to \infty_1, \\ \frac{t-\bar{\alpha}}{z+\bar{\gamma}} & \text{as} \quad z \to -\bar{\gamma}_2, \\ \left(\bar{\beta} - \frac{\bar{\alpha}}{z}\right) & \text{as} \quad z \to \infty_2, \end{cases}$$

where, by 1 and 2 we indicate the sheets.

Poles and residues of the Schwarz function determine all the coefficients of the curve $f(z,\bar{z}) = a(z)(\bar{z} - S^{(1)}(z))(\bar{z} - S^{(2)}(z)) = \overline{a(z)}(z - \bar{S}^{(1)}(\bar{z}))(z - \bar{S}^{(2)}(\bar{z}))$ except one. The behavior at ∞ of $z, \bar{z}$ gives $k_1 = \gamma - \bar{\beta}$, $k_2 = -\gamma\bar{\beta}$. Hereafter we choose the origin by setting $\gamma = 0$. The equation of the curve then reads $Q_t(z,\bar{z}) = 0$, where $Q_t(z,\bar{z})$ is given by

$$z^2\bar{z}^2 - z^2\bar{z}\bar{\beta} - z\bar{z}^2\beta + \left(|\bar{\beta}|^2 + \alpha + \bar{\alpha} - t\right)z\bar{z} + z\bar{\beta}(t-\alpha) + \bar{z}\beta(t-\bar{\alpha}) + h_t, \quad (6.61)$$

where the free term h_t is to be determined by the area of the unphysical droplet.

If this solution is chosen, the exterior of the physical droplet can be mapped to the exterior of the unit disk by the Joukowsky map

$$z(w) = rw + u_0 + \frac{u}{w-a}, \quad |w| > 1, \ |a| < 1. \qquad (6.62)$$

The inverse map is given by the branch $w_1(z)$ (such that $w_1 \to \infty$ as $z \to \infty$) of the double-valued function

$$w_{1,2}(z) = \frac{z - u_0 + ar \pm \sqrt{(z-z_1)(z-z_2)}}{2r}, \quad z_{1,2} = u_0 + ar \mp 2\sqrt{r(u + au_0)}.$$

The function
$$\bar{z}(w^{-1}) = rw^{-1} + \bar{u}_0 + \frac{\bar{u}}{w^{-1} - \bar{a}} \qquad (6.63)$$

is a meromorphic function of w with two simple poles at $w = 0$ and $w = \bar{a}^{-1}$. Treated as a function of z, it covers the z-plane twice. Two branches of the Schwarz function are $S^{(1,2)}(z) = \bar{z}(w_{1,2}^{-1}(z))$. On the physical sheet, $S^{(1)}(z) = \bar{z}(w_1(z))$ is the analytic continuation of $\bar{z}$ away from the boundary. This function is meromorphic outside the droplet. Apart from a cut between the branching points $z_{1,2}$, the sheets also meet at the double point $z_* = -\bar{\gamma} + a^{-1}re^{2i\phi}$, where $S^{(1)}(z_*) = S^{(2)}(z_*)$, $\phi = \arg\left(ar + \frac{u\bar{a}}{1-|a|^2}\right)$.

A direct way to obtain the complex curve from the conformal map is the following. First, rewrite (6.62) and (6.63) as

$$\begin{cases} z - u_0 + ar = rw + a(z + \bar{\gamma})w^{-1} \\ \bar{z} - \bar{u}_0 + \bar{a}r = rw^{-1} + \bar{a}(\bar{z} + \gamma)w, \end{cases} \qquad (6.64)$$

and treat w and $1/w$ as independent variables. Then impose the condition $w \cdot w^{-1} = 1$. One obtains

$$\left| \det \begin{bmatrix} z - u_0 + ar & a(z + \gamma) \\ \bar{z} - \bar{u}_0 + \bar{a}r & r \end{bmatrix} \right|^2 = \left(\det \begin{bmatrix} r & a(z + \bar{\gamma}) \\ \bar{a}(\bar{z} + \gamma) & r \end{bmatrix} \right)^2 .$$

This gives the equation of the curve and in particular h, in terms of u, u_0, r, a and eventually through the deformation parameters α, β, γ and t.

Analyzing singularities of the Schwarz function, one connects parameters of the conformal map with the deformation parameters:

$$\begin{cases} \gamma = \frac{\bar{u}}{\bar{a}} - \bar{u}_0, \\ t - \bar{\alpha} = r^2 - \frac{ur}{a^2}, \\ \beta = \frac{r}{a} + u_0 + \frac{u\bar{a}}{1 - |a|^2} \end{cases} \tag{6.65}$$

$$\frac{\text{Area of the droplet}}{\pi} = t = r^2 - \frac{|u|^2}{(1 - |a|^2)^2}.$$

A critical degeneration occurs when the double point merges with a branching point located inside the droplet ($z_* = z_2$) to form a triple point z_{**}. This may happen on the boundary only. At this point, the boundary has a 3/2 cusp. In local coordinates, it is $x^2 \sim y^3$. This is a critical point of the conformal map: $w'(z_{**}) = \infty$. A critical point may form under Laplacian growth with these data, if the parameter $\alpha \in \mathbb{R}$ is negative. In that case, the classical solution breaks down when the critical point is reached, and a 3/2 cusp forms (like in the Polubarinova cardioid). For real and positive values, however, the solution can be continued beyond the critical time, and the droplet undergoes a change in connectivity, becoming doubly-connected, see [33].

Stochastic Laplacian growth and resolution of singularities

Since in the limit of infinte-size normal matrices, the equilibrium distribution of eigenvalues is given by the characteristic function of the classical Laplacian growth problem, it is interesting to derive the corresponding generalization of the spectral curve, and use it for the purpose of producing a stochastic weak resolution of the classical singularities discussed above. This is achieved by identifying the stochastic variant of the spectral curve with a canonical object from the theory of asymptotic expansions for ordinary differential equations on $\mathbb{C}$.

According to the general theory of linear differential equations, the semiclassical (known either as Wentzell–Kramers–Brillouin, or WKB, or as Liouville–Green) asymptotic expansion of solutions to equation (6.58), as $\hbar \to 0$, is found by solving the eigenvalue problem for the matrix $\mathcal{L}_n(z)$ [577]. More precisely, the basic object of the WKB approach is the spectral curve [577] of the matrix $\mathcal{L}_n$, which is defined, for every integer $n > 0$, by the secular equation $\det(\mathcal{L}_n(z) - \tilde{z}) = 0$ (here

$\tilde{z}$ means $\tilde{z} \cdot \mathbf{1}$, where $\mathbf{1}$ is the unit $d \times d$ matrix). It is clear that the left-hand side of the secular equation is a polynomial in $\tilde{z}$ of degree d. We define the spectral curve by an equivalent equation

$$f_n(z, \tilde{z}) = a(z) \det(\mathcal{L}_n(z) - \tilde{z}) = 0, \qquad (6.66)$$

where the factor $a(z)$ is added to make $f_n(z, \tilde{z})$ a polynomial in z as well. The factor $a(z)$ then has zeros at the points where poles of the matrix function $\mathcal{L}(z)$ are located. It does not depend on n. We will soon see that the degree of the polynomial $a(z)$ is equal to d. Assume that all poles of $A(z)$ are simple, then zeros of the $a(z)$ are just the $d - 1$ poles of $A(z)$ and another simple zero at the point $\overline{A(\infty)}$. Therefore, we conclude that the matrix $\mathcal{L}_n(z)$ is rather special. For a general $d \times d$ matrix function with the same d poles, the factor $a(z)$ would be of degree d^2.

Note that the matrix $\mathcal{L}_n(z) - \tilde{z}$ enters the differential equation

$$\hbar \partial_z |\psi(n)|^2 = \bar{\psi}(n)(\mathcal{L}_n(z) - \tilde{z})\psi(n) \qquad (6.67)$$

for the squared amplitude $|\psi(n)|^2 = \psi^\dagger(n)\psi(n) = e^{-\frac{|z|^2}{\hbar}}|\chi(n)|^2$ of the vectors $\psi(n)$ built from the orthonormal wave functions (6.39).

The equation of the curve can be interpreted as a "resultant" of the non-commutative polynomials $K_2 - zK_1$ and $M_2^\dagger - \tilde{z}M_1^\dagger$ (cf. [56]). Indeed, the point $(z, \tilde{z})$ belongs to the curve if and only if the linear system

$$\begin{cases} (K_2 c)_k = z(K_1 c)_k & n - d \le k \le n - 1 \\ (M_2^\dagger c)_k = \tilde{z}(M_1^\dagger c)_k & n \le k \le n + d - 1 \end{cases} \qquad (6.68)$$

has non-trivial solutions. The system contains $2d$ equations for $2d$ variables c_{n-d}, $\ldots$, c_{n+d-1}. Vanishing of the $2d \times 2d$ determinant yields the equation of the spectral curve. Below we use this method to find the equation of the curve in the examples. It appears to be much easier than determination of the matrix $\mathcal{L}_n(z)$.

The spectral curve (6.66) possesses an important property: it admits an anti-holomorphic involution. In the coordinates z, $\tilde{z}$ the involution reads $(z, \tilde{z}) \mapsto (\bar{\tilde{z}}, \bar{z})$. This simply means that the secular equation $\det(\bar{\mathcal{L}}_n(\tilde{z}) - z) = 0$ for the matrix $\bar{\mathcal{L}}_n(\tilde{z}) \equiv \overline{\mathcal{L}_n(\bar{\tilde{z}})}$ defines the same curve. Therefore, the polynomial f_n takes real values for $\tilde{z} = \bar{z}$:

$$f_n(z, \bar{z}) = \overline{f_n(z, \bar{z})}. \qquad (6.69)$$

Points of the real section of the curve ($\tilde{z} = \bar{z}$) are fixed points of the involution.

The curve (6.66) was discussed in papers [56, 55] in the context of Hermitian two-matrix models with polynomial potentials. The dual realizations of the curve pointed out in [56] correspond to the antiholomorphic involution in our case. The involution can be proven along the lines of these works. The proof is rather technical and we omit it, restricting ourselves to the examples below. We simply note that the involution relies on the fact that the squared modulus of the wave function is real.

We will give a concrete example for construction of the spectral curve, comparing it to the classical Schottky double discussed earlier, and describing the classical limit of the problem.

Schwarz function for the stochastic spectral curve

The polynomial $f_n(z, \bar{z})$ can be factorized in two ways:

$$f_n(z, \bar{z}) = a(z)(\bar{z} - S_n^{(1)}(z)) \cdots (\bar{z} - S_n^{(d)}(z)), \tag{6.70}$$

where $S_n^{(i)}(z)$ are eigenvalues of the matrix $\mathcal{L}_n(z)$, or

$$f_n(z, \bar{z}) = \overline{a(z)}(z - \bar{S}_n^{(1)}(\bar{z})) \cdots (z - \bar{S}_n^{(d)}(\bar{z})), \tag{6.71}$$

where $\bar{S}_n^{(i)}(\bar{z})$ are eigenvalues of the matrix $\bar{\mathcal{L}}_n(\bar{z})$. One may understand them as different branches of a multi-valued function $S(z)$ (respectively, $\bar{S}(z)$) on the plane (here we do not indicate the dependence on n, for simplicity of the notation). It then follows that $S(z)$ and $\bar{S}(z)$ are mutually inverse functions:

$$\bar{S}(S(z)) = z. \tag{6.72}$$

As we recall, an algebraic function with this property is called *the Schwarz function*. By the equation $f(z, S(z)) = 0$, it defines a complex curve with an antiholomorphic involution. An upper bound for genus of this curve is $g = (d-1)^2$, where d is the number of branches of the Schwarz function. The real section of this curve is a set of all fixed points of the involution. It consists of a number of contours on the plane (and possibly a number of isolated points, if the curve is not smooth). The structure of this set is known to be complicated. Depending on coefficients of the polynomial, the number of disconnected contours in the real section may vary from 0 to $g + 1$. If the contours divide the complex curve into two disconnected "halves", or sides (related by the involution), then the curve can be realized as the *Schottky double* of one of these sides. Each side is a Riemann surface with a boundary.

Let us come back to equation (6.58). It has d independent solutions. They are functions on the spectral curve. One of them is a physical solution corresponding to biorthogonal polynomials. The physical solution defines the "physical sheet" of the curve.

The Schwarz function on the physical sheet is a particular root, say $S_n^{(1)}(z)$, of the polynomial $f_n(z, \bar{z})$ (see (6.70)). It follows from (6.49) that this root is selected by the requirement that it has the same poles and residues as the potential A.

The semiclassical analysis gives a guidance for the form of the recurrence relations. Let us use an ansatz for the L-operator, which resembles the conformal map (6.62):

$$L = r_n \hat{w} + u_n^{(0)} + (\hat{w} - a_n)^{-1} u_n,$$

so that

$$(\hat{w} - a_n)L = (\hat{w} - a_n)r_n\hat{w} + (\hat{w} - a_n)u_n^{(0)} + u_n, \tag{6.73}$$

$$L^\dagger(\hat{w}^{-1} - \bar{a}_n) = \hat{w}^{-1}r_n(\hat{w}^{-1} - \bar{a}_n) + \bar{u}_n^{(0)}(\hat{w}^{-1} - \bar{a}_n) + \bar{u}_n, \tag{6.74}$$

where $\hat{w}$ is the shift operator $n \to n + 1$.

Now we follow the procedure of the previous section. Since the potential has only one pole, $\mathcal{L}_n$ can be cast into 2×2 matrix form. Let us apply the lines (6.73), (6.74) to an eigenvector (c_n, c_{n+1}) of the operator $\mathcal{L}_n$, and set the eigenvalue to be $\tilde{z}$:

$$\begin{cases} (z + r_{n-1}a_{n-1} - u_n^{(0)})c_n = r_n c_{n+1} + a_{n-1}(z + \bar{\gamma}_{n-1})c_{n-1} \\ (\tilde{z} + r_n\bar{a}_n - \bar{u}_{n+1}^{(0)})c_n = \bar{a}_{n+1}(\tilde{z} + \gamma_{n+1})c_{n+1} + r_n c_{n-1}. \end{cases} \tag{6.75}$$

We have defined $\bar{\gamma}_n = \frac{u_n}{a_n} - u_n^0$. The equations are compatible if c_{n-1} and c_{n+1} found through c_n differ by the shift $n \to n + 2$. We have

$$c_{n+1} = \frac{c_n}{d_n} \det \begin{vmatrix} z + r_{n-1}a_{n-1} - u_n^{(0)} & a_{n-1}(z + \bar{\gamma}_{n-1}) \\ \tilde{z} + r_n\bar{a}_n - \bar{u}_{n+1}^{(0)} & r_n \end{vmatrix} = c_n \frac{\widetilde{\mathcal{D}}_n}{d_n},$$

$$c_{n-1} = \frac{c_n}{d_n} \det \begin{vmatrix} r_n & z + r_{n-1}a_{n-1} - u_n^{(0)} \\ \bar{a}_{n+1}(\tilde{z} + \gamma_{n+1}) & \tilde{z} + r_n\bar{a}_n - \bar{u}_{n+1}^{(0)} \end{vmatrix} = c_n \frac{\mathcal{D}_n}{d_n},$$

where

$$d_n = \det \begin{vmatrix} r_n & a_{n-1}(z + \bar{\gamma}_{n-1}) \\ \bar{a}_{n+1}(\tilde{z} + \gamma_{n+1}) & r_n. \end{vmatrix}.$$

This yields the curve

$$\widetilde{\mathcal{D}}_n \cdot \mathcal{D}_{n+1} = d_n d_{n+1}. \tag{6.77}$$

Comparing the two forms of the curve (1.14) and (6.77), we obtain the conservation laws of growth:

$$\gamma = \gamma_n = \frac{\bar{u}_n}{\bar{a}_n} - \bar{u}_n^0, \qquad \beta = \frac{r_n}{\bar{a}_{n+1}} + u_{n+1}^{(0)} + \frac{u_{n+1}^{(0)}a_n\bar{a}_{n+1}}{1 - a_n\bar{a}_{n+1}},$$

$$n\hbar - \bar{\alpha} = r_n r_{n+1} - \frac{r_{n+1}u_{n+1}}{a_n a_{n+1}}.$$

They are the discrete version of (6.65). Analyzing the solutions of this system of nonlinear recurrence relations, for generic initial conditions, presents a possible method for regularization of finite-time singularities of classical Laplacian growth, and constitutes a generalization of the weak solutions discussed in Chapter 3, since it corresponds to entropy-maximization solutions of discretized (atomic) distributions whose limit gives the classical case.

Chapter 7

Integrability and Moments

7.1 The Hele-Shaw directional derivative

Laplacian growth is a special kind of domain variation. Following [370] we introduce the notation $\nabla(a)$ for the directional derivative corresponding to Hele-Shaw injection at the point a. This derivative can be regarded as a tangent vector in the infinite-dimensional space $\mathcal{M}^{(a)}$ of bounded domains $\Omega \subset \mathbb{C}$ containing the point a and having analytic boundary, and it acts on smooth functionals

$$\mathcal{F} : \mathcal{M}^{(a)} \to \mathbb{C}$$

as follows. Given a domain $\Omega \in \mathcal{M}^{(a)}$, let $\{\Omega(t) : 0 \leq t < \varepsilon\}$ be the Hele-Shaw evolution with a source at a and initial domain $\Omega(0) = \Omega$. We normalize the source to be of strength $Q = 1$. This means that

$$\frac{d}{dt} \int_{\Omega(t)} \varphi d\sigma = - \int_{\partial\Omega(t)} \varphi(z) \frac{\partial G(z, a)}{\partial n} ds$$

for every smooth function φ, and where $G(z, \zeta) = G_{\Omega(t)}(z, \zeta)$ is the Green function (1.18) of the domain. Then we define the differential operator $\nabla(a)$ by

$$(\nabla(a)\mathcal{F})(\Omega) = \frac{d}{dt}\bigg|_{t=0} \mathcal{F}(\Omega(t)).$$

One can show that any infinitesimal domain variation of a domain $\Omega \in \mathcal{M}^{(a)}$ can be approximated by linear combinations of $\nabla(a)$ with $a \in \Omega$. See [496] for a general theorem in this direction. Taking a close to the boundary, the variation corresponding to $\nabla(a)$, i.e., Hele-Shaw injection at a, produces a 'bump' on the boundary at a (cf. [374], [372]).

Example 7.1.1. Let h be a fixed harmonic function defined in a neighbourhood of the closure of Ω and define $\mathcal{F}$ by

$$\mathcal{F}(\Omega) = \int_{\Omega} h d\sigma.$$

Then

$$(\nabla(a)\mathcal{F})(\Omega) = \frac{d}{dt}\Big|_{t=0} \int_{\Omega(t)} h d\sigma = -\int_{\partial\Omega} h(z)\frac{\partial G_\Omega(z,a)}{\partial n} ds = h(a).$$

Taking $h(z) = \frac{1}{\pi}z^k$ we get the harmonic moments M_k as special cases of the functional $\mathcal{F}$, with

$$\nabla(a)M_k(\Omega) = \frac{a^k}{\pi} \tag{7.1}$$

for $k \geq 0$. Note that this is zero, i.e., the moments are conserved, only when $a = 0$ and $k > 0$.

Example 7.1.2. As a second example, choose two points $b, c \in \Omega$ and take $\mathcal{F}$, defined in $\mathcal{M}^{(b)} \cap \mathcal{M}^{(c)}$, to be

$$\mathcal{F}(\Omega) = G_\Omega(b,c).$$

For a general domain variation, the change of this functional is given by the *Hadamard variational formula*. Following classical tradition, let δn denote an infinitesimal change of the boundary in the outward normal direction and let δG denote the corresponding change of the Green function. Then the Hadamard formula is

$$\delta G(b,c) = \int_{\partial\Omega} \delta n \frac{\partial G(\cdot,b)}{\partial n} \frac{\partial G(\cdot,c)}{\partial n} ds.$$

On dividing with a corresponding infinitesimal time interval δt we may view $\delta n/\delta t$ as the normal velocity of the boundary. In the case of Hele-Shaw injection at a we have

$$\frac{\delta n}{\delta t} = -\frac{\partial G(\cdot,a)}{\partial n}.$$

Thus, we get the beautiful formula

$$\nabla(a)G_\Omega(b,c) = -\int_{\partial\Omega} \frac{\partial G(\cdot,a)}{\partial n} \frac{\partial G(\cdot,b)}{\partial n} \frac{\partial G(\cdot,c)}{\partial n} ds. \tag{7.2}$$

In particular, $\nabla(a)G_\Omega(b,c)$ is completely symmetric in a, b and c.

The symmetry in the above formula (7.2) calls for an explanation. To this end, we first remark that the Richardson conservation of moments for Hele-Shaw evolutions, which in a stronger form is built into the concept of weak solution and into the balayage formulation, contains an integrability statement: Hele-Shaw injection at a point a followed by injection at a point b gives the same result as if the injections had been performed in the opposite order. In infinitesimal form this statement becomes

$$\nabla(a)\nabla(b) = \nabla(b)\nabla(a).$$

Therefore, the symmetry in (7.2) would be explained by exhibiting a functional $\mathcal{E}$ such that

$$G_\Omega(b, c) = \nabla(b)\nabla(c)\mathcal{E}(\Omega) + \text{constant}.$$

Such an $\mathcal{E}$ can indeed be found. Let $\mathbb{D}_R = \mathbb{D}(0, R)$ be a disk containing the closure of Ω. We define $\mathcal{E}(\Omega)$ to be (minus) the electrostatic energy of the exterior domain, cut off at radius R:

$$\mathcal{E}(\Omega) = \frac{1}{4\pi} \int_{\mathbb{D}_R \setminus \Omega} \int_{\mathbb{D}_R \setminus \Omega} \log |z - \zeta| \, d\sigma_z d\sigma_\zeta. \tag{7.3}$$

This functional of course depends on the choice of R (actually on the entire choice of the ambient space $\mathbb{D}_R$), but the essential dependence is that on Ω. With respect to this dependence $\mathcal{E}(\Omega)$ is sometimes referred to as a *prepotential* (cf. [373], [371]). It is related to the τ-function [321] appearing in the theory of integrable hierarchies, and discussed also in Chapters 6 and 8, by

$$\log \tau(\Omega) = -\frac{4}{\pi}\mathcal{E}(\Omega). \tag{7.4}$$

Several important domain functions (like the Green function) can be produced by letting $\nabla(a)$ act on $\mathcal{E}(\Omega)$, one or more times. This is the contents of the following proposition, where the last item is our "explanation" for the symmetry (7.2). For the formulation we decompose the Green function $G(z, \zeta)$ as

$$G(z, \zeta) = -\frac{1}{2\pi} \log |z - \zeta| + H(z, \zeta), \tag{7.5}$$

where $H(z, \zeta)$ is the regular part, harmonic in each variable and symmetric. Recall also (see after (3.23)) that, in general, U^D denotes the logarithmic potential of the set D, considered as a body with density one.

Proposition 7.1.1. *The result of single and repeated actions of ∇ on $\mathcal{E}(\Omega)$ is given by*

$$\nabla(a)\mathcal{E}(\Omega) = -\frac{1}{2\pi} \int_{\mathbb{D}_R \setminus \Omega} \log |z - a| d\sigma_z = U^{\mathbb{D}_R \setminus \Omega}(a), \tag{7.6}$$

$$\nabla(a)\nabla(b)\mathcal{E}(\Omega) = H(a, b) = G(a, b) + \frac{1}{2\pi} \log |a - b|, \tag{7.7}$$

$$\nabla(a)\nabla(b)\nabla(c)\mathcal{E}(\Omega) = \nabla(a)G(b, c). \tag{7.8}$$

Note that, in (7.7), the term $\frac{1}{2\pi} \log |a - b|$ does not depend on Ω, hence is just a "constant" with respect to variations of Ω.

Proof. In the definition (7.3) of $\mathcal{E}(\Omega)$ the domain Ω occurs symmetrically at two places. Thus, the result of $\nabla(a)$ acting on Ω in $\mathcal{E}(\Omega)$ will be twice the result of its action on only one of the occurences. This gives

$$\nabla(a)\mathcal{E}(\Omega) = \frac{1}{2\pi}\frac{d}{dt}\Big|_{t=0} \int_{\mathbb{D}_R\setminus\Omega} \int_{\mathbb{D}_R\setminus\Omega(t)} \log|z-\zeta|d\sigma_\zeta d\sigma_z$$

$$= \frac{1}{2\pi} \int_{\mathbb{D}_R\setminus\Omega} \int_{\partial\Omega} \log|z-\zeta|\frac{\partial G(\zeta,a)}{\partial n} ds_\zeta d\sigma_z$$

$$= -\frac{1}{2\pi} \int_{\mathbb{D}_R\setminus\Omega} \log|z-a|d\sigma_z = U^{\mathbb{D}_R\setminus\Omega}(a).$$

Here we used that $\log|z-\zeta|$ is harmonic in Ω in the variable ζ when $z\in\mathbb{D}_R\setminus\Omega$. This proves (7.6).

Applying next $\nabla(b)$ to the above and using that $G(z,a)=0$ on $\partial\Omega$ we obtain

$$\nabla(b)\nabla(a)\mathcal{E}(\Omega) = -\frac{1}{2\pi} \int_{\partial\Omega} \log|z-a|\frac{\partial G(z,b)}{\partial n} ds_z$$

$$= \int_{\partial\Omega} (G(z,a) - H(z,a))\frac{\partial G(z,b)}{\partial n} ds_z$$

$$= H(b,a) = G(b,a) + \frac{1}{2\pi}\log|b-a|.$$

This gives (7.7) (with reversed roles of a and b), and (7.8) finally is immediate by another application of ∇. □

7.2 Moment coordinates

Above, infinitesimal Laplacian growth was considered as a directional derivative, equivalently as a vector field, on the infinite-dimensional space $\mathcal{M}^{(a)}$. As an alternative to a directional derivative one may consider it as a partial derivative, by introducing a full set of coordinates on the space of domains. In this section (and further on) we shall choose $a=0$, and then write $\mathcal{M}=\mathcal{M}^{(0)}$. In addition, we shall consider only simply connected domains, for which a natural set of (local) coordinates are the harmonic moments $M_0, M_1, M_2, \ldots$. As is clear from the Richardson moment law (1.35), Laplacian growth with a source at the origin will correspond to evolution along the vector field $\partial/\partial M_0$. One may then ask what kind of evolutions the vector fields $\partial/\partial M_k$, for $k>0$, correspond to. A slight complication arising here is that the M_k for $k>0$ are actually complex coordinates, so one needs to consider also the conjugate derivatives $\partial/\partial \bar{M}_k$, and only real combinations of these (Wirtinger) derivatives will correspond to directions within the space of domains. Recall that

$$\frac{\partial}{\partial M_j} = \frac{1}{2}\left(\frac{\partial}{\partial\operatorname{Re} M_j} - i\frac{\partial}{\partial\operatorname{Im} M_j}\right), \tag{7.9}$$

$$\frac{\partial}{\partial \bar{M}_j} = \frac{1}{2}\left(\frac{\partial}{\partial\operatorname{Re} M_j} + i\frac{\partial}{\partial\operatorname{Im} M_j}\right). \tag{7.10}$$

We shall below connect such derivatives to Laplacian evolution driven by multipole singularities at the origin.

By *Laplacian evolution* in general we understand domain evolution driven by harmonic gradients, i.e., such that the normal velocity V_n of the boundary of the domain equals (minus) the normal derivative of a harmonic function p in the domain, vanishing on the boundary and having prescribed singularities inside the domain (or possibly satisfying given boundary conditions on some fixed inner boundaries). Thus $V_n = -\frac{\partial p}{\partial n}$, and if p has singularities in Ω represented by a given distribution μ in Ω (so that $-\Delta p = \mu$), then

$$\frac{d}{dt}\int_{\Omega(t)} h d\sigma = \langle \mu, h \rangle$$

for harmonic functions h defined in some neighborhood of the closure of $\Omega(t)$. With $p = G_\Omega(\cdot, a)$ we get ordinary Laplacian growth with a source of strength $Q = 1$ at a.

Since V_n in general need not have any particular sign we call it evolution rather than growth. If one allows the singularities to depend on time, every kind of evolution can in fact be approximated by Laplacian evolution. This is because every harmonic function p defined near the boundary can be approximated by harmonic functions in $\Omega \setminus \{0\}$ with only polar singularities at the origin (cf. [496]).

As for coordinates, the most obvious set of coordinates on the space of simply connected domains is of course the coefficients of the power series expansion

$$f(\zeta) = \sum_{k=1}^{\infty} a_k \zeta^k \quad (a_1 > 0)$$

of the normalized conformal map $f : \mathbb{D} \to \Omega$. Recall that the harmonic moments can be written as boundary integrals,

$$M_n = \frac{1}{2\pi i}\int_{\partial\Omega} z^n \bar{z} dz = \frac{1}{2\pi i}\int_{\partial\mathbb{D}} f(\zeta)^n f^*(\zeta) f'(\zeta) d\zeta, \tag{7.11}$$

and that evaluating the latter integral gives Richardson's transition formula, see (1.37) (or (5.22)), for the variable change $(M_0, M_1, M_2, \dots) \leftrightarrow (a_1, a_2, a_3, \dots)$.

The boundary integral form (7.11) of the moments has the additional advantage that it gives a meaning to the negative moments M_n, $n < 0$, which we shall also consider below. Further on, from (1.37) it is clear that if we restrict ourselves to polynomial mapping functions, say

$$f(\zeta) = \sum_{k=1}^{N+1} a_k \zeta^k \tag{7.12}$$

for a fixed $N \geq 0$, then there will only be finitely many non-zero moments with positive index, namely $(M_0, M_1, \dots, M_N)$ (but still infinitely many M_n with $n < 0$).

Conversely, if $M_{N+1} = M_{N+2} = \cdots = 0$ then one can show that the conformal map f necessarily is a polynomial of degree at most $N+1$. Indeed, if the moment sequence is finite then Ω is a quadrature domain (see Section 3.3.1) with the origin as the only quadrature node, and the assertion made follows from a detailed version of statement (iv) in Theorem 3.3.1 (see after the proof of Theorem 3.3.1, and also [560], [517]).

The above observation gives the possibility to work rigorously within a finite-dimensional setting by considering only those domains which are images of the unit disk under polynomial (of a fixed degree) conformal maps. For simplicity we shall require the map f to be univalent on the full closure of the unit disk. This makes the space of domains a manifold (without boundary) in the ordinary sense. (Actually local univalence, i.e., non-vanishing of f' on the closed disk, would be enough for most purposes.) Let $\mathcal{M}_{2N+1}$ denote the so obtained manifold, of real dimension $2N+1$, namely

$$
\mathcal{M}_{2N+1} = \left\{ \Omega = f(\mathbb{D}) : f(\zeta) = \sum_{k=1}^{N+1} a_k \zeta^k, f \text{ is univalent on } \overline{\mathbb{D}}, \quad a_1 > 0 \right\}.
$$

It is known that the change of variables $(M_0, M_1, M_2, \dots) \leftrightarrow (a_1, a_2, a_3, \dots)$ is at least locally one-to-one on $\mathcal{M}_{2N+1}$, i.e., that the appropriate Jacobi determinant is non-zero. An explicit expression for this determinant has been found by O. Kuznetsova and V. Tkachev [338], [556], based on a conjecture of C. Ullemar [558]. It can be expressed in terms of the ratio between the volume forms, as follows.

$$
d\bar{M}_N \wedge \cdots d\bar{M}_1 \wedge dM_0 \wedge dM_1 \wedge \cdots \wedge dM_N
$$
$$
= 2a_1^{(N+1)^2} \prod_{k=1}^{N} f'^*(\omega_k) \cdot d\bar{a}_{N+1} \wedge \cdots d\bar{a}_2 \wedge da_1 \wedge da_2 \wedge \cdots \wedge da_{N+1}. \tag{7.13}
$$

Here $\omega_1, \dots, \omega_N$ are the zeros of f', and we recall that $f'^*(\zeta) = \sum_{k=1}^{N+1} k\bar{a}_k \zeta^{1-k}$. Note that $\prod_{k=1}^{N} f'^*(\omega_k) \neq 0$ when $|\omega_k| > 1$ for all k, i.e., whenever f is locally univalent on $\overline{\mathbb{D}}$.

Any smooth variation within $\mathcal{M}_{2N+1}$ can be thought of as Laplacian evolution with time-dependent multipole singularities at the origin. Such a deformation of a domain Ω corresponds to a smooth curve $t \mapsto \Omega(t)$ with $\Omega(0) = \Omega$ (equivalently, $t \mapsto f(\zeta, t)$) in $\mathcal{M}_{2N+1}$, and its velocity at $t = 0$ is a vector in the real tangent space of $\mathcal{M}_{2N+1}$. This tangent space has, in terms of the coordinates introduced above, two natural bases, namely

$$
\frac{\partial}{\partial a_1}, \quad \frac{\partial}{\partial \mathrm{Re}\, a_k}, \quad \frac{\partial}{\partial \mathrm{Im}\, a_k},
$$
$$
\frac{\partial}{\partial M_0}, \quad \frac{\partial}{\partial \mathrm{Re}\, M_j}, \quad \frac{\partial}{\partial \mathrm{Im}\, M_j}.
$$

$(2 \leq k \leq N+1, 1 \leq j \leq N)$, respectively. It is natural to consider also the corresponding complexified tangent space, obtained by allowing complex coefficients in front of the above basis vectors. For this tangent space we also have the bases:

$$\frac{\partial}{\partial a_1}, \quad \frac{\partial}{\partial a_k}, \quad \frac{\partial}{\partial \bar{a}_k},$$
$$\frac{\partial}{\partial M_0}, \quad \frac{\partial}{\partial M_j}, \quad \frac{\partial}{\partial \overline{M}_j}.$$

The dependence of the conformal map f on the coefficients $a_1, \ldots, a_{N+1}$ is certainly analytic by (7.12), but as can be understood from the appearance of conjugations in (1.37), the dependence on $M_0, \ldots, M_N$ is no longer analytic. Therefore we prefer to write this dependence as

$$f(\zeta) = f(\zeta; \overline{M}_N, \ldots, \overline{M}_1, M_0, M_1, \ldots, M_N), \qquad (7.14)$$

or just $f(\zeta; M)$, with M shorthand for the list of moments:

$$M = (\overline{M}_N, \ldots, \overline{M}_1, M_0, M_1, \ldots, M_N). \qquad (7.15)$$

It should be said that this kind of dependence in principle is only local, as there are examples of different domains having the same moments (see [498]).

The directional derivative $\nabla(z)$ ($z \in \Omega$) in Section 7.1 can of course be expressed in the coordinates $(\overline{M}_N, \ldots, \overline{M}_1, M_0, M_1, \ldots, M_N)$. Indeed, using the chain rule and (7.1) we have

$$\nabla(z) = (\nabla(z)M_0)\frac{\partial}{\partial M_0} + \sum_{k=1}^{N}\left((\nabla(z)M_k)\frac{\partial}{\partial M_k} + (\nabla(z)\overline{M}_k)\frac{\partial}{\partial \overline{M}_k}\right)$$
$$= \frac{1}{\pi}\left[\frac{\partial}{\partial M_0} + \sum_{k=1}^{N}\left(z^k \frac{\partial}{\partial M_k} + \bar{z}^k \frac{\partial}{\partial \overline{M}_k}\right)\right].$$

Of interest in itself is the analytic part (without constant term) of $\nabla(z)$:

$$D(z) = \frac{1}{\pi}\sum_{k=1}^{N} z^k \frac{\partial}{\partial M_k}.$$

Then

$$\nabla(z) = \frac{1}{\pi}\frac{\partial}{\partial M_0} + D(z) + \overline{D(z)} = \nabla(0) + D(z) + \overline{D(z)}.$$

In the above notation, (7.7) gives the following formula for the Green function for Ω:

$$G_\Omega(z, \zeta) = -\frac{1}{2\pi}\log|z - \zeta| + (\nabla(0) + D(z) + \overline{D(z)})(\nabla(0) + D(\zeta) + \overline{D(\zeta)})\mathcal{E}(\Omega).$$

Another expression for the same Green function is obtained via the inverse conformal map $f^{-1} : \Omega \to \mathbb{D}$ and the Green function for $\mathbb{D}$:

$$G_\Omega(z, \zeta) = -\frac{1}{2\pi} \log \left| \frac{f^{-1}(z) - f^{-1}(\zeta)}{1 - f^{-1}(z)\overline{f^{-1}(\zeta)}} \right|.$$

By extracting and identifying, in these two expressions, the parts which are analytic in both z and ζ (special care is needed for the constant term) one arrives at the following formula for the logarithmic difference quotient of the inverse map.

$$\log \frac{f^{-1}(z) - f^{-1}(\zeta)}{z - \zeta} = 4\pi \left[\frac{1}{2}\nabla(0)^2 - (\nabla(0) + D(z))(\nabla(0) + D(\zeta)) \right] \mathcal{E}(\Omega)$$

$$= \frac{2}{\pi} \left(\frac{\partial^2 \mathcal{E}(\Omega)}{\partial M_0^2} - 2 \sum_{k,j=0}^{N} z^k \zeta^j \frac{\partial^2 \mathcal{E}(\Omega)}{\partial M_k \partial M_j} \right). \qquad (7.16)$$

This equation is related to the dispersionless Hirota equation and the dispersionless Fay identity, see [321].

Choosing in particular $\zeta = 0$ in (7.16) one gets a formula for the inverse conformal map f^{-1} in terms of $\mathcal{E}(\Omega)$:

$$f^{-1}(z) = z \exp \left[-\frac{2}{\pi} \frac{\partial^2 \mathcal{E}(\Omega)}{\partial M_0^2} - \frac{4}{\pi} \sum_{k=1}^{N} z^k \frac{\partial^2 \mathcal{E}(\Omega)}{\partial M_0 \partial M_k} \right].$$

And inserting this, and the corresponding formula for $f^{-1}(\zeta)$, in (7.16) gives after some simplifications the following *Fay identity* (cf. [184] and [321]):

$$(z - \zeta) \exp \left[\sum_{k,j=0}^{N} z^k \zeta^j \frac{\partial^2 \log \tau}{\partial M_k \partial M_j} \right]$$

$$= z \exp \left[\sum_{k=0}^{N} z^k \frac{\partial^2 \log \tau}{\partial M_0 \partial M_k} \right] - \zeta \exp \left[\sum_{k=0}^{N} \zeta^k \frac{\partial^2 \log \tau}{\partial M_0 \partial M_k} \right].$$

Here we have used $\log \tau$ in place of $\mathcal{E}(\Omega)$, see (7.4), in order to avoid some ugly constants.

As another application of (7.16), one gets a linear representation of the Schwarzian derivative of f^{-1}. Recall that the Schwarzian derivative of a holomorphic function $w = w(z)$ in general is defined by

$$S_w(z) = (\log w')'' - \frac{1}{2}((\log w')')^2 = \frac{w'''}{w'} - \frac{3}{2}\left(\frac{w''}{w'}\right)^2,$$

and that it also can be represented as

$$S_w(z) = 6 \lim_{\zeta \to z} \frac{\partial^2}{\partial z \partial \zeta} \log \frac{w(\zeta) - w(z)}{\zeta - z}.$$

For $w = f^{-1}(z)$ we get, using the latter representation in combination with (7.16),

$$S_{f^{-1}}(z) = -\frac{24}{\pi} \sum_{k,j=1}^{N} kj z^{k+j-2} \frac{\partial^2 \mathcal{E}(\Omega)}{\partial M_k \partial M_j}.$$

Further identities of this kind can be found in papers by M. Mineev-Weinstein, P. Wiegmann, A. Zabrodin and others, for example [327, 374, 393, 580].

7.3 Poisson bracket and string equation

We shall now derive formulas for derivatives such as $\partial f/\partial M_k$, $\partial M_{-j}/\partial M_k$, and also find Hamiltonian formulations for Laplacian growth. The equation for $\partial f/\partial M_0$ will be equivalent to the Polubarinova–Galin equation, but it will be derived in a completely different way, and it is in the present context known as the *string equation*.

We define a *Poisson bracket* between any two functions $\varphi = \varphi(\zeta; M)$ and $\psi = \psi(\zeta; M)$, which are analytic for ζ in a neighborhood of the unit circle and which moreover depend on the moments M, by

$$\{\varphi, \psi\} = \zeta \frac{\partial \varphi}{\partial \zeta} \frac{\partial \psi}{\partial M_0} - \zeta \frac{\partial \psi}{\partial \zeta} \frac{\partial \varphi}{\partial M_0}. \tag{7.17}$$

Clearly, $\{\varphi, \psi\}$ will itself be analytic in a neighborhood of the unit circle and depend on the moments. The bracket is uniquely determined by its restriction to the unit circle, which on writing $\zeta = e^{i\theta}$ takes the form

$$\{\varphi, \psi\} = i \frac{\partial(\varphi, \psi)}{\partial(M_0, \theta)} \quad \text{on } \partial\mathbb{D}. \tag{7.18}$$

As a first choice we shall take $\varphi = f$, the conformal map in (7.12), and $\psi = f^*$, the reflection of f in the unit circle. We can connect f and f^* via the Schwarz function S of $\partial\Omega$: we have $f^* = S \circ f$ identically since this relation holds on $\partial\mathbb{D}$. Since also $S = S(z; M)$ depends on the moments the relationship spells out to

$$f^*(\zeta; M) = S(f(\zeta; M); M).$$

Using hence the chain rule when computing $\frac{\partial f^*}{\partial M_0}$ one arrives at

$$\{f, f^*\} = \zeta \frac{\partial f}{\partial \zeta} \cdot \left(\frac{\partial S}{\partial M_0} \circ f\right). \tag{7.19}$$

Recall from Section 1.7 (equation (1.51)) that the formal Laurent series of the Schwarz function for $\partial\Omega$ has the complete set of moments M_k ($k \in \mathbb{Z}$) as coefficients:

$$S(z; M) = \sum_{k=-\infty}^{\infty} \frac{M_k}{z^{k+1}}. \tag{7.20}$$

Even though the full series here need not converge anywhere it has the following precise meaning: the negative part of the series defines a germ of an analytic function at the origin and the positive part a germ of an analytic function at infinity. When $\partial\Omega$ is analytic the domains of analyticity of these two analytic functions overlap, and the overlap region contains $\partial\Omega$. What is important in the present context is that

$$S(z; M) = \sum_{k=0}^{N} \frac{M_k}{z^{k+1}} + S_+(z; M) = S_-(z) + S_+(z),$$

where $S_+(z; M)$ is holomorphic in Ω (recall (1.48)). Since the coefficients M_1, $M_2, \ldots$ are independent of M_0 it follows that

$$\frac{\partial S}{\partial M_0}(z; M) = \frac{1}{z} + \frac{\partial S_+}{\partial M_0}(z; M),$$

where the last term is holomorphic in Ω.

Combining this with (7.19) one sees that $\{f, f^*\}$ is holomorphic in $\mathbb{D}$ and equals one at the origin. Moreover, by (7.18) for example, $\{f, f^*\}$ is real-valued on $\partial\mathbb{D}$. Hence it must be identically one. Thus we have proved the following result.

Proposition 7.3.1. *The normalized conformal map $f : \mathbb{D} \to \Omega$ satisfies the string equation,*

$$\{f, f^*\} = 1. \tag{7.21}$$

Next we shall connect the derivatives $\frac{\partial}{\partial M_j}$, $\frac{\partial}{\partial \overline{M}_j}$ to Laplacian evolutions. Let L denote a real-valued differential operator with constant coefficients acting on functions in $\mathbb{C}$, for example $L = \frac{\partial^k}{\partial x^k}$. Then, given $\Omega(0)$ with analytic boundary, there exists locally an evolution $t \mapsto \Omega(t)$ such that

$$\frac{d}{dt} \int_{\Omega(t)} h \, d\sigma = (Lh)(0) \tag{7.22}$$

for every function h which is harmonic in a region containing the closure of $\Omega(t)$. This is an instance of Laplacian evolution, in fact the one driven by the harmonic function

$$p(z) = \{L_a G(z, a)\}_{a=0}, \tag{7.23}$$

where $G(z, a) = -\frac{1}{2\pi} \log|z - a| + \text{harmonic}$ is the Green function with pole at a, and the differential operator $L_a = L$ acts on that variable (the location of the pole). Thus p is harmonic in Ω, has a multipole singularity at the origin and vanishes on $\partial\Omega$.

It is actually not easy to prove that evolutions $\Omega(t)$ driven by a given harmonic gradient really exist, but it can be done for $\Omega \in \mathcal{M}$ using tools as in [179, 181, 467], for example, or otherwise by using tools of partial balayage, as described in Section 3.2.2 with Remark 3.2.1 taken into account. In the algebraic framework with $\Omega \in \mathcal{M}_{2N+1}$, the full evolution will stay in $\mathcal{M}_{2N+1}$ provided the order k of the differential operator L is at most N, and the existence proof is much easier (cf. [233]). In fact, the existence and uniqueness can even be read off from (7.13).

To prove the assertion that p in (7.23) really achieves (7.22), recall that the evolution $t \mapsto \Omega(t)$ is defined by the outward normal velocity of the boundary being $V_n = -\frac{\partial p}{\partial n}$. Since $-\Delta G(\cdot, a) = \delta_a$ we have

$$-\Delta p = -\{\Delta L_a G(\cdot, a)\}_{a=0} = -\{L_a \Delta G(\cdot, a)\}_{a=0} = \{L_a \delta_a\}_{a=0}.$$

With h harmonic this gives

$$\frac{d}{dt} \int_{\Omega(t)} h \, d\sigma = \int_{\partial \Omega(t)} V_n h \, ds = -\int_{\partial \Omega(t)} h \frac{\partial p}{\partial n} \, ds$$

$$= -\int_{\Omega(t)} h \Delta p \, d\sigma = \{L_a h(a)\}_{a=0} = (Lh)(0),$$

proving (7.22).

Useful particular choices of L above are, for $1 \le k \le N$,

$$L^{(1)} = \frac{\partial^k}{\partial x^k}, \quad L^{(2)} = \frac{\partial^k}{\partial x^{k-1} \partial y}. \tag{7.24}$$

We may allow h in (7.22) to be complex-valued, and choosing $h(z) = z^j$ and evaluating at $z = 0$ one gets, for $j = k$,

$$\{L^{(1)}(z^k)\}_{z=0} = k!, \quad \{L^{(2)}(z^k)\}_{z=0} = i \cdot k!, \tag{7.25}$$

and $\{L^{(i)}(z^j)\}_{z=0} = 0$ in all remaining cases. Similarly, with $h(z) = \bar{z}^j$ we have

$$\{L^{(1)}(\bar{z}^k)\}_{z=0} = k!, \quad \{L^{(2)}(\bar{z}^k)\}_{z=0} = -i \cdot k!, \tag{7.26}$$

and $\{L^{(i)}(\bar{z}^j)\}_{z=0} = 0$ in the remaining cases.

In what follows, we will denote the time coordinates for the two evolutions corresponding to $L^{(1)}$ and $L^{(2)}$ by different letters, namely t_1 and t_2, respectively. Let p_i be the harmonic function given by (7.23) for $L^{(i)}$ ($i = 1, 2$). It is a consequence of (7.22) and (7.25), (7.26) that the evolution $t_1 \mapsto \Omega(t_1)$ generated by p_1 has the property that

$$\frac{d}{dt_1} M_k = \frac{d}{dt_1} \bar{M}_k = \frac{k!}{\pi}, \quad \frac{d}{dt_1} M_j = \frac{d}{dt_1} \bar{M}_j = 0 \quad (j \ne k).$$

Similarly, for the evolution generated by p_2:

$$\frac{d}{dt_2} M_k = -\frac{d}{dt_2} \bar{M}_k = \frac{ik!}{\pi}, \quad \frac{d}{dt_2} M_j = \frac{d}{dt_2} \bar{M}_j = 0 \quad (j \ne k).$$

In other words, the tangent vector giving the velocity for the evolution driven by p_1 is

$$\frac{d}{dt_1} = \frac{k!}{\pi}\left(\frac{\partial}{\partial M_k} + \frac{\partial}{\partial \bar{M}_k}\right) = \frac{k!}{\pi}\frac{\partial}{\partial \mathrm{Re}\, M_k},$$

and that for p_2 is

$$\frac{d}{dt_2} = \frac{ik!}{\pi}\left(\frac{\partial}{\partial M_k} - \frac{\partial}{\partial \bar{M}_k}\right) = \frac{k!}{\pi}\frac{\partial}{\partial \mathrm{Im}\, M_k}.$$

Now we wish to extract $\frac{\partial}{\partial M_k}$ and $\frac{\partial}{\partial \bar{M}_k}$ from the above relations. We then see that these partial derivatives will correspond to linear combinations with complex (and non-real) coefficients of the two different evolutions, or tangent vectors, hence they will not themselves correspond to evolutions of domains. Still they make sense, of course, as vectors in the complexified tangent space of $\mathcal{M}_{2N+1}$. Precisely, in view of (7.9), (7.10) we get

$$\frac{\partial}{\partial M_k} = \frac{\pi}{2k!}\left(\frac{d}{dt_1} - i\frac{d}{dt_2}\right), \tag{7.27}$$

$$\frac{\partial}{\partial \bar{M}_k} = \frac{\pi}{2k!}\left(\frac{d}{dt_1} + i\frac{d}{dt_2}\right). \tag{7.28}$$

It follows, for example, that for harmonic functions h in Ω,

$$\frac{\partial}{\partial M_k}\left(\frac{1}{\pi}\int_\Omega h d\sigma\right) = \frac{1}{k!}\frac{\partial^{k-1}}{\partial x^{k-1}}\frac{\partial h}{\partial z}(0) = \frac{1}{k!}\frac{\partial^k h}{\partial z^k}(0). \tag{7.29}$$

Recall the prepotential (7.3), which we now consider as a functional of the moments via $\Omega = \Omega(M)$. Using (7.27), (7.28) we get

$$\frac{\partial \mathcal{E}(\Omega)}{\partial M_k} = \frac{1}{4\pi}\frac{\partial}{\partial M_k}\int_{\mathbb{D}_R\setminus\Omega}\int_{\mathbb{D}_R\setminus\Omega}\log|z-\zeta|d\sigma_z d\sigma_\zeta$$

$$= \frac{1}{8k!}\left(\frac{d}{dt_1} - i\frac{d}{dt_2}\right)\int_{\mathbb{D}_R\setminus\Omega}\int_{\mathbb{D}_R\setminus\Omega}\log|z-\zeta|d\sigma_z d\sigma_\zeta.$$

When computing the $\frac{d}{dt_1}$, for example, Ω is to be replaced with the evolution $\Omega(t_1)$, and we evaluate the derivative at $t_1 = 0$ ($\Omega(0) = \Omega$). There are two occurrences of Ω in the expression for $\mathcal{E}(\Omega)$, and $\mathcal{E}(\Omega)$ symmetric in them, so it is enough to differentiate one of the occurrences and then multiply the result by two. Using (7.22), (7.24) and (7.11) this gives

$$\frac{d}{dt_1}\Big|_{t_1=0}\int_{\mathbb{D}_R\setminus\Omega(t_1)}\int_{\mathbb{D}_R\setminus\Omega(t_1)}\log|z-\zeta|\, d\sigma_z d\sigma_\zeta$$

$$= \frac{i(k-1)!}{2}\int_{\partial\Omega}\left(\frac{1}{\zeta^k} + \frac{1}{\bar{\zeta}^k}\right)\bar{\zeta}d\zeta = -\pi k!\left(\frac{1}{k}M_{-k} + \frac{1}{k}\bar{M}_{-k}\right).$$

Similarly, the corresponding derivative with respect to t_2 gives

$$\frac{d}{dt_2}\Big|_{t_2=0} \int_{\mathbb{D}_R\setminus\Omega(t_1)} \int_{\mathbb{D}_R\setminus\Omega(t_1)} \log|z-\zeta|\, d\sigma_z d\sigma_\zeta = -i\pi k!\left(\frac{1}{k}M_{-k} - \frac{1}{k}\bar{M}_{-k}\right).$$

Hence, in the combination $\frac{d}{dt_1} - i\frac{d}{dt_2}$, $\bar{M}_{-k}$ will cancel out, and we get

$$\frac{\partial\mathcal{E}(\Omega)}{\partial M_k} = -\frac{\pi}{4k}M_{-k}.$$

As for $\frac{\partial\mathcal{E}(\Omega)}{\partial M_0}$, it follows from (7.6) that

$$\frac{\partial\mathcal{E}(\Omega)}{\partial M_0} = \pi\nabla(0)\mathcal{E}(\Omega) = -\frac{1}{2}\int_{\mathbb{D}_R\setminus\Omega} \log|z|d\sigma_z.$$

We thus arrive at the following integrability formulas due to I. Krichever, A. Marshakov, M. Mineev-Weinstein, P. Wiegmann, A. Zabrodin, see for example [327, 374, 393, 580].

Proposition 7.3.2. *With notations as above, and using (7.4), we have*

$$\frac{\partial\mathcal{E}(\Omega)}{\partial M_k} = -\frac{\pi}{4k}M_{-k}(\Omega) \quad (1\le k \le N), \tag{7.30}$$

$$\frac{\partial\mathcal{E}(\Omega)}{\partial M_0} = -\frac{1}{2}\int_{\mathbb{D}_R\setminus\Omega} \log|z|d\sigma_z. \tag{7.31}$$

As a consequence,

$$\frac{1}{k}\frac{\partial M_{-k}}{\partial M_j} = \frac{1}{j}\frac{\partial M_{-j}}{\partial M_k} \quad (1\le k,j\le N). \tag{7.32}$$

7.4 Hamiltonians

As a final topic, we wish to explain, in our notations and settings, the Hamiltonian descriptions of the evolution of the mapping function f presented in [580] and related articles. Recall that we work within $\mathcal{M}_{2N+1}$, and we use M as a shorthand notation for the moments, which serve as coordinates on $\mathcal{M}_{2N+1}$, see (7.15).

Let $\mathcal{W} = \mathcal{W}(z; M)$ be a primitive function (with respect to z) of the Schwarz function $S = S(z; M)$. By (7.20) the power series expansion of $\mathcal{W}$ is

$$\mathcal{W}(z; M) = -\sum_{k=1}^{N}\frac{M_k}{k}\frac{1}{z^k} + M_0\log z + \sum_{j=1}^{\infty}\frac{M_{-j}}{j}z^j + C(M), \tag{7.33}$$

where $C(M)$ is a constant (with respect to z), which may be allowed to depend on the moments M. We shall be more precise about $C(M)$ later on (see (7.41)), for now we just declare that we choose $C(M)$ to be real.

Because of the logarithm, $\mathcal{W}$ is multi-valued, but the real part $w = \operatorname{Re} \mathcal{W}$ is perfectly well defined (after choice of $C(M)$). It is a harmonic function in Ω with a multipole singularity at $z = 0$, and it satisfies easily recognized boundary conditions. In fact, $S(z) = 2\frac{\partial w}{\partial z}$, and since $S(z) = \bar{z}$ on $\partial\Omega$ it follows that the real-valued function

$$u(z) = \frac{1}{4}|z|^2 - \frac{1}{2}w(z) \tag{7.34}$$

satisfies $\frac{\partial u}{\partial z} = 0$ on $\partial\Omega$. In other words, the gradient of u vanishes on $\partial\Omega$, and in particular u itself is constant on $\partial\Omega$. We shall fix the free additive constant in u, and hence that in w, by requiring that $u = 0$ on $\partial\Omega$. It follows that u then satisfies

$$\begin{cases} \Delta u = 1 & \text{in a neighborhood of } \partial\Omega, \\ u = \frac{\partial u}{\partial z} = 0 & \text{on } \partial\Omega. \end{cases}$$

Clearly, u and w depend on the moments M, and $\frac{\partial u}{\partial M_k} = -\frac{1}{2}\frac{\partial w}{\partial M_k}$ by (7.34). Since u vanishes on $\partial\Omega$ together with its gradient, u will vanish on $\partial\Omega$ also after an 'infinitesimal variation' of $\partial\Omega$. In other words, we have

$$\frac{\partial u}{\partial M_k} = \frac{\partial w}{\partial M_k} = 0 \quad \text{on } \partial\Omega \quad (0 \le k \le N). \tag{7.35}$$

From (7.33) we see that

$$\frac{\partial \mathcal{W}}{\partial M_0} = \log z + \sum_{j=1}^{\infty} \frac{1}{j}\frac{\partial M_{-j}}{\partial M_0} z^j + \frac{\partial C}{\partial M_0}. \tag{7.36}$$

Similarly, for $k = 1, \dots, N$,

$$\frac{\partial \mathcal{W}}{\partial M_k} = -\frac{1}{kz^k} + \sum_{j=1}^{\infty} \frac{1}{j}\frac{\partial M_{-j}}{\partial M_k} z^j + \frac{\partial C}{\partial M_k}. \tag{7.37}$$

Thus the derivatives $\frac{\partial \mathcal{W}}{\partial M_k}$ are analytic functions in Ω with specific logarithmic or polar singularities at $z = 0$.

As $\frac{\partial w}{\partial M_0} = \log|z| + $ harmonic by (7.36) we conclude from (7.35) that $\frac{\partial w}{\partial M_0}$ essentially coincides with the Green function $G = G_\Omega(z, 0)$ of Ω with pole at the origin. More precisely we have

$$\frac{\partial w}{\partial M_0} = -2\pi G. \tag{7.38}$$

For the regular part H of G (see (7.5)) this gives

$$\operatorname{Re}\left(\sum_{j=1}^{\infty} \frac{1}{j}\frac{\partial M_{-j}}{\partial M_0} z^j\right) + \frac{\partial C(M)}{\partial M_0} = -2\pi H(z, 0), \tag{7.39}$$

and by choosing $z = 0$,

$$\frac{\partial C(M)}{\partial M_0} = -2\pi H(0,0).$$
(7.40)

The right-hand member in (7.40) is the *Robin constant* in geometric function theory (see, e.g., [404], and also Chapter 5 above).

By (7.7) with $a = b = 0$ we have on the other hand that

$$H(0,0) = \frac{1}{\pi^2} \frac{\partial^2 \mathcal{E}(\Omega)}{\partial M_0^2}.$$

Combining this with (7.31) shows (with (7.40)) that

$$\frac{\partial}{\partial M_0} \left(C(M) - \frac{1}{\pi} \int_{\mathbb{D}_R \setminus \Omega} \log |z| d\sigma_z \right) = 0.$$

More generally, one shows (in a similar way) that actually

$$\frac{\partial}{\partial M_k} \left(C(M) - \frac{1}{\pi} \int_{\mathbb{D}_R \setminus \Omega} \log |z| d\sigma_z \right) = 0$$

for all $k = 0, 1, \ldots, N$. Using that

$$\int_{\mathbb{D}_R \setminus \Omega} \log |z| d\sigma_z + \int_{\Omega} \log |z| d\sigma_z = \int_{\mathbb{D}_R} \log |z| d\sigma_z,$$

where the right-hand member depends only on R, we thus see that we can choose $C(M)$ to be a logarithmic moment:

$$C(M) = -\frac{1}{\pi} \int_{\Omega} \log |z| d\sigma_z + \text{constant},$$
(7.41)

where the constant now depends on neither z nor M, and hence can be chosen to be zero, for example.

Next, let $\mathcal{G} = G + i^* G$ be the analytic completion of $G = G_{\Omega}(z,0)$. This Green function $\mathcal{G} = \mathcal{G}(z; M)$ is (like G) a function of $z \in \Omega$ and the moments M. By (7.38),

$$2\pi \mathcal{G}(z; M) = -\frac{\partial \mathcal{W}(z; M)}{\partial M_0},$$
(7.42)

at least up to an additive imaginary local constant (the conjugate Green function being multi-valued and not uniquely determined). We define the (complex-valued) *Hamiltonian function* $\mathcal{H}_0 = \mathcal{H}_0(\zeta; M)$ of order zero to be exactly the above quantity, but considered as a function of $\zeta \in \mathbb{D}$:

$$\mathcal{H}_0(\zeta; M) = -\frac{\partial \mathcal{W}(z; M)}{\partial M_0}.$$
(7.43)

Here $z = f(\zeta; M)$, and in the right-hand member the derivative $\frac{\partial}{\partial M_0}$ acts only on the M_0 which appears in $\mathcal{W}(z; M)$, not that in $f(\zeta; M)$. Similarly for other places (e.g., (7.45) below).

Explicitly we have

$$\mathcal{H}_0(\zeta; M) = 2\pi\mathcal{G}(f(\zeta; M); M) = -\log\zeta. \tag{7.44}$$

where we used that $\mathcal{G} \circ f$ is the Green function of $\mathbb{D}$. Thus $\mathcal{H}_0$ is a multi-valued analytic function defined in $\mathbb{D}$, and moreover it is independent of the moments. Since $\frac{\partial \mathcal{H}_0}{\partial M_0} = 0$, $\zeta\frac{\partial \mathcal{H}_0}{\partial \zeta} = -1$ by (7.44), we trivially arrive at the evolution equation

$$\frac{\partial\varphi}{\partial M_0} = \{\varphi, \mathcal{H}_0\},$$

valid for any function $\varphi = \varphi(\zeta; M)$ for which the Poisson bracket makes sense.

The higher-order Hamiltonians are defined by

$$\mathcal{H}_k(\zeta; M) = -\frac{\partial \mathcal{W}(z; M)}{\partial M_k}, \quad k \geq 1. \tag{7.45}$$

These are again analytic functions in $\mathbb{D}$, and (unlike $\mathcal{H}_0$) they do depend on the moments. We may summarize the definitions of the above Hamiltonians by writing up the total differential of $\mathcal{W} = \mathcal{W}(z; M)$, with respect to all variables:

$$d\mathcal{W}(z; M) = S(z; M)dz - \mathcal{H}_0(\zeta)\,dM_0 - \sum_{k=1}^{\infty}(\mathcal{H}_k(\zeta; M)dM_k + \overline{\mathcal{H}_k(\zeta; M)}d\bar{M}_k).$$

Here, z and ζ are related by $z = f(\zeta; M)$.

In terms of the above Hamiltonians we have the following evolution equations for f, sometimes referred to as the *Lax–Sato equations* (see [580] for example).

Proposition 7.4.1. *For the conformal map $f : \mathbb{D} \to \Omega$ we have*

$$\frac{\partial f}{\partial M_k} = \{f, \mathcal{H}_k\} \quad (1 \leq k \leq N). \tag{7.46}$$

Proof. Differentiating first (7.44) with respect to ζ and M_k gives

$$2\pi\frac{\partial\mathcal{G}}{\partial z}(z; M)\frac{\partial f}{\partial \zeta}(\zeta; M) = -\frac{1}{\zeta},$$

$$\frac{\partial\mathcal{G}}{\partial z}(z; M)\frac{\partial f}{\partial M_k}(\zeta; M) + \frac{\partial\mathcal{G}}{\partial M_k}(z; M)) = 0,$$

which after elimination of $\frac{\partial\mathcal{G}}{\partial z}$ reduces to

$$\frac{\partial f}{\partial M_k}(\zeta; M) = 2\pi\zeta\frac{\partial f}{\partial \zeta}(\zeta; M)\frac{\partial\mathcal{G}}{\partial M_k}(z; M). \tag{7.47}$$

Here we wish to remove also $\frac{\partial \mathcal{G}}{\partial M_k}$, in favor of partial derivatives of $\mathcal{H}_k$. From (7.45) we get

$$\zeta \frac{\partial \mathcal{H}_k}{\partial \zeta}(\zeta; M) = -\frac{\partial^2 \mathcal{W}}{\partial z \partial M_k}(z; M) \cdot \zeta \frac{\partial f}{\partial \zeta}(\zeta; M),$$

and, using also (7.42),

$$\zeta \frac{\partial f}{\partial \zeta}(\zeta; M) \frac{\partial \mathcal{H}_k}{\partial M_0}(\zeta; M)$$

$$= -\zeta \frac{\partial f}{\partial \zeta}(\zeta; M) \frac{\partial^2 \mathcal{W}}{\partial z \partial M_k}(z; M) \frac{\partial f}{\partial M_0}(\zeta; M) - \zeta \frac{\partial f}{\partial \zeta}(\zeta; M) \frac{\partial^2 \mathcal{W}}{\partial M_0 \partial M_k}(z; M)$$

$$= \zeta \frac{\partial \mathcal{H}_k}{\partial \zeta}(\zeta; M) \frac{\partial f}{\partial M_0}(\zeta; M) + 2\pi \zeta \frac{\partial f}{\partial \zeta}(\zeta; M) \frac{\partial \mathcal{G}}{\partial M_k}(z; M).$$

The last term coincides with the right member in (7.47). Thus substituting into (7.47) we get

$$\frac{\partial f}{\partial M_k} = \{f, \mathcal{H}_k\},$$

as desired. $\square$

By (7.34), the real part of the potential $\mathcal{W}$ has on $\partial \Omega$ the explicit form $w(z) = \frac{1}{2}|z|^2$. Similarly, the imaginary part $\operatorname{Im} W$ has a simple geometric interpretation on $\partial \Omega$. In fact, fix a point $z_0 \in \partial \Omega$ and consider points $z \in \partial \Omega$ close to z_0. Let $D = D(z)$ be the region bounded by the two radii from the origin to the points z_0 and z together with the part γ of $\partial \Omega$ from z_0 to z. We may assume that Ω is star-shaped with respect to the origin, for simplicity. Also, let $A = A(z)$ denote the area of D.

We make $\mathcal{W}$ uniquely determined in a neighborhood z_0 by setting $\operatorname{Im} \mathcal{W}(z_0) = 0$. Noting that the differential $\bar{z}dz - zd\bar{z}$ vanishes along any radius from the origin we then get

$$\operatorname{Im} \mathcal{W}(z) = \operatorname{Im} \int_\gamma d\mathcal{W} = \operatorname{Im} \int_\gamma dS(z)dz = \operatorname{Im} \int_\gamma \bar{z}dz$$

$$= \frac{1}{2i} \int_\gamma (\bar{z}dz - zd\bar{z}) = \frac{1}{2i} \int_{\partial D} (\bar{z}dz - zd\bar{z}) = \frac{1}{i} \int_D d\bar{z}dz = 2A(z).$$

In conclusion, the potential $\mathcal{W}$ is given on $\partial \Omega$ by

$$\mathcal{W}(z) = \frac{1}{2}|z|^2 + 2iA(z) \quad (z \in \partial \Omega).$$

Finally we wish to write up, for the classical Laplacian growth, a formulation within a traditional Hamiltonian framework (with a real-valued Hamiltonian function). In the moment coordinates the situation is really trivial (in view of the explicit solution (1.34)), but once the equations are written up one may make

transformations to other coordinates (for example the coefficients of the conformal map (7.12)). One possibility (out of many) is to choose the phase space to
be $\partial\mathbb{D} \times \mathcal{M}_{2N+1} \subset \partial\mathbb{D} \times \mathbb{R}^{2N+1}$, with real coordinates (θ, M_0, Re M_1, Im $M_1 \ldots$,
Re M_N, Im M_N), where θ parametrizes $\partial\mathbb{D}$ by $e^{i\theta} \in \partial\mathbb{D}$, and symplectic form

$$
\omega = \frac{1}{2}d\theta \wedge dM_0 + d\text{Re}\,M_1 \wedge d\text{Im}\,M_1 + \cdots + d\text{Re}\,M_N \wedge d\text{Im}\,M_N
$$

$$
= \frac{1}{2i}\left[id\theta \wedge dM_0 + d\bar{M}_1 \wedge dM_1 + \cdots + d\bar{M}_N \wedge dM_N\right].
$$

The Hamiltonian function will then be, in case of a source of strength Q,

$$
H(\theta, M_0, \text{Re}\,M_1, \ldots, \text{Im}\,M_N) = \frac{Q}{2\pi}\theta.
$$

The Hamilton equations are, generally speaking (see [26]),

$$
dH = \omega(\cdot, \xi), \tag{7.48}
$$

where $\xi = \frac{d}{dt}$ is the velocity vector for the evolution, a vector in the tangent space
of the phase space. In our case we have (using dot for time derivative)

$$
\xi = \dot{\theta}\frac{\partial}{\partial\theta} + \dot{M}_0\frac{\partial}{\partial M_0} + \text{Re}\,\dot{M}_1\frac{\partial}{\partial\text{Re}\,M_1} + \cdots + \text{Im}\,\dot{M}_N\frac{\partial}{\partial\text{Im}\,M_N},
$$

giving in (7.48)

$$
\dot{\theta} = 0, \quad \dot{M}_0 = \frac{Q}{\pi},
$$

$$
\text{Re}\,\dot{M}_j = \text{Im}\,\dot{M}_j = 0 \quad (1 \leq j \leq N),
$$

as expected (cf. (1.34) and remarks after (1.39)).

It may be interesting to observe that the Liouville form, ω^{N+1}, for the above
symplectic structure equals, up to a constant factor, the wedge product of $d\theta$ with
the volume form in (7.13).

7.5   Exterior domains

Much of the work on Laplacian growth and Hele-Shaw flow concerns the case that
the active domain (i.e., the fluid domain, or the domain where the harmonic function p lives) is a domain Ω in the Riemann sphere containing the point of infinity.
In particular, the work of M. Mineev-Weinstein, P. Wiegmann, A. Zabrodin et al.,
referred to earlier, is of this kind. Of course the formulas will be quite similar to
those for the bounded case, but let us still spell out some of the details.

So assume that Ω is a domain in the Riemann sphere such that $\infty \in \Omega$ and
such that $\partial\Omega$ is an analytic curve (in particular, Ω is simply connected). Then

$D = \mathbb{C} \setminus \overline{\Omega}$ is a bounded domain and we may still define the moments by (7.11). Thus the moments for D and Ω are by definition the same, and expressed in terms of Ω they are

$$M_n = -\frac{1}{2\pi i} \int_{\partial\Omega} z^n \bar{z} dz,$$

with $\partial\Omega$ oriented as the boundary of Ω. When we think of the M_n ($n \in \mathbb{Z}$) as moments for Ω it is more natural to consider $M_0, M_{-1}, M_{-2}, \ldots$ as independent variables and to view M_j, $j > 0$, as functions of these. Then the integrability identities (7.32) become

$$\frac{1}{k}\frac{\partial M_k}{\partial M_{-j}} = \frac{1}{j}\frac{\partial M_j}{\partial M_{-k}} \quad (k, j \geq 1), \tag{7.49}$$

$$\frac{1}{k}M_k = \frac{\partial \log \tau}{\partial M_{-k}} \quad (k \geq 1), \tag{7.50}$$

where now

$$\log \tau = \frac{1}{\pi^2} \int_D \int_D \log \frac{1}{|z - \zeta|} d\sigma_z d\sigma_\zeta. \tag{7.51}$$

Independent on whether the interior or exterior of the curve $\partial\Omega$ is considered to be the active domain, the Schwarz function $S(z)$ and its potential $\mathcal{W}(z)$ will always be the same. This gives a way to identify notations between different sources. For example, in [580, 393, 370, 327] the potential $\mathcal{W}(z) = \mathcal{W}(z; M)$, in the mentioned sources denoted $\Omega(z) = \Omega(z, t)$ with t shorthand for "(generalized) times", is expanded as

$$\mathcal{W}(z) = \sum_{k=1}^{\infty} t_k z^k + t \log z - \frac{1}{2}v_0(t) - \sum_{k=1}^{\infty} \frac{v_k(t)}{k} z^{-k}.$$

Comparing with our expansion (7.33) this gives the following list for translating notations, our notations appearing to the right:

$C_k = \pi M_{-k}$ 'moments'   ($k \in \mathbb{Z}$),

$v_0 = -2C(M) = \dfrac{2}{\pi} \displaystyle\int_D \log|z| d\sigma_z$ 'logarithmic moment',

$v_k = M_k$ 'interior moments'   ($k \geq 1$),

$t_0 = M_0$ 'ordinary time',

$t_k = \dfrac{1}{k}M_{-k}$ 'generalized times'   ($k \geq 1$).

The word 'times' for the exterior moments refer to their role as independent variables in a certain dispersionless 2D Toda hierarchy (compare related discussions in Chapters 6 and 8). The integrability equations become, for example,

$$\frac{\partial C_{-k}}{\partial t_j} = \frac{\partial C_{-j}}{\partial t_k} \quad (k, j \geq 1).$$

Chapter 8

Shape Evolution and Integrability

The Hele-Shaw (strong or classical) advancing evolution in the plane is an example of evolution in the infinite-dimensional manifold of smooth shapes. It is a typical 'field' problem, i.e., given an initial shape, the further evolution is well defined at least for a short time. By *shape* we understand a simple closed curve in the complex plane dividing it into two simply connected domains. The study of 2D shapes is one of the central problems in the field of applied sciences. A program of such study and its importance was summarized by Mumford at ICM 2002 in Beijing [397]. The harmonic (Richardson's) moments of the Hele-Shaw evolution (or of the Laplacian growth) are conserved (see (1.34)) under this evolution and serve as the evolution parameters or generalized times. The infinite number of evolution parameters constitutes the infinite number of degrees of freedom of the system, and clearly suggests to apply field theory methods as a natural tool of study, which logically lead to integrable systems, the dispersionless Toda hierarchies, in particular.

Another group of models, in which the evolution is governed by an infinite number of parameters, can be observed in controllable dynamical systems, where the infinite number of degrees of freedom follows from the infinite number of driving terms. Surprisingly, a similar algebraic structural background appears for this group. We develop this viewpoint in the present section.

One of the general approaches to the homotopic evolution of shapes starting from a canonical shape, the unit disk in our case, was provided by Löwner and Kufarev [328, 362, 446]. A shape evolution is described by a time-dependent conformal parametric map from the canonical domain onto the domain bounded by the shape at any fixed instant. In fact, these one-parameter conformal maps satisfy the Löwner–Kufarev differential equation, or an infinite-dimensional controllable system, for which the infinite number of conservation laws is given by the *Virasoro generators* in their covariant form.

We start with the further comparison of the Hele-Shaw evolution and the Löwner and Kufarev evolution, which having certain similarities differ quite much both in analytic and field theoretical aspects.

Let us consider the solutions to the Polubarinova–Galin equation (1.22) in the case of injection. The fluid is advancing in the normal direction and the solutions form subordination chains of conformal univalent maps (and corresponding chain of hyperbolic univalent domains). This particular case of subordination chains has been considered in the preceding chapters. The existence theorem makes it natural to assume that at least the initial domain Ω_0 of the Hele-Shaw dynamics $\Omega(t)$ is bounded by a smooth analytic curve. The equation (1.22), being extended inside the unit disk $\mathbb{D}$ with the help of the Schwarz–Poisson kernel, looks like (1.24), which we rewrite here again

$$\dot{f}(\zeta, t) = \zeta f'(\zeta, t) P(\zeta, t), \quad \zeta \in \mathbb{D}, \tag{8.1}$$

where

$$P(\zeta, t) = \frac{Q}{4\pi^2} \int\limits_0^{2\pi} \frac{1}{|f'(e^{i\theta}, t)|^2} \frac{e^{i\theta} + \zeta}{e^{i\theta} - \zeta} d\theta, \quad \zeta \in \mathbb{D}. \tag{8.2}$$

Equation (8.1) is a quite complex integro-differential equation, which was discussed in Chapters 1 and 2. In contrast, the Löwner–Kufarev equation for subordinating domains is the same as (8.1) but the *feed-back term* $P(\zeta, t)$ is changed to an arbitrary holomorphic function $p(\zeta, t)$ with $\operatorname{Re} p(\zeta, t) > 0$ in the unit disk $\zeta \in \mathbb{D}$. In this case, the function p becomes a *driving term* and the equation (8.1) becomes a linear equation but with an infinite-dimensional control.

The results of this section were obtained in joint works with Irina Markina [366, 367]. We are also thankful to Roland Friedrich for inspiring conversations, and some ideas used here, first of all, related to Grassmannians, were independently used in [196]. We will discuss in brief in the next chapter the stochastic approach to the Löwner theory which received much attention during recent decades, see, e.g., [349, 350, 351, 518].

Let us also mention here an interesting paper by Gibbons and Tsarëv [206], the first who noticed that the chordal Löwner equation plays an essential role in classification of reductions of the Benney equations. We discuss this at the end of the chapter. Later Takebe, Teo, and Zabrodin [537] showed that the chordal and radial Löwner PDE served as consistency conditions for one-variable reductions of dispersionless KP and Toda hierarchies, respectively. Although our approach is different, it is remarkable and common that the Löwner–Kufarev equation leads to relations with the KP, dKP and Toda hierarchies.

8.1   Löwner–Kufarev evolution

Recall that we denote the class of normalized univalent maps $f \colon \mathbb{D} \to \mathbb{C}$, $f(z) = z + a_2 z^2 + \cdots$ by S. A t-parameter family $\Omega(t)$ of simply connected hyperbolic univalent domains forms a *Löwner subordination chain* in the complex plane $\mathbb{C}$, for $0 \leq$

$t < \tau$ (where τ may be ∞), if $\Omega(t) \subseteq \Omega(s)$, whenever $t < s$, and the family is continuous in the Carathéodory sense. We suppose that the origin is a point of $\Omega(0)$.

A Löwner subordination chain $\Omega(t)$ is described by a t-dependent family of conformal maps $z = f(\zeta, t)$ from $\mathbb{D}$ onto $\Omega(t)$, normalized by $f(\zeta, t) = a_1(t)\zeta + a_2(t)\zeta^2 + \cdots$, $a_1(t) > 0$, $\dot{a}_1(t) > 0$. Pommerenke [446] described governing evolution equations in partial and ordinary derivatives, known now as the Löwner–Kufarev equations.

One can normalize the growth of evolution of a subordination chain by the conformal radius of $\Omega(t)$ with respect to the origin setting $a_1(t) = e^t$.

We say that the function p is from the Carathéodory class if it is analytic in $\mathbb{D}$, normalized as $p(\zeta) = 1 + p_1\zeta + p_2\zeta^2 + \cdots$, $\zeta \in \mathbb{D}$, and such that $\operatorname{Re} p(\zeta) > 0$ in $\mathbb{D}$. Given a Löwner subordination chain of domains $\Omega(t)$ defined for $t \in [0, \tau)$, there exists a function $p(\zeta, t)$, measurable in $t \in [0, \tau)$ for any fixed $z \in \mathbb{D}$, and from the Carathéodory class for almost all $t \in [0, \tau)$, such that the conformal mapping $f \colon \mathbb{D} \to \Omega(t)$ solves the equation

$$\frac{\partial f(\zeta, t)}{\partial t} = \zeta \frac{\partial f(\zeta, t)}{\partial \zeta} p(\zeta, t), \tag{8.3}$$

for $\zeta \in \mathbb{D}$ and for almost all $t \in [0, \tau)$. Equation (8.3) is called the Löwner–Kufarev equation due to two seminal papers: by Löwner [362] who considered the case when

$$p(\zeta, t) = \frac{e^{iu(t)} + \zeta}{e^{iu(t)} - \zeta}, \tag{8.4}$$

where $u(t)$ is a continuous function regarding to $t \in [0, \tau)$, and by Kufarev [328] who proved differentiability of f with respect to t for all ζ in the case of general p from the Carathéodory class.

Let us consider a reverse process. We are given an initial domain $\Omega(0) \equiv \Omega_0$ (and therefore, the initial mapping $f(\zeta, 0) \equiv f_0(\zeta)$), and an analytic function $p(\zeta, t)$ of positive real part normalized by $p(\zeta, t) = 1 + p_1\zeta + \cdots$. Let us solve the equation (8.3) and ask ourselves, whether the solution $f(\zeta, t)$ defines a subordination chain of simply connected univalent domains $f(\mathbb{D}, t)$. The initial condition $f(\zeta, 0) = f_0(\zeta)$ is not given on the characteristics of the partial differential equation (8.3), hence the solution exists and is unique but not necessarily univalent. Assuming s as a parameter along the characteristics we have

$$\frac{dt}{ds} = 1, \quad \frac{d\zeta}{ds} = -\zeta p(\zeta, t), \quad \frac{df}{ds} = 0,$$

with the initial conditions $t(0) = 0$, $\zeta(0) = z$, $f(\zeta, 0) = f_0(\zeta)$, where z is in $\mathbb{D}$. Obviously, we can assume $t = s$. Observe that the domain of ζ is the entire unit disk. However, the solutions to the second equation of the characteristic system range within the unit disk but do not fill it. Therefore, introducing another letter

w (in order to distinguish the function $w(z,t)$ from the variable ζ) we arrive at the Cauchy problem for the Löwner–Kufarev equation in ordinary derivatives

$$\frac{dw}{dt} = -wp(w,t), \qquad (8.5)$$

for a function $\zeta = w(z,t)$ with the initial condition $w(z,0) = z$. The equation (8.5) is a non-trivial characteristic equation for (8.3). Unfortunately, this approach requires the extension of $f_0(w^{-1}(\zeta,t))$ into the whole $\mathbb{D}$ (here w^{-1} means the inverse function in ζ) because the solution to (8.3) is the function $f(\zeta,t)$ given as $f_0(w^{-1}(\zeta,t))$, where $\zeta = w(z,s)$ is a solution of the initial value problem for the characteristic equation (8.5) that maps $\mathbb{D}$ into $\mathbb{D}$. Therefore, the solution of the initial value problem for the equation (8.3) may be non-univalent.

On the other hand, solutions to equation (8.5) are holomorphic univalent functions $w(z,t) = e^{-t}(z + a_2(t)z^2 + \cdots)$ in the unit disk that map $\mathbb{D}$ into itself. Every function f from the class S can be represented by the limit

$$f(z) = \lim_{t \to \infty} e^t w(z,t), \qquad (8.6)$$

where $w(z,t)$ is a solution to (8.5) with some function $p(z,t)$ of positive real part for almost all $t \geq 0$ (see [446, pages 159–163]). Each function $p(z,t)$ generates a unique function from the class S. The reciprocal statement is not true. In general, a function $f \in S$ can be obtained using different functions $p(\cdot,t)$.

Now we are ready to formulate the condition of univalence of the solution to the equation (8.3) in terms of the limiting function (8.6), which can be obtained by a combination of known results of [446].

Theorem 8.1.1 ([446], [455]). *Given a function $p(\zeta,t)$ of positive real part normalized by $p(\zeta,t) = 1 + p_1\zeta + \cdots$, the solution to the equation (8.3) is unique, analytic and univalent with respect to ζ for almost all $t \geq 0$, if and only if, the initial condition $f_0(\zeta)$ is taken in the form (8.6), where the function $w(\zeta,t)$ is the solution to the equation (8.5) with the same driving function p.*

Geometrically, the Löwner–Kufarev evolution readily corresponds to the normal motion of the boundary $\partial\Omega(t)$ similarly to the Hele-Shaw evolution. Indeed, supposing an analytic boundary $\partial\Omega(t)$ the normal vector in the outward direction is $\mathbf{n} = \zeta f'(\zeta,t)/|f'(\zeta,t)|$, $|\zeta| = 1$, and defining $p(\zeta,t)$ as $(\dot{f}(\zeta,t))/(\zeta f'(\zeta,t))$ the normal velocity V_n is given by

$$V_n = \operatorname{Re}\left(\frac{\dot{f}(\zeta,t)\overline{\zeta f'(\zeta,t)}}{|f'(\zeta,t)|}\right) = \operatorname{Re}\left[p(\zeta,t)|f'(\zeta,t)|\right], \qquad |\zeta| = 1,$$

and is positive. Therefore, $\operatorname{Re} p(\zeta,t) > 0$ as stated in (8.1). Of course, the general case of nonanalytic boundary requires finer argumentation.

Since we will work with the Löwner subordination chains and with Löwner–Kufarev evolution with smooth boundaries, we will need the following lemmas:

Lemma 8.1.1. *Let the function $w(z,t)$ be a solution to the Cauchy problem (8.5). If the driving function $p(\cdot,t)$, being from the Carathéodory class for almost all $t \geq 0$, is C^∞ smooth in the closure $\overline{\mathbb{D}}$ of the unit disk $\mathbb{D}$ and summable with respect to t, then the boundaries of the domains $\Omega(t) = w(\mathbb{D},t) \subset \mathbb{D}$ are smooth for all t and $w(\cdot,t)$ extended to S^1 is injective on S^1.*

Proof. Observe that the continuous and differentiable dependence of the solution of a differential equation $\dot{x} = F(t,x)$ on the initial condition $x(0) = x_0$ is a classical problem. One can refer, e.g., to [574] in order to assure that summability of $F(\cdot,x)$ regarding to t for each fixed x and continuous differentiability (C^1 with respect to x for almost all t) imply that the solution $x(t,x_0)$ exists, is unique, and is C^1 with respect to x_0. In our case, the solution to (8.5) exists, is unique and analytic in $\mathbb{D}$, and moreover, C^1 on its boundary S^1. Let us differentiate (8.5) inside the unit disk $\mathbb{D}$ with respect to z and write

$$\log w' = -\int_0^t (p(w(z,\tau),\tau) + w(z,\tau)p'(w(z,\tau),\tau))d\tau,$$

choosing the branch of the logarithm such as $\log w'(0,t) = -t$. This equality is extendable onto S^1 because the right-hand side is, and therefore, $w' \in C^1(S^1)$ and $w \in C^2(S^1)$. We continue analogously and write the formula

$$w'' = -w'\int_0^t (2w'(z,\tau)p'(w(z,\tau),\tau) + w(z,\tau)w'(z,\tau)p''(w(z,\tau),\tau))d\tau,$$

which guarantees that $w \in C^3(S^1)$. Finally, we come to the conclusion that w is C^∞ on S^1. $\qquad\square$

Let us denote by $f(z,\infty)$ the final point of the trajectory $f(z,t) = e^t w(z,t)$, $t \in [0,\infty)$, where $w(z,t)$ is a solution to the Cauchy problem (8.5) with the driving function $p(z,t)$ satisfying the conditions of Lemma 8.1.1. Then $f(z,t) \in \mathcal{F}_0$ for all $t \in [0,\infty)$ but not necessarily for $t = \infty$. One can formulate a stronger reciprocal statement.

Lemma 8.1.2. *With the above notations let $f(z) \in \mathcal{F}_0$. Then there exists a function $p(\cdot,t)$ from the Carathéodory class for almost all $t \geq 0$, and C^∞ smooth in $\overline{\mathbb{D}}$, such that $f(z) = \lim_{t\to\infty} f(z,t)$ is the final point of the Löwner–Kufarev trajectory with the driving term $p(z,t)$.*

Proof. Indeed, the domain $\Omega^+ = f(\mathbb{D})$ has a complement Ω^- which is a simply connected domain with infinity ∞ as an internal point of Ω^- and $\partial\Omega^+ = \partial\Omega^-$. Let us construct a subordination chain $\Omega^+(t)$ such that $\partial\Omega^+(t)$ is a level line of the Green function of the domain Ω^- with a singularity at ∞, and such that the conformal radius of $\Omega^+(t)$ with respect to the origin is equal to e^t. This can be always achieved, see [446]. Then we can construct a one-parameter subordination chain of univalent maps $F(z,t) = e^t(z + \cdots)$, $F(\cdot,t): \mathbb{D} \to \Omega^+(t)$ that exists for the time interval $[0,\infty)$, $f(z) = F_0(z) = F(z,0)$ and $f(\mathbb{D}) = \Omega^+ = \Omega^+(0)$, and such that $\Omega^+(\infty) = \mathbb{C}$. Set up the function $p(z,t) = \dot{F}/zF'$, where $\dot{F}$ and F' are the real t-derivative and the complex z-derivative respectively. It is obviously smooth on the boundary and belongs to the Carathéodory class. The function $w(z,t) = F(F^{-1}(z,t),0)$ is defined in the whole unit disk (as an analytic continuation from $F^{-1}(F_0(z),t) \subset \mathbb{D}$), satisfies the Löwner–Kufarev equation (8.5), and $f(z,t) = e^t w(z,t)$ has the limit $f(z) = f(z,\infty)$. The latter statement can be found in [446, 454].   □

Concluding this section we remark that the Löwner and Löwner–Kufarev equations are described in several monographs [9, 17, 103, 156, 213, 223, 259, 446, 482].

8.2 Structures of Vir , Diff S^1, and Diff S^1/S^1

8.2.1 Witt and Virasoro algebras

The complex *Witt algebra* is the Lie algebra of holomorphic vector fields defined on $\mathbb{C}^* = \mathbb{C} \setminus \{0\}$ acting by derivation over the ring of Laurent polynomials $\mathbb{C}[z, z^{-1}]$. It is spanned by the basis $L_n = z^{n+1}\frac{\partial}{\partial z}$, $n \in \mathbb{Z}$. The Lie bracket of two basis vector fields is given by the commutator $[L_n, L_m] = (m - n)L_{n+m}$. Its central extension is the complex *Virasoro algebra* $\mathfrak{vir}_\mathbb{C}$ with the central element c commuting with all L_n, $[L_n, c] = 0$, and with the Virasoro commutation relation

$$[L_n, L_m] = (m - n)L_{n+m} + \frac{c}{12}n(n^2 - 1)\delta_{n,-m}, \quad n, m \in \mathbb{Z},$$

where $c \in \mathbb{C}$ is the central charge denoted by the same character. These algebras play important roles in conformal field theory. In order to construct their representations one can use an analytic realization.

8.2.2 Group of diffeomorphisms

Let us denote by Diff S^1 the group of orientation preserving C^∞ diffeomorphisms of the unit circle S^1, where the group operation is given by the superposition of diffeomorphisms, the identity element of the group is the identity map on the circle, and the inverse element is the inverse diffeomorphism. Topologically the group Diff S^1 is an open subset of the space of smooth functions on the unit circle $C^\infty(S^1 \to S^1)$, endowed with the C^∞-topology. This allows us to consider the group Diff S^1 as a Lie–Fréchet group. The corresponding Lie–Fréchet algebra $\mathfrak{diff}\, S^1$ is identified with the tangent space $T_{\mathrm{id}}\mathrm{Diff}\, S^1$ at the identity id, and it inherits the Fréchet topology from $C^\infty(S^1 \to S^1)$. In its turn $T_{\mathrm{id}}\mathrm{Diff}\, S^1$ can be thought of as the set of all velocity vectors of smooth curves at time zero passing through id. Every such velocity vector is just a smooth real vector field on S^1. Denote by Vect $S^1 = \{\phi = \phi(\theta)\frac{d}{d\theta} \mid \phi \in C^\infty(S^1 \to \mathbb{R})\}$ the space of smooth real vector fields on the circle. This construction allows us to identify the Lie–Fréchet algebra $\mathfrak{diff}\, S^1$ of Diff S^1 with the space Vect S^1 equipped with the Lie brackets $[\phi_1(\theta)\frac{d}{d\theta}, \phi_2(\theta)\frac{d}{d\theta}]$, see, e.g., [389].

The Virasoro–Bott group Vir is the central extension of the group Diff S^1 by the group of real numbers $\mathbb{R}$. This central extension is given by the Bott continuous cocycle [61], which is a map Diff $S^1 \times$ Diff $S^1 \to S^1$ of the form

$$(\varphi_1, \varphi_2) \mapsto \frac{1}{2}\int_{S^1} \log(\varphi_1 \circ \varphi_2)' d\log \varphi_2'.$$

The Lie algebra $\mathfrak{vir}$ for Vir is called the (real) Virasoro algebra and it is given by the central extension of the Lie–Fréchet algebra Vect S^1 by the algebra of real numbers. The central extension consists of unique nontrivial modulo isomorphisms and is given by the Gelfand–Fuchs 2-cocycle [204]

$$\omega(\phi_1, \phi_2) = \int_{S_1} \phi_1'(\theta)\phi_2''(\theta)d\theta.$$

Both groups Diff S^1 and Vir are modeled over a real Fréchet space.

Let us denote by [id] the equivalence class in Diff S^1/S^1 of the identity element id $\in$ Diff S^1. Then $T_{[\mathrm{id}]}$Diff S^1/S^1 is associated with the quotient Vect $_0 S^1 =$ Vect S^1/const of the algebra Vect S^1 by the constant vector fields and can be realized as the space of vector fields $\phi(\theta)\frac{d}{d\theta}$ from Vect S^1 with vanishing mean value over S^1. All constant vector fields form the equivalence class [0].

8.2.3 CR and complex structures

In Section 8.3 we shall describe relations between the groups Vir, Diff S^1, the homogeneous manifold Diff S^1/S^1 and different spaces of univalent functions. The algebraic objects are essentially real, meanwhile the spaces of univalent functions carry natural complex structures as well as the algebraic definition of the Witt and Virasoro algebras in Subsection 8.2.1 considers vector fields over the field of complex numbers. Therefore, we need to complexify the real objects in order to present these relations. Structures and mappings on infinite-dimensional manifolds are more general than for finite-dimensional ones, however, being restricted to the latter they coincide with the standard ones. For the completeness we give some necessary definitions mostly based on [64, 358].

Given a smooth manifold $\mathcal{M}$, we consider the tangent space $T_p\mathcal{M}$ at each point $p \in \mathcal{M}$ as a real vector space. After tensoring with $\mathbb{C}$ and splitting $T_p\mathcal{M} \otimes \mathbb{C} = T_p^{(1,0)}\mathcal{M} \oplus T_p^{(0,1)}\mathcal{M}$, we form the holomorphic $T^{(1,0)}\mathcal{M}$ and antiholomorphic $T^{(0,1)}\mathcal{M}$ tangent bundles. The pair $(\mathcal{M}, T^{(1,0)}\mathcal{M})$ is an almost complex manifold which becomes complex in the integrable case meaning that any commutator of vector fields from $T^{(1,0)}\mathcal{M}$ remains in $T^{(1,0)}\mathcal{M}$, and similarly, the commutators of vector fields from $T^{(0,1)}\mathcal{M}$ remain in $T^{(0,1)}\mathcal{M}$.

A Lie group $\mathbb{G}$ with a neutral element e and with a Lie algebra $\mathfrak{g}$ possesses a left invariant complex structure $(\mathbb{G}, \mathfrak{g}^{(1,0)})$ if one can construct a complexification $\mathfrak{g}_{\mathbb{C}} = (T_e\mathbb{G})_{\mathbb{C}}$ of the Lie algebra $\mathfrak{g}$, such that the decomposition $\mathfrak{g}_{\mathbb{C}} = \mathfrak{g}^{(1,0)} \oplus \mathfrak{g}^{(0,1)}$ is integrable, that is equivalent to saying that $\mathfrak{g}^{(1,0)}$ is a subalgebra.

Let us recall the definition of the Cauchy–Riemann (CR) structure on a manifold $\mathcal{N}$. Given a smooth manifold $\mathcal{N}$ and its complexified tangent bundle $T\mathcal{N} \otimes \mathbb{C}$ we find a complex corank one subbundle H of $T\mathcal{N} \otimes \mathbb{C}$. The splitting $H = H^{(1,0)} \oplus H^{(0,1)}$ defines an almost CR structure. If it is integrable, then the pair $(\mathcal{N}, H^{(1,0)})$ is called a CR manifold. Roughly speaking the holomorphic part of a CR structure represents a maximal subbundle of the real tangent bundle that admits a complex structure. The left-invariant CR-structure $(\mathbb{G}, \mathfrak{h}^{(1,0)})$ is defined similarly to the left-invariant complex structure above.

As an example of CR manifold we can consider an embedded real hypersurface (that is an embedded real corank 1 submanifold) into a complex manifold. Namely, let $\mathcal{N}$ be a real hypersurface of the complex manifold $(\mathcal{M}, T^{(1,0)}\mathcal{M})$. Then the CR manifold $(\mathcal{N}, H^{(1,0)})$ is defined by setting $H^{(1,0)} = T^{(1,0)}\mathcal{M}\big|_{\mathcal{N}} \bigcap(T\mathcal{N} \otimes \mathbb{C})$.

A CR manifold $(\mathcal{N}, H^{(1,0)})$ is called pseudoconvex if $[X, \bar{X}] \notin H^{(1,0)} \oplus H^{(0,1)}$ for any non-vanishing vector field $X \in H^{(1,0)}$.

A smooth mapping F from a complex manifold $(\mathcal{M}_1, T^{(1,0)}\mathcal{M}_1)$ to a complex manifold $(\mathcal{M}_2, T^{(1,0)}\mathcal{M}_2)$ is called holomorphic if the holomorphic part ∂F of its differential $dF = \partial F + \bar{\partial} F$ is the mapping $\partial F : T^{(1,0)}\mathcal{M}_1 \to T^{(1,0)}\mathcal{M}_2$ and $\bar{\partial} F = 0$. The problem of solving the equation $\bar{\partial} F = 0$ is quite difficult. Some results in this direction have been found, e.g., [460].

Analogously, a smooth mapping F from a CR manifold $(\mathcal{N}_1, H_1^{(1,0)})$ to a CR manifold $(\mathcal{N}_2, H_2^{(1,0)})$ is called CR if its holomorphic differential is a map $\partial F : H_1^{(1,0)} \to H_2^{(1,0)}$ and $\bar{\partial} F = 0$.

Given a non-trivial representative ϕ of the equivalence class $[\phi]$ of Vect $_0S^1$,

$$\phi(\theta) = \sum_{n=1}^{\infty} a_n \cos n\theta + b_n \sin n\theta,$$

let us define an almost complex structure J by the operator

$$J(\phi)(\theta) = \sum_{n=1}^{\infty} -a_n \sin n\theta + b_n \cos n\theta.$$

Then $J^2 = -id$. On Vect $_{0\mathbb{C}} :=$ Vect $_0S^1 \otimes \mathbb{C}$, the operator J diagonalizes and we have the isomorphism

$$\text{Vect}_{0}S^1 \ni \phi \leftrightarrow v := \frac{1}{2}(\phi - iJ(\phi)) = \sum_{n=1}^{\infty}(a_n - ib_n)e^{in\theta} \in H^{(1,0)}$$

$$:= (\text{Vect}_{0}S^1 \otimes \mathbb{C})^{(1,0)},$$

and the latter series extends into the unit disk as a holomorphic function. So Diff $_{\mathbb{C}}S^1/S^1 = (\text{Diff } S^1/S^1, H^{(1,0)})$ becomes a complex manifold and (Diff S^1, $H^{(1,0)}$) becomes a CR manifold where $H^{(1,0)}$ is isomorphic to Vect $_0S^1$. Thus, the group Diff S^1 possesses the left-invariant CR-structure (Diff $S^1, H^{(1,0)}$), and C^* forms a Cartan subalgebra of Vect $S^1 \otimes \mathbb{C} = (H^{(1,0)} \oplus H^{(0,1)}) \oplus \mathbb{C}^*$. Taking the complex Fourier basis $v_n = e^{in\theta}\frac{d}{d\theta}$, $n \in \mathbb{Z}$, in Vect $S^1 \otimes \mathbb{C}$ we arrive at the Witt commutation relations $[v_n, v_m] = (m - n)v_{n+m}$, where the commutators $[v_n, v_m]$ remain in $H^{(1,0)}$ for $n, m > 0$ and in $H^{(0,1)}$ for $n, m < 0$, however the Lie hull $\text{Lie}(H^{(1,0)}, H^{(0,1)}) \not\subset H^{(1,0)} \oplus H^{(0,1)}$.

8.3 Relations between Vir, Diff S^1, and Diff S^1/S^1 and spaces of univalent functions

Let us introduce necessary classes of univalent functions in order to formulate our main statements. Let $\mathcal{A}_0$ and $\widetilde{\mathcal{A}}_0$ denote the classes of holomorphic functions in the unit disk $\mathbb{D}$ defined by

$$\mathcal{A}_0 = \{f \in C^{\infty}(\overline{\mathbb{D}}) \mid f \in \text{Hol}(\mathbb{D}), f(0) = 0\}, \quad \widetilde{\mathcal{A}}_0 = \{f \in \mathcal{A}_0 \mid f'(0) = 0\},$$

where $\overline{\mathbb{D}}$ is the closure of the unit disk $\mathbb{D}$. The classes $\mathcal{A}_0$ and $\widetilde{\mathcal{A}}_0$ are complex Fréchet vector spaces, where the topology is defined by the seminorms

$$\|f\|_m = \sup\{|f^{(m)}(z)| \mid z \in \overline{\mathbb{D}}\},$$

which is equivalent to the uniform convergence of all derivatives in $\overline{\mathbb{D}}$. Notice that both $\mathcal{A}_0$ and $\widetilde{\mathcal{A}}_0$ can be considered as complex manifolds where the real tangent space is naturally isomorphic to the holomorphic part of the splitting. Then we define

$$\mathcal{F} = \{f \in \mathcal{A}_0 \mid f \text{ is univalent in } \mathbb{D}, \text{ injective and smooth on } \partial\mathbb{D}\}.$$

Geometrically, class $\mathcal{F}$ defines all differentiable embeddings of the closed disk $\overline{\mathbb{D}}$ to $\mathbb{C}$ and analytically it is represented by functions $f = cz(1 + \sum_{n=1}^{\infty} c_n z^n)$, $c, c_n \in \mathbb{C}$. As a subset of $\mathcal{A}_0$ the space of univalent functions $\mathcal{F}$ forms an open subset inheriting the Fréchet topology of complex vector space $\mathcal{A}_0$. Next we consider the class

$$\mathcal{F}_1 = \{f \in \mathcal{F} \mid |f'(0)| = 1\},$$

whose elements can be written as $f = e^{i\phi}z(1 + \sum_{n=1}^{\infty} c_n z^n)$, $\phi \in \mathbb{R}$ mod 2π. The set $\mathcal{F}_1$ is the pseudo-convex surface of real codimension 1 in the complex open set $\mathcal{F} \subset \mathcal{A}_0$.

The last class of functions is

$$\mathcal{F}_0 = \{f \in \mathcal{F} \mid f'(0) = 1\}.$$

The elements of this class have the form $f = z(1 + \sum_{n=1}^{\infty} c_n z^n)$. It is obvious that $\mathcal{F}_0$ can be considered both as the quotient $\mathcal{F}_1/S^1$ and as the quotient $\mathcal{F}/\mathbb{C}^*$, $\mathbb{C}^* = \mathbb{C} \setminus \{0\}$. In the latter case, $\mathcal{F}$ is the holomorphic trivial $\mathbb{C}^*$-principal bundle over the base space $\mathcal{F}_0$. Since the set $\mathcal{F}_0$ can be also considered as an open subset of the affine space $v + \widetilde{\mathcal{A}}_0$, where $v(z) = z$, the tangent space $T_f \mathcal{F}_0$ inherits the natural complex structure of complex vector space $\widetilde{\mathcal{A}}_0$ [12]. The tangent space $T_f \mathcal{F}_0$ with the induced complex structure from $\widetilde{\mathcal{A}}_0$ is isomorphic to the complex vector space $T_f^{(1,0)} \mathcal{F}_0$ of the complexification $T\mathcal{F}_0 \otimes \mathbb{C} = T^{(1,0)}\mathcal{F}_0 \oplus T^{(0,1)}\mathcal{F}_0$. Moreover, the affine coordinates can be introduced so that to every $f \in \mathcal{F}_0$, written in the form $f(z) = z(1 + \sum_{n=1}^{\infty} c_n z^n)$, there will correspond the sequence $\{c_n\}_{n=1}^{\infty}$.

Theorem 8.3.1 ([358]). *The Virasoro–Bott group* Vir *has a left invariant complex structure, and as a complex manifold* Vir$_{\mathbb{C}}$, *it is biholomorphic to* $\mathcal{F}$.

Theorem 8.3.2 ([358]). *The group* Diff S^1 *has a left invariant CR structure and with this CR structure it is isomorphic to the hypersurface* $\mathcal{F}_1$.

The last theorem concerns the homogeneous space Diff S^1/S^1, where S^1 is considered as a subgroup of Diff S^1. The group S^1 acts transversally to the CR structure of Diff S^1, leaving it invariant.

Theorem 8.3.3 ([314], [358]). *The homogeneous space* Diff S^1/S^1 *has a complex structure, and as a complex manifold* Diff $_{\mathbb{C}}S^1/S^1$, *is biholomorphic to* $\mathcal{F}_0$.

It can be shown that Diff S^1/S^1 admits not only complex but even Kählerian structure. The entire necessary background for construction of the theory of unitary representations of Diff S^1 is found in [12], [313]–[316].

It was mentioned that $\mathcal{F}$ is the holomorphic trivial $\mathbb{C}^*$-principal bundle over $\mathcal{F}_0$. In order to prove Theorem 8.3.1, Lempert showed [358] that the complexification Vir$_{\mathbb{C}}$ of the Virasoro–Bott group Vir is also a holomorphic trivial $\mathbb{C}^*$-principal bundle over Diff $_{\mathbb{C}}S^1/S^1$. This implies the existence of a biholomorphic map between $\mathcal{F}$ and Vir$_{\mathbb{C}}$.

We will assign the same character $\mathcal{F}_0$ to both, the class of univalent functions defined in the closure unit disk $\mathcal{F}_0(\overline{\mathbb{D}})$, and the class of functions restricted to the unit circle $\mathcal{F}_0(S^1)$. Obviously both classes are isomorphic.

The right action of the group Diff S^1 over the manifold Diff S^1/S^1 is well defined and it gives the right action Diff S^1 over the class $\mathcal{F}_0(S^1)$ due to Theorem 8.3.3, which is technically impossible to write explicitly because the Riemann mapping theorem gives no explicit formulas. However, it is possible [314] to write the infinitesimal generator making use of the Schaeffer and Spencer variation [508, page 32]

$$L[f, \phi](z) := \frac{f(z)^2}{2\pi} \int\limits_{S^1} \left(\frac{w f'(w)}{f(w)} \right)^2 \frac{\phi(w)\, dw}{w(f(w) - f(z))} \in T_f \mathcal{F}_0,$$

defined for $f \in \mathcal{F}_0$, $\phi \in \text{Vect}\, S^1$. It extends by linearity to a map $L[f, \cdot]$: Vect $_\mathbb{C} S^1 \to T_f \mathcal{F}_0 \otimes \mathbb{C} = T_f^{(1,0)} \mathcal{F}_0 \oplus T_f^{(0,1)} \mathcal{F}_0$. The variation $L[f, \cdot]$ defines the isomorphism of vector spaces $H^{(1,0)} \leftrightarrow T_f^{(1,0)} \mathcal{F}_0$, which is given explicitly by (8.7). At the same time $L[f, \cdot]$ defines an isomorphism of the Lie algebras $H^{(1,0)} \leftrightarrow T_f^{(1,0)} \mathcal{F}_0$, where $H^{(1,0)}$ is considered as a subalgebra of the Witt algebra Vect $_\mathbb{C} S^1$ and $T_f^{(1,0)} \mathcal{F}_0$ is endowed with the usual commutator of vectors. In order to obtain a homomorphism of the entire Witt algebra we extend $L[f, \cdot]$ to $H^{(1,0)} \oplus H^{(0,1)} \oplus \mathbb{C}^* \to T_f^{(1,0)} \mathcal{F}_0$.

Explicitly, this homomorphism $L[f, \cdot]$ is given by the residue calculus, see, e.g., [12], [316]. Taking the holomorphic part of the Fourier basis $v_k = -iz^k$, $k = 1, 2, \ldots$, for Vect $S^1 \otimes \mathbb{C}$, we obtain

$$L[f, v_k](z) = L_k[f](z) = z^{k+1} f'(z) \quad L_k[f] \in T_f^{(1,0)} \mathcal{F}_0, \tag{8.7}$$

and taking the antiholomorphic part of the basis $v_{-k} = -iz^{-k}$, $k = 1, 2, \ldots$, we obtain expressions for $L_{-k}[f] \in T_f^{(1,0)} \mathcal{F}_0$ that are rather difficult. The first two of them are

$$L_{-1}[f](z) = f'(z) - 2c_1 f(z) - 1, \quad L_{-2}[f](z) = \frac{f'(z)}{z} - \frac{1}{f(z)} - 3c_1 + (c_1^2 - 4c_2) f(z),$$

and others can be obtained by the commutation relations [12, 314]

$$[L_k, L_n] = (n - k) L_{k+n}, \quad k, n \in \mathbb{Z}. \tag{8.8}$$

The constant vector $v_0 = -i$ is mapped to $L_0[f](z) = z f'(z) - f(z)$. The vector fields L_k, $k \in \mathbb{Z}$ were obtained in [314] and received the name of Kirillov's vector fields, see also [12]. We have

$$T_{\text{id}}^{(1,0)} \mathcal{F}_0 = \text{span}\, \{L_0[\text{id}], L_1[\text{id}], L_2[\text{id}], \ldots\} = \text{span}\, \{z^2, z^3, \ldots\}.$$

Let us recall that id $\in \mathcal{F}_0$ is the image of an equivalence class of the identity diffeomorphism from Diff S^1/S^1.

Summarizing, the group Diff S^1 acts transitively on the homogeneous manifold Diff S^1/S^1 defining an action on the manifold $\mathcal{F}_0$. The infinitesimal generator of this action produces the left-invariant section of the tangent bundle $T\mathcal{F}_0$ by the Schaeffer–Spencer linear map. We get the isomorphism

$$T_f \mathcal{F}_0 \simeq T_f^{(1,0)} \mathcal{F}_0 = \text{span}\, \{L_1[f], L_2[f], \ldots\},$$

at a point $f \in \mathcal{F}_0$. The vector $L_0[f]$ is the image of the constant unit vector i under the Schaeffer–Spencer linear map at an arbitrary point $f \in \mathcal{F}_0$ with value 0 at $\mathrm{id} \in \mathcal{F}_0$.

The vector fields L_k, $k \in \mathbb{Z}$, at $f(z) = z\left(1 + \sum_{n=1}^{\infty} c_n z^n\right) \in \mathcal{F}_0$ can be written in the affine coordinates $\{c_n\}_{n=1}^{\infty}$ by making use of the isomorphism $z^{n+1} \mapsto \partial_n$, where $\partial_n = \frac{\partial}{\partial c_n}$, as the following first-order differential operators

$$L_k[f] = \partial_k + \sum_{n=1}^{\infty}(n+1)c_n\partial_{k+n}, \quad k > 0,$$

$$L_0[f] = \sum_{n=1}^{\infty} nc_n\partial_n,$$

$$L_{-1}[f] = \sum_{n=1}^{\infty}\left((n+2)c_{n+1} - 2c_1c_n\right)\partial_n, \tag{8.9}$$

$$L_{-2}[f] = \sum_{n=1}^{\infty}\left((n+3)c_{n+2} + (c_1^2 - 4c_2)c_n - \alpha_{(n+2)}\right)\partial_n,$$

where α_n can be found from the recurrent relations $\alpha_n = -\sum_{k=1}^{n} c_k\alpha_{n-k}$, $\alpha_0 = 1$. Here, for example,

$$\alpha_1 = -c_1, \quad \alpha_2 = c_1^2 - c_2, \quad \alpha_3 = -c_1^3 + 2c_1c_2 - c_3, \quad \dots.$$

For other negative values of k the expressions of $L_k[f]$ are more complicated but can be found by an algebraic procedure, see, e.g., [12], [14].

8.4   Segal–Wilson Grassmannian

Sato's (universal) Grassmannian appeared first in 1982 in [504] as an infinite-dimensional generalization of the classical finite-dimensional Grassmannian manifolds and it is described as 'the topological closure of the inductive limit of' a finite-dimensional Grassmanian as the dimensions of the ambient vector space and its subspaces tend to infinity. It turned out to be a very important infinite-dimensional manifold, being related to the representation theory of loop groups, integrable hierarchies, microlocal analysis, conformal and quantum field theories, the second quantization of fermions, and to many other topics [128, 398, 512, 582]. In the Segal and Wilson approach [512] the infinite-dimensional Grassmannian $\mathrm{Gr}\,(H)$ is taken over the separable Hilbert space H. The first systematic description of the infinite-dimensional Grassmannian can be found in [451].

We present here a general definition of the infinite-dimensional smooth Grassmannian $\mathrm{Gr}_{\infty}(H)$. As a separable Hilbert space we take the space $L^2(S^1)$ and consider its dense subspace $H = C_{\|\cdot\|_2}^{\infty}(S^1)$ of smooth complex-valued functions defined on the unit circle endowed with $L^2(S^1)$ inner product $\langle f, g \rangle = \frac{1}{2\pi}\int\limits_{S^1} f\bar{g}\,ds$,

$f, g \in H$. The orthonormal basis of H is $\{z^k\}_{k \in \mathbb{Z}} = \{e^{ik\theta}\}_{k \in \mathbb{Z}}$, $e^{i\theta} \in S^1$. Let us split all integers $\mathbb{Z}$ into two sets

$$\mathbb{Z}^+ = \{0, 1, 2, 3, \ldots\} \quad \text{and} \quad \mathbb{Z}^- = \{\ldots, -3, -2, -1\},$$

and let us define a polarization by

$$H_+ = \operatorname{span}_H \{z^k, \ k \in \mathbb{Z}^+\}, \quad H_- = \operatorname{span}_H \{z^k, \ k \in \mathbb{Z}^-\}.$$

Here and further span is taken in the appropriate space indicated as a subscription. The Grassmanian is thought of as the set of closed linear subspaces W of H, which are commensurable with H_+ in the sense that they have finite codimension in both H_+ and W. This can be defined by means of the descriptions of the orthogonal projections of the subspace $W \subset H$ to H_+ and H_-.

Definition 8.4.1. The infinite-dimensional smooth Grassmannian $\operatorname{Gr}_\infty(H)$ over the space H is the set of subspaces W of H, such that

1. the orthogonal projection $pr_+ : W \to H_+$ is a Fredholm operator,
2. the orthogonal projection $pr_- : W \to H_-$ is a compact operator.

The requirement that pr_+ is Fredholm means that the kernel and cokernel of pr_+ are finite dimensional. The reader can find more information about Fredholm operators in [147]. It was proved in [451], that $\operatorname{Gr}_\infty(H)$ is a dense submanifold in a Hilbert manifold modeled over the space $\mathcal{L}_2(H_+, H_-)$ of Hilbert–Schmidt operators from H_+ to H_-, that itself has the structure of a Hilbert space, see [464]. Any $W \in \operatorname{Gr}_\infty(H)$ can be thought of as a graph W_T of a Hilbert–Schmidt operator $T : W \to W^\perp$, and points of a neighborhood U_W of $W \in \operatorname{Gr}_\infty(H)$ are in one-to-one correspondence with operators from $\mathcal{L}_2(W, W^\perp)$.

Let us denote by $\mathfrak{S}$ the set of all collections $\mathbb{S} \subset \mathbb{Z}$ of integers such that $\mathbb{S} \setminus \mathbb{Z}^+$ and $\mathbb{Z}^+ \setminus \mathbb{S}$ are finite. Thus, any sequence $\mathbb{S}$ of integers is bounded from below and contains all positive numbers starting from some number. It is clear that the sets $H_\mathbb{S} = \operatorname{span}_H \{z^k, \ k \in \mathbb{S}\}$ are elements of the Grassmanian $\operatorname{Gr}_\infty(H)$ and they are usually called *special points*. The collection of neighborhoods $\{U_\mathbb{S}\}_{\mathbb{S} \in \mathfrak{S}}$,

$$U_\mathbb{S} = \{W \mid \exists \text{ an orthogonal projection } \pi : W \to H_\mathbb{S} \text{ that is an isomorphism}\}$$

forms an open cover of $\operatorname{Gr}_\infty(H)$. The virtual cardinality of $\mathbb{S}$ defines the *virtual dimension* (v.d.) of $H_\mathbb{S}$, namely:

$$\operatorname{virtcard}(\mathbb{S}) = \operatorname{virtdim}(H_\mathbb{S}) = \dim(\mathbb{N} \setminus \mathbb{S}) - \dim(\mathbb{S} \setminus \mathbb{N}) = \operatorname{ind}(pr_+). \quad (8.10)$$

The expression $\operatorname{ind}(pr_+) = \dim \ker(pr_+) - \dim \operatorname{coker}(pr_-)$ is called the index of the Fredholm operator pr_+. According to their virtual dimensions the points of $\operatorname{Gr}_\infty(H)$ belong to different components of connectivity. The Grassmannian is the disjoint union of connected components parametrized by their virtual dimensions.

8.5   Hamiltonian formalism

Let the driving term $p(z,t)$ in the Löwner–Kufarev ODE (8.5) be from the Carathéodory class for almost all $t \geq 0$, C^∞-smooth in $\mathbb{D}$, and summable with respect
to t as in Lemma 8.1.1. Then the domains $\Omega(t) = f(\mathbb{D}, t) = e^t w(\mathbb{D}, t)$ have smooth
boundaries $\partial\Omega(t)$ and the function f is injective on S^1, i.e.; $f \in \mathcal{F}_0$. So the Löwner–
Kufarev equation can be extended to the closed unit disk $\overline{\mathbb{D}} = \mathbb{D} \cup S^1$.

Let us consider functions $\psi \in H = C^\infty_{\|\cdot\|_2}$ from $T_f^*\mathcal{F}_0 \otimes \mathbb{C}$, $f \in \mathcal{F}_0$,

$$\psi(z) = \sum_{k \in \mathbb{Z}} \psi_k z^{k-1}, \quad |z| = 1,$$

and the space of observables on $T^*\mathcal{F}_0 \otimes \mathbb{C}$, given by integral functionals

$$\mathcal{R}(f, \bar{\psi}, t) = \frac{1}{2\pi} \int_{z \in S^1} r(f(z), \bar{\psi}(z), t) \frac{dz}{iz},$$

where the function $r(\xi, \eta, t)$ is smooth in variables ξ, η and measurable in t.

We define a special observable, the time-dependent pseudo-Hamiltonian $\mathcal{H}$, by

$$\mathcal{H}(f, \bar{\psi}, p, t) = \frac{1}{2\pi} \int_{z \in S^1} \bar{z}^2 f(z, t)(1 - p(e^{-t} f(z, t), t)) \bar{\psi}(z, t) \frac{dz}{iz}, \qquad (8.11)$$

with the driving function (control) $p(z, t)$ satisfying the above properties. The
Poisson structure on the space of observables is given by the canonical brackets

$$\{\mathcal{R}_1, \mathcal{R}_2\} = 2\pi \int_{z \in S^1} z^2 \left(\frac{\delta \mathcal{R}_1}{\delta f} \frac{\delta \mathcal{R}_2}{\delta \bar{\psi}} - \frac{\delta \mathcal{R}_1}{\delta \bar{\psi}} \frac{\delta \mathcal{R}_2}{\delta f} \right) \frac{dz}{iz},$$

where $\frac{\delta}{\delta f}$ and $\frac{\delta}{\delta \bar{\psi}}$ are the variational derivatives, $\frac{\delta}{\delta f}\mathcal{R} = \frac{1}{2\pi}\frac{\partial}{\partial f}r$, $\frac{\delta}{\delta \bar{\psi}}\mathcal{R} = \frac{1}{2\pi}\frac{\partial}{\partial \bar{\psi}}r$.

Representing the coefficients c_n and $\bar{\psi}_m$ of f and $\bar{\psi}$ as integral functionals

$$c_n = \frac{1}{2\pi} \int_{z \in S^1} \bar{z}^{n+1} f(z, t) \frac{dz}{iz}, \quad \bar{\psi}_m = \frac{1}{2\pi} \int_{z \in S^1} z^{m-1} \bar{\psi}(z, t) \frac{dz}{iz},$$

$n \in \mathbb{N}$, $m \in \mathbb{Z}$, we obtain $\{c_n, \bar{\psi}_m\} = \delta_{n,m}$, $\{c_n, c_k\} = 0$, and $\{\bar{\psi}_l, \bar{\psi}_m\} = 0$, where
$n, k \in \mathbb{N}$, $l, m \in \mathbb{Z}$.

The infinite-dimensional Hamiltonian system is written as

$$\frac{dc_k}{dt} = \{c_k, \mathcal{H}\}, \qquad (8.12)$$

$$\frac{d\bar{\psi}_k}{dt} = \{\bar{\psi}_k, \mathcal{H}\}, \qquad (8.13)$$

where $k \in \mathbb{Z}$ and $c_0 = c_{-1} = c_{-2} = \cdots = 0$, or equivalently, multiplying by corresponding powers of z and summing up,

$$\frac{df(z,t)}{dt} = f(1 - p(e^{-t}f, t)) = 2\pi \frac{\delta \mathcal{H}}{\delta \bar{\psi}} z^2 = \{f, \mathcal{H}\}, \qquad (8.14)$$

$$\frac{d\bar{\psi}}{dt} = -(1 - p(e^{-t}f, t) - e^{-t}fp'(e^{-t}f, t))\bar{\psi} = -2\pi \frac{\delta \mathcal{H}}{\delta f} z^2 = \{\bar{\psi}, \mathcal{H}\}, \quad (8.15)$$

where $z \in S^1$. So the phase coordinates $(f, \bar{\psi})$ play the role of the canonical Hamiltonian pair. Observe that the equation (8.14) is the Löwner–Kufarev equation (8.5) for the function $f = e^t w$.

Let us set up the *generating function* $\mathcal{G}(z) = \sum_{k \in \mathbb{Z}} \mathcal{G}_k z^{k-1}$, such that

$$\bar{\mathcal{G}}(z) := f'(z, t)\bar{\psi}(z, t).$$

Consider the 'non-positive' $(\bar{\mathcal{G}}(z))_{\leq 0}$ and 'positive' $(\bar{\mathcal{G}}(z))_{>0}$ parts of the Laurent series for $\bar{\mathcal{G}}(z)$:

$$(\bar{\mathcal{G}}(z))_{\leq 0} = (\bar{\psi}_1 + 2c_1\bar{\psi}_2 + 3c_2\bar{\psi}_3 + \cdots) + (\bar{\psi}_2 + 2c_1\bar{\psi}_3 + \cdots)z^{-1} + \cdots$$

$$= \sum_{k=0}^{\infty} \bar{\mathcal{G}}_{k+1} z^{-k},$$

$$(\bar{\mathcal{G}}(z))_{>0} = (\bar{\psi}_0 + 2c_1\bar{\psi}_1 + 3c_2\bar{\psi}_2 + \cdots)z + (\bar{\psi}_{-1} + 2c_1\bar{\psi}_0 + 3c_2\bar{\psi}_1 \cdots)z^2 + \cdots$$

$$= \sum_{k=1}^{\infty} \bar{\mathcal{G}}_{-k+1} z^{k}.$$

Proposition 8.5.1. *Let the driving term $p(z,t)$ in the Löwner–Kufarev ODE be from the Carathéodory class for almost all $t \geq 0$, C^{∞}-smooth in $\overline{\mathbb{D}}$, and summable with respect to t. The functions $\mathcal{G}(z)$, $(\mathcal{G}(z))_{<0}$, $(\mathcal{G}(z))_{\geq 0}$, and all coefficients $\mathcal{G}_n$ are time-independent for all $z \in S^1$.*

Proof. It is sufficient to check the equality $\dot{\bar{\mathcal{G}}} = \{\bar{\mathcal{G}}, \mathcal{H}\} = 0$ for the function $\mathcal{G}$, and then, the same holds for the coefficients of the Laurent series for $\mathcal{G}$. $\qquad \square$

Proposition 8.5.2. *The conjugates $\bar{\mathcal{G}}_k$, $k = 1, 2, \ldots$, to the coefficients of the generating function satisfy the Witt commutation relation $\{\bar{\mathcal{G}}_m, \bar{\mathcal{G}}_n\} = (n - m)\bar{\mathcal{G}}_{n+m}$ for $n, m \geq 1$, with respect to our Poisson structure.*

The *proof* is straightforward.

The isomorphism $\iota : \bar{\psi}_k \to \partial_k = \frac{\partial}{\partial c_k}$, $k > 0$, is a Lie algebra isomorphism $(T_f^{*(0,1)}\mathcal{F}_0, \{,\}) \to (T_f^{(1,0)}\mathcal{F}_0, [,])$. It makes a correspondence between the conjugates $\bar{\mathcal{G}}_n$ of the coefficients $\mathcal{G}_n$ of $(\mathcal{G}(z))_{\geq 0}$ at the point $(f, \bar{\psi})$ and the Kirillov vectors $L_n[f] = \partial_n + \sum_{k=1}^{\infty} (k+1)c_k \partial_{n+k}$, $n \in \mathbb{N}$. Both satisfy the Witt commutation relations (8.8).

8.6   Curves in Grassmannian

Let us recall, that the underlying space for the universal smooth Grassmannian $\mathrm{Gr}_\infty(H)$ is $H = C^\infty_{\|\cdot\|_2}(S^1)$ with the canonical L^2 inner product of functions defined on the unit circle. Its natural polarization

$$H_+ = \operatorname{span}_H\{1, z, z^2, z^3, \dots\}, \qquad H_- = \operatorname{span}_H\{z^{-1}, z^{-2}, \dots\},$$

was introduced before. The pseudo-Hamiltonian $\mathcal{H}(f, \bar\psi, t)$ is defined for an arbitrary $\psi \in L^2(S^1)$, but we consider only smooth solutions of the Hamiltonian system, therefore, $\psi \in H$. We identify this space with the dense subspace of $T_f^*\mathcal{F}_0 \otimes \mathbb{C}$, $f \in \mathcal{F}_0$. The generating function $\mathcal{G}$ defines a linear map $\bar{\mathcal{G}}$ from the dense subspace of $T_f^*\mathcal{F}_0 \otimes \mathbb{C}$ to H, which being written in a matrix form becomes

$$
\begin{pmatrix} \cdots \\ \bar{\mathcal{G}}_{-2} \\ \bar{\mathcal{G}}_{-1} \\ \bar{\mathcal{G}}_0 \\ \bar{\mathcal{G}}_1 \\ \bar{\mathcal{G}}_2 \\ \bar{\mathcal{G}}_3 \\ \cdots \end{pmatrix}
=
\left(\begin{array}{ccccc|ccccc}
\ddots & \ddots & \ddots & \ddots & \ddots & \ddots & \ddots & \ddots & \ddots & \ddots \\
\cdots & 0 & 1 & 2c_1 & 3c_2 & 4c_3 & 5c_4 & 6c_5 & 7c_6 & \cdots \\
\cdots & 0 & 0 & 1 & 2c_1 & 3c_2 & 4c_3 & 5c_4 & 6c_5 & \cdots \\
\cdots & 0 & 0 & 0 & 1 & 2c_1 & 3c_2 & 4c_3 & 5c_4 & \cdots \\
\cdots & 0 & 0 & 0 & 0 & 1 & 2c_1 & 3c_2 & 4c_3 & \cdots \\
\cdots & 0 & 0 & 0 & 0 & 0 & 1 & 2c_1 & 3c_2 & \cdots \\
\cdots & 0 & 0 & 0 & 0 & 0 & 0 & 1 & 2c_1 & \cdots \\
\ddots & \ddots & \ddots & \ddots & \ddots & \ddots & \ddots & \ddots & \ddots & \ddots
\end{array}\right)
\begin{pmatrix} \cdots \\ \bar\psi_{-2} \\ \bar\psi_{-1} \\ \bar\psi_0 \\ \bar\psi_1 \\ \bar\psi_2 \\ \bar\psi_3 \\ \cdots \end{pmatrix}
$$

$$\tag{8.16}$$

or in the matrix block form as

$$
\begin{pmatrix} \bar{\mathcal{G}}_{>0} \\ \bar{\mathcal{G}}_{\le 0} \end{pmatrix}
=
\begin{pmatrix} C_{1,1} & C_{1,2} \\ 0 & C_{1,1} \end{pmatrix}
\begin{pmatrix} \bar\psi_{>0} \\ \bar\psi_{\le 0} \end{pmatrix}.
\tag{8.17}
$$

The proof of the following proposition is obvious.

Proposition 8.6.1. *The operator $C_{1,1}\colon H_+ \to H_+$ is invertible.*

The generating function also defines a map $\mathcal{G}\colon T^*\mathcal{F}_0 \otimes \mathbb{C} \to H$ by

$$T^*\mathcal{F}_0 \otimes \mathbb{C} \ni (f(z), \psi(z)) \mapsto \mathcal{G} = \bar f'(z)\psi(z) \in H.$$

Observe that any solution $(f(z,t), \bar\psi(z,t))$ of the Hamiltonian system is mapped into a single point of the space H, since all $\mathcal{G}_k$, $k \in \mathbb{Z}$ are time-independent by Proposition 8.5.1.

Consider a bundle $\pi\colon \mathcal{B} \to T^*\mathcal{F}_0 \otimes \mathbb{C}$ with a typical fiber isomorphic to $\mathrm{Gr}_\infty(H)$. We are aimed at construction of a curve $\Gamma\colon [0, T] \to \mathcal{B}$ that is traced by the solutions to the Hamiltonian system, or in other words, by the Löwner–Kufarev evolution. The curve Γ will have the form

$$\Gamma(t) = \left(f(z,t), \psi(z,t), W_{T_n}(t) \right)$$

in the local trivialization.

Here W_{T_n} is the graph of a finite rank operator $T_n \colon H_+ \to H_-$, such that W_{T_n} belongs to the connected component of U_{H_+} of virtual dimension 0. In other words, we build an hierarchy of finite rank operators $T_n \colon H_+ \to H_-$, $n \in \mathbb{Z}^+$, whose graphs in the neighborhood U_{H_+} of the point $H_+ \in \mathrm{Gr}_\infty(H)$ are

$$T_n((\mathcal{G}(z))_{>0}) = T_n(\mathcal{G}_1, \mathcal{G}_2, \ldots, \mathcal{G}_k, \ldots) = \begin{cases} G_0(\mathcal{G}_1, \mathcal{G}_2, \ldots, \mathcal{G}_k, \ldots) \\ G_{-1}(\mathcal{G}_1, \mathcal{G}_2, \ldots, \mathcal{G}_k, \ldots) \\ \cdots \\ G_{-n+1}(\mathcal{G}_1, \mathcal{G}_2, \ldots, \mathcal{G}_k, \ldots), \end{cases}$$

with $G_0 z^{-1} + G_{-1} z^{-2} + \cdots + G_{-n+1} z^{-n} \in H_-$. Let us write $G_k = \mathcal{G}_k$, $k \in \mathbb{N}$. The elements $G_0, G_{-1}, G_{-2}, \ldots$ are constructed so that all $\{\bar{G}_k\}_{k=-n+1}^\infty$ satisfy the truncated Witt commutation relations

$$\{\bar{G}_k, \bar{G}_l\}_n = \begin{cases} (l-k)\bar{G}_{k+l}, & \text{for } k+l \geq -n+1, \\ 0, & \text{otherwise}, \end{cases}$$

and are related to Kirillov's vector fields under the isomorphism ι. The projective limit as $n \leftarrow \infty$ recovers the whole Witt algebra and the Witt commutation relations. We present an explicit algorithm consisting of two steps in order to define the coefficients G_{-k}, $k = 0, 1, 2, \ldots, n-1$.

STEP 1. In the first step we remove the dependence of $\bar{\mathcal{G}}_{>0} = \{\bar{\mathcal{G}}_{-k}\}_{k=0}^\infty$ on $\bar{\psi}_{>0} = \{\bar{\psi}_{-k}\}_{k=0}^\infty$ defining

$$\tilde{\mathcal{G}}_{>0} = \bar{\mathcal{G}}_{>0} - C_{1,1}\bar{\psi}_{>0}, \tag{8.18}$$

where $C_{1,1}$ is the upper triangular block in the matrix (8.17). Thus, $\tilde{\mathcal{G}}_{>0} = \tilde{\mathcal{G}}_{>0}(\bar{\psi}_{\leq 0})$. Since the matrix $C_{1,1}$ is invertible we can write $\bar{\psi}_{\leq 0} = C_{1,1}^{-1}\bar{\mathcal{G}}_{\leq 0}$, that implies

$$\tilde{\mathcal{G}}_{>0} = \tilde{\mathcal{G}}_{>0}(C_{1,1}^{-1}\bar{\mathcal{G}}_{\leq 0}) = \tilde{\mathcal{G}}_{>0}(\bar{\mathcal{G}}_{\leq 0}).$$

Let us denote by $\tilde{T}_n$ the operator that maps a vector $\sum_{k=0}^\infty \mathcal{G}_{k+1} z^k$ from H_+ to a finite-dimensional vector $\sum_{k=1}^n \tilde{\mathcal{G}}_{-k+1} z^k \in H_-$. These operators can be written as the superpositions $\tilde{T}_n = C_{1,2}^{(n)} \circ C_{1,1}^{-1} \colon H_+ \to H_-$, where $C_{1,2}^{(n)}$ is equal to the nth cut of the block $C_{1,2}$ in (8.17) of the first lower n-rows and with vanishing others. The operators $\tilde{T}_n \colon H_+ \to H_-$ are of finite rank, and therefore, compact. Their graphs $W_{\tilde{T}_n} = (\mathrm{id} + \tilde{T}_n)(H_+) \in \mathrm{Gr}_\infty(H)$ belong to the connected component of virtual dimension 0.

STEP 2. Observe that up to now there is no clear relation of operators $\tilde{T}_n$, or their graphs with the Kirillov vector fields L_k and L_{-k}. However, it is not hard to see that the quantities $\bar{\mathcal{G}}_k$, considered as functions of $\bar{\psi}$, are mapped to $L_k[f]$ under the isomorphism ι for $k > 0$. In Step 2 we aim to modify $\tilde{\mathcal{G}}_{-k}$, defined in (8.18) to G_{-k} in such a way that the isomorphism ι maps $\bar{G}_{-k}$ to the 'non-positive' Kirillov vector fields L_{-k}. We will construct only $\bar{G}_0, \bar{G}_{-1}, \bar{G}_{-2}$, and then we extend the isomorphism ι to the Lie algebra isomorphism by defining $\bar{G}_{-(n+m)}(m-n) = \{\bar{G}_{-n}, \bar{G}_{-m}\}, n, m \geq 0$.

Let us recall that the first three Virasoro generators written in affine coordinates are

- $L_0[f](z) = \sum_{n=1}^{\infty} nc_n \partial_n$;
- $L_{-1}[f](z) = \sum_{n=1}^{\infty} \left((n+2)c_{n+1} - 2c_1 c_n\right)\partial_n$;
- $L_{-2}[f](z) = \sum_{n=1}^{\infty} \left((n+3)c_{n+2} + (c_1^2 - 4c_2)c_n - a_{(n+2)}\right)\partial_n$, where the coefficient a_n can be found from the recurrent relations

$$a_n = -\sum_{k=1}^{n} c_k a_{n-k}, \qquad a_0 = 1. \tag{8.19}$$

In order to construct $\bar{G}_k = \iota^{-1}(L_k)$, $k = 0, -1$ we consider the coefficients $\tilde{\mathcal{G}}$ from (8.18) as functions of $\bar{\psi}_{>0}$, and write $\bar{\psi}_0^* = \sum_{k=1}^{\infty} c_k \bar{\psi}_k$. We deduce that

$$G_0 = \tilde{\mathcal{G}}_0 - \psi_0^*, \qquad G_{-1} = \tilde{\mathcal{G}}_{-1} - 2\bar{c}_1 \psi_0^*.$$

Since $\tilde{\mathcal{G}}_{-2} = \sum_{k=1}^{\infty} (k+3)\bar{c}_{k+2} \psi_k$, we have

$$G_{-2} = \tilde{\mathcal{G}}_{-2} + \sum_{k=1}^{\infty} \left((\bar{c}_1^2 - 4\bar{c}_2)\bar{c}_k - \bar{a}_{(k+2)}\right)\psi_k.$$

Let us write this in terms of operators. Let

$$\tilde{B}^{(0)} = \begin{pmatrix} \cdots & \cdots & \cdots & \cdots & \cdots \\ \cdots & 0 & 0 & 0 & 0 \\ \cdots & 0 & 0 & 0 & 0 \\ \cdots & -c_4 & -c_3 & -c_2 & -c_1 \end{pmatrix},$$

$$\tilde{B}^{(1)} = \begin{pmatrix} \cdots & \cdots & \cdots & \cdots & \cdots \\ \cdots & 0 & 0 & 0 & 0 \\ \cdots & -2c_1 c_4 & -2c_1 c_3 & -2c_1 c_2 & -2c_1 c_1 \\ \cdots & -c_4 & -c_3 & -c_2 & -c_1 \end{pmatrix},$$

$$\widetilde{B}^{(2)} = \begin{pmatrix} \cdots & \cdots & \cdots & \cdots & \cdots \\ \cdots & 0 & 0 & 0 & 0 \\ \cdots & (c_1^2 - 4c_2)c_4 - \alpha_6 & (c_1^2 - 4c_2)c_3 - \alpha_5 & (c_1^2 - 4c_2)c_2 - \alpha_4 & 2c_1 - 6c_1c_2 + c_3 \\ \cdots & -2c_1c_4 & -2c_1c_3 & -2c_1c_2 & -2c_1c_1 \\ \cdots & -c_4 & -c_3 & -c_2 & -c_1 \end{pmatrix},$$

$$C_{1,2}^{(0)} = \begin{pmatrix} \cdots & \cdots & \cdots & \cdots & \cdots \\ \cdots & 0 & 0 & 0 & 0 \\ \cdots & 0 & 0 & 0 & 0 \\ \cdots & 5c_4 & 4c_3 & 3c_2 & 2c_1 \end{pmatrix}, \quad C_{2,1}^{(1)} = \begin{pmatrix} \cdots & \cdots & \cdots & \cdots & \cdots \\ \cdots & 0 & 0 & 0 & 0 \\ \cdots & 6c_5 & 5c_4 & 4c_3 & 3c_2 \\ \cdots & 5c_4 & 4c_3 & 3c_2 & 2c_1 \end{pmatrix},$$

$$C_{2,1}^{(2)} = \begin{pmatrix} \cdots & \cdots & \cdots & \cdots & \cdots \\ \cdots & 0 & 0 & 0 & 0 \\ \cdots & 7c_6 & 6c_5 & 5c_4 & 4c_3 \\ \cdots & 6c_5 & 5c_4 & 4c_3 & 3c_2 \\ \cdots & 5c_4 & 4c_3 & 3c_2 & 2c_1 \end{pmatrix},$$

where a_n are given by (8.19). Then the operators T_n such that their conjugates $\bar{T}_n = (\widetilde{B}^{(n)} + C_{2,1}^{(n)}) \circ C_{1,1}^{-1}$ are operators from H_+ to H_- of finite rank and their graphs $W_{T_n} = (\mathrm{id} + T_n)(H_+)$ are elements of the component of virtual dimension 0 in $\mathrm{Gr}_\infty(H)$. We can choose a basis $\{e_0, e_1, e_2, \dots\}$ in W_{T_n} as a set of Laurent polynomials constructed by means of operators T_n and $\bar{C}_{1,1}$ as

$$\{\psi_1, \psi_2, \dots\} \xrightarrow{\bar{C}_{1,1}} \{G_1, G_2, \dots\} \xrightarrow{\mathrm{id} + T_n} \{G_{-n+1}, G_{-n+2}, \dots, G_0, G_1, G_2, \dots\},$$

projecting the canonical basis $\{1, 0, 0, \dots\}$, $\{0, 1, 0, \dots\}$, $\{0, 0, 1, \dots\}$,…:

$$e_0 = 1 + \bar{c}_1 \frac{1}{z} + (3\bar{c}_2 - 2\bar{c}_1^2)\frac{1}{z^2} + (5\bar{c}_3 + 2\bar{c}_1^3 - 6\bar{c}_1\bar{c}_2)\frac{1}{z^3} + \cdots$$
$$+ G_{-n+1}(\bar{C}_{1,1}(1, 0, 0, \dots))\frac{1}{z^n},$$

$$e_1 = z + 2\bar{c}_1 + 2\bar{c}_2 \frac{1}{z} + (4\bar{c}_3 - 2\bar{c}_1\bar{c}_2)\frac{1}{z^2}$$
$$+ (6\bar{c}_4 - 5\bar{c}_2^2 - 2\bar{c}_1\bar{c}_3 + 4\bar{c}_1^2\bar{c}_2 - \bar{c}_1^4)\frac{1}{z^3} + \cdots$$
$$+ G_{-n+1}(\bar{C}_{1,1}(0, 1, 0, \dots))\frac{1}{z^n},$$

$$e_2 = z^2 + 2\bar{c}_1 z + 3\bar{c}_2 + 3\bar{c}_3 \frac{1}{z} + (5\bar{c}_4 - 2\bar{c}_1\bar{c}_3)\frac{1}{z^2}$$
$$+ (7\bar{c}_5 - 6\bar{c}_2\bar{c}_3 + 3\bar{c}_1\bar{c}_2^2 - 2\bar{c}_1\bar{c}_4 + 4\bar{c}_1^2\bar{c}_3 - 4\bar{c}_1^3 c_2 + \bar{c}_1^5)\frac{1}{z^3} + \cdots$$
$$+ G_{-n+1}(\bar{C}_{1,1}(0, 0, 1, \dots))\frac{1}{z^n},$$

....

Let us formulate the result as the following *main statement* of this section.

Proposition 8.6.2. *The operator* $(\mathrm{id}+T_n)$ *defines a graph* $W_{T_n}= \mathrm{span}\{e_0, e_1, e_2, \ldots\}$ *in the Grassmannian* Gr_∞ *of virtual dimension 0. Given any* $\psi = \sum_{k=0}^\infty \psi_{k+1} z^k \in H_+ \subset H$, *the function*

$$G(z) = \sum_{k=-n}^\infty G_{k+1} z^k = \sum_{k=0}^\infty \psi_{k+1} e_k,$$

is an element of W_{T_n}.

Proposition 8.6.3. *In the autonomous case of the Cauchy problem* (8.15), *when the function* $p(z, t)$ *does not depend on* t, *the pseudo-Hamiltonian* $\mathcal{H}$ *plays the role of time-dependent energy and* $\mathcal{H}(t) = \bar{G}_0(t) + const$, *where* $\bar{G}_0\big|_{t=0} = 0$. *The constant is defined as* $\sum_{n=1}^\infty p_k \bar{\psi}_k(0)$.

Proof. In the autonomous case we have $\frac{d}{dt}\mathcal{H} = \frac{\partial}{\partial t}\mathcal{H}$. By straightforward calculation we assure that $\frac{d}{dt}\bar{G}_0 = \frac{\partial}{\partial t}\mathcal{H}$, which leads to conclusion of the proposition. The constant is calculated by substituting $t = 0$ in $\mathcal{H}$.   □

Remark 8.6.1. The Virasoro generator L_0 plays the role of an energy functional in CFT. In view of the isomorphism ι, the observable $\bar{G}_0 = \iota^{-1}(L_0)$ plays an analogous role.

As a *conclusion*, we constructed a countable family of curves $\Gamma_n : [0, T] \to \mathcal{B}$ in the trivial bundle $\mathcal{B} = T^*\mathcal{F}_0 \otimes \mathbb{C} \times Gr_\infty(H)$, such that the curve Γ_n admits the form $\Gamma_n(t) = \big(f(z, t), \psi(z, t), W_{T_n}(t)\big)$, for $t \in [0, T]$ in the local trivialization. Here $\big(f(z, t), \bar{\psi}(z, t)\big)$ is the solution of the Hamiltonian system (8.12)–(8.13). Each operator $T_n(t): H_+ \to H_-$ that maps $\mathcal{G}_{>0}$ to $\big(G_0(t), G_{-1}(t), \ldots, G_{-n+1}(t)\big)$ defined for any $t \in [0, T]$, $n = 1, 2, \ldots$, is of finite rank and its graph $W_{T_n}(t)$ is a point in $Gr_\infty(H)$ for any t. The graphs W_{T_n} belong to the connected component of the virtual dimension 0 for every time $t \in [0, T]$ and for fixed n. Each coordinate $(G_{-n+1}, \ldots, G_{-2}, G_{-1}, G_0, G_1, G_2, \ldots)$ of a point in the graph W_{T_n} considered as a function of ψ is isomorphic to the Kirilov vector fields

$$(L_{-n+1}, \ldots, L_{-2}, L_{-1}, L_0, L_1, L_1, L_2, \ldots)$$

under the isomorphism ι.

Remark 8.6.2. Although we performed all constructions for the operators T_n of finite rank, the limiting operator T_∞ is Hilbert–Schmidt because of the embedding of the conformal welding into the Grassmannian, see, e.g., [314, 542].

8.7 τ-function

τ- and Baker–Akhiezer functions were discussed in Section 6.2.5. Here we define them in relation with the Grassmannian. Remember that any function g holomorphic in the unit disk, non-vanishing on the boundary and normalized by $g(0) = 1$, defines the multiplication operator $g\varphi$, $\varphi(z) = \sum_{k\in\mathbb{Z}} \varphi_k z^k$, that can be written in the matrix form

$$
\begin{pmatrix} a & b \\ 0 & d \end{pmatrix} \begin{pmatrix} \varphi_{\geq 0} \\ \varphi_{<0} \end{pmatrix}. \tag{8.20}
$$

All these upper triangular matrices form a subgroup GL_{res}^+ of the group of automorphisms GL_{res} of the Grassmannian $\mathrm{Gr}_\infty(H)$.

With any function g and any graph W_{T_n} constructed in the previous section (which is transverse to H_-) we can relate the τ-function $\tau_{W_{T_n}}(g)$ by the formula

$$
\tau_{W_{T_n}}(g) = \det(1 + a^{-1} b T_n),
$$

where a, b are the blocks in the multiplication operator generated by g^{-1}. If we write the function g in the form $g(z) = \exp(\sum_{n=1}^\infty t_n z^n) = 1 + \sum_{k=1}^\infty S_k(\mathbf{t}) z^k$, where the coefficients $S_k(\mathbf{t})$ are the homogeneous elementary Schur polynomials, then the coefficients t_n are called generalized times. For any fixed W_{T_n} we get an orbit in $\mathrm{Gr}_\infty(H)$ of curves Γ constructed in the previous section under the action of the elements of the subgroup GL_{res}^+ defined by the function g. On the other hand, the τ-function defines a section in the determinant bundle over $\mathrm{Gr}_\infty(H)$ for any fixed $f \in \mathcal{F}_0$ at each point of the curve Γ.

8.8   Baker–Akhiezer function, KP flows, and KP equation

Let us consider the component Gr^0 of the Grassmannian Gr_∞ of virtual dimension 0, and let g be a holomorphic function in $\mathbb{D}$ considered as an element of GL_{res}^+ analogously to the previous section. Then g is an upper triangular matrix with 1s on the principal diagonal. Observe that $g(0) = 1$ and g does not vanish on S^1. Given a point $W \in \mathrm{Gr}^0$ let us define a subset $\Gamma^+ \subset GL_{res}^+$ as $\Gamma^+ = \{g \in GL_{res}^+ : g^{-1}W$ is transverse to $H_-\}$. Then there exists [512] a unique function $\Psi_W[g](z)$ defined on S^1, such that for each $g \in \Gamma^+$, the function $\Psi_W[g]$ is in W, and it admits the form

$$
\Psi_W[g](z) = g(z) \left(1 + \sum_{k=1}^\infty \omega_k(g, W) \frac{1}{z^k} \right).
$$

The coefficients $\omega_k = \omega_k(g, W)$ depend both on $g \in \Gamma^+$ and on $W \in \mathrm{Gr}^0$, besides they are holomorphic on Γ^+ and extend to meromorphic functions on GL_{res}^+. The function $\Psi_W[g](z)$ is called the *Baker–Akhiezer function* of W or the *wave function*. It plays a crucial role in the definition of the KP (Kadomtsev–Petviashvili) hierarchy which we will present in what follows. We are going to construct the Baker–Akhiezer function explicitly in our case.

Let $W = W_{T_n}$ be a point of Gr^0 defined in Proposition 8.6.2. This point corresponds to the *adèlic Grassmannian* introduced by Wilson [588, 589] in the

study of the bispectral problem and the rational solutions of the KP equation. We
remark that the rational solutions to KP were studied earlier by Matveev [381] and
Krichever [323]. It is possible to consider W_{T_∞} as well because of Remark 8.6.2.
Take a function $g(z) = 1+a_1z+a_2z^2+\cdots \in \Gamma^+$, and let us write the corresponding
bi-infinite series for the Baker–Akhiezer function $\Psi_W[g](z)$ explicitly as

$$\Psi_W[g](z) = \sum_{k\in\mathbb{Z}} \mathcal{W}_k z^k = (1\ +a_1z+a_2z^2+\cdots)\left(1+\frac{\omega_1}{z}+\frac{\omega_2}{z^2}+\cdots\right)$$

$$= \cdots + (a_2 + a_3\omega_1 + a_4\omega_2 + a_5\omega_3 + \cdots)z^2$$
$$+ (a_1 + a_2\omega_1 + a_3\omega_2 + a_4\omega_3 + \cdots)z$$
$$+ (1 + a_1\omega_1 + a_2\omega_2 + a_3\omega_3 + \cdots)$$
$$+ (\omega_1 + a_1\omega_2 + a_2\omega_3 + \cdots)\frac{1}{z}$$
$$+ (\omega_2 + a_1\omega_3 + a_2\omega_4 + \cdots)\frac{1}{z^2} + \cdots$$
$$\cdots + (\omega_k + a_1\omega_{k+1} + a_2\omega_{k+2} + \cdots)\frac{1}{z^k} + \cdots$$

The Baker–Akhiezer function for g and W_{T_n} must be of the form

$$\Psi_{W_{T_n}}[g](z) = g(z)\left(1 + \sum_{k=1}^n \omega_k(g)\frac{1}{z^k}\right) = \sum_{k=-n}^\infty \mathcal{W}_k z^k,$$

and for W_{T_∞},

$$\Psi_{W_{T_\infty}}[g](z) = g(z)\left(1 + \sum_{k=1}^\infty \omega_k(g)\frac{1}{z^k}\right) = \sum_{k=-\infty}^\infty \mathcal{W}_k z^k.$$

In the first case, for a fixed $n \in \mathbb{N}$, we truncate the bi-infinite series by putting
$\omega_k = 0$ for all $k > n$. The Baker–Akhiezer function must belong to W_{T_n}. In order to
satisfy the definition of W_{T_n}, and determine the coefficients $\omega = (\omega_1, \omega_2, \ldots, \omega_n)$,
we must check that there exists a vector $\{\psi_1, \psi_2, \ldots\}$, such that $\Psi_{W_{T_n}}[g](z) = \sum_{k=0}^\infty e_k\psi_{k+1} \in W_{T_n}$. We define ψ_k by means of the coefficients $\mathcal{W}_k$ at the positive
powers of z in the expansion of $\Psi_{W_{T_n}}[g](z)$, and then, recover ω_k by means of the
coefficients $\mathcal{W}_k$ at the negative powers of z. First we express ψ_k as linear functions
$\psi_k = \psi_k(\omega_1, \omega_2, \ldots, \omega_n) = \psi_k(\omega)$ by

$$(\psi_1, \psi_2, \psi_3, \ldots) = \bar{C}_{1,1}^{-1}\Big(\mathcal{W}_0(\omega), \mathcal{W}_1(\omega), \ldots\Big). \tag{8.21}$$

Using Wronski's formula we can write

$$\psi_1 = \mathcal{W}_0 - 2\bar{c}_1\mathcal{W}_1 - (3\bar{c}_2 - 4\bar{c}_1^2)\mathcal{W}_2 - (4\bar{c}_3 - 12\bar{c}_2\bar{c}_1 + 8\bar{c}_1^3)\mathcal{W}_3 + \cdots,$$
$$\psi_2 = \mathcal{W}_1 - 2\bar{c}_1\mathcal{W}_2 - (3\bar{c}_2 - 4\bar{c}_1^2)\mathcal{W}_3 - (4\bar{c}_3 - 12\bar{c}_2\bar{c}_1 + 8\bar{c}_1^3)\mathcal{W}_4 + \cdots,$$
$$\psi_3 = \mathcal{W}_2 - 2\bar{c}_1\mathcal{W}_3 - (3\bar{c}_2 - 4\bar{c}_1^2)\mathcal{W}_4 - (4\bar{c}_3 - 12\bar{c}_2\bar{c}_1 + 8\bar{c}_1^3)\mathcal{W}_5 + \cdots.$$
$$\cdots\cdots\cdots$$

Next we define $\omega_1, \omega_2, \ldots, \omega_n$ as functions of g and W_{T_n}, or in other words, as functions of $a_k, \bar{c}_k$ by solving linear equations

$$\omega_1 = \bar{c}_1 \psi_1(a, \omega) + 2\bar{c}_2 \psi_2(a, \omega) + \cdots k \bar{c}_k \psi_k(a, \omega) + \cdots ,$$

$$\omega_2 = \sum_{k=1}^{\infty} \left((k+2)\bar{c}_{k+1} - 2\bar{c}_1 \bar{c}_k \right) \psi_k(a, \omega),$$

$$\ldots\ldots\ldots$$

where ψ_k are taken from (8.21). The above procedure is valid for W_{T_∞} as well. The solution exists and is unique because of the general fact of the existence of the Baker–Akhiezer function. It is a quite difficult task in general, however, in the case $n = 1$, it is possible to write the solution explicitly in matrix form.

In what follows, the procedure is rather standard but we would like to present it in order to have some solution in a closed form. If

$$A = \begin{pmatrix} \cdots \\ 3\bar{c}_3 \\ 2\bar{c}_2 \\ \bar{c}_1 \end{pmatrix}^T \bar{C}_{1,1}^{-1} \begin{pmatrix} \cdots \\ a_3 \\ a_2 \\ a_1 \end{pmatrix}, \quad B = \begin{pmatrix} \cdots \\ 3\bar{c}_3 \\ 2\bar{c}_2 \\ \bar{c}_1 \end{pmatrix}^T \bar{C}_{1,1}^{-1} \begin{pmatrix} \cdots \\ a_2 \\ a_1 \\ 1 \end{pmatrix},$$

then $\omega_1 = \frac{B}{1-A}$.

In order to apply further theory of integrable systems we need to change variables $a_n \to a_n(\mathbf{t})$, $n > 0$, $\mathbf{t} = \{t_1, t_2, \ldots\}$ in the following way:

$$a_n = a_n(t_1, \ldots, t_n) = S_n(t_1, \ldots, t_n),$$

where S_n is the nth elementary homogeneous Schur polynomial

$$1 + \sum_{k=1}^{\infty} S_k(\mathbf{t}) z^k = \exp\left(\sum_{k=1}^{\infty} t_k z^k \right) = e^{\xi(\mathbf{t}, z)}.$$

In particular,

$$S_1 = t_1, \quad S_2 = \frac{t_1^2}{2} + t_2, \quad S_3 = \frac{t_1^3}{6} + t_1 t_2 + t_3,$$

$$S_4 = \frac{t_1^4}{24} + \frac{t_2^2}{2} + \frac{t_1^2 t_2}{2} + t_1 t_3 + t_4.$$

Then the Baker–Akhiezer function corresponding to the graph W_{T_n} is written as

$$\Psi_{W_{T_n}}[g](z) = \sum_{k=-n}^{\infty} \mathcal{W}_k z^k = e^{\xi(\mathbf{t}, z)} \left(1 + \sum_{k=1}^{n} \frac{\omega_k(\mathbf{t}, W_{T_n})}{z^k} \right),$$

and $\mathbf{t} = \{t_1, t_2, \ldots\}$ is called the vector of generalized times. It is easy to see that

$$\partial_{t_k} a_m = 0, \quad \text{for all } m = 1, 2 \ldots, k-1,$$

$\partial_{t_k} a_m = 1$ and

$$\partial_{t_k} a_m = a_{m-k}, \quad \text{for all } m > k.$$

In particular, $B = \partial_{t_1} A$. Let us write $\partial := \partial_{t_1}$. Then in the case $n = 1$ we have

$$\omega_1 = \frac{\partial A}{1 - A}. \tag{8.22}$$

Now we consider the associative algebra of pseudo-differential operators $\mathcal{A} = \sum_{k=-\infty}^{n} a_k \partial^k$ over the space of smooth functions, where the derivation symbol ∂ satisfies the Leibniz rule and the integration symbol and its powers satisfy the algebraic rules $\partial^{-1}\partial = \partial\partial^{-1} = 1$ and $\partial^{-1}a$ is the operator

$$\partial^{-1}a = \sum_{k=0}^{\infty} (-1)^k (\partial^k a) \partial^{-k-1}$$

(see, e.g., [143]). The action of the operator ∂^m, $m \in \mathbb{Z}$, is well defined over the function $e^{\xi(t,z)}$, where $\xi(\mathbf{t}, z) = \sum_{k=1}^{\infty} t_k z^k$, so that the function $e^{\xi(t,z)}$ is the eigenfunction of the operator ∂^m for any integer m, i.e., it satisfies the equation

$$\partial^m e^{\xi(t,z)} = z^m e^{\xi(t,z)}, \quad m \in \mathbb{Z}, \tag{8.23}$$

see, e.g., [28], [143]. As usual, we identify $\partial = \partial_{t_1}$, and $\partial^0 = 1$.

Let us introduce the dressing operator $\Lambda = \phi \partial \phi^{-1} = \partial + \sum_{k=1}^{\infty} \lambda_k \partial^{-k}$, where ϕ is a pseudo-differential operator $\phi = 1 + \sum_{k=1}^{\infty} w_k(\mathbf{t}) \partial^{-k}$. The operator Λ is defined up to the multiplication on the right by a series $1 + \sum_{k=1}^{\infty} b_k \partial^{-k}$ with constant coefficients b_k. The mth KP flow is defined by making use of the vector field

$$\partial_m \phi := -\Lambda^m_{<0} \phi, \quad \partial_m = \frac{\partial}{\partial t_m},$$

and the flows commute. In the Lax form the KP flows are written as

$$\partial_m \Lambda = [\Lambda^m_{\geq 0}, \Lambda]. \tag{8.24}$$

If $m = 1$, then $\partial \Lambda = [\partial, \Lambda] = \sum_{k=1}^{\infty} (\partial \lambda_k) \partial^{-k}$, which justifies the identification $\partial = \partial_{t_1}$.

Thus, the Baker–Akhiezer function $\Psi_{W_{T_n}}[g](z)$ admits the form

$$\Psi_{W_{T_n}}[g](z) = \phi \exp(\xi(\mathbf{t}, z))$$

where ϕ is a pseudo-differential operator $\phi = 1 + \sum_{k=1}^{n} \omega_k(\mathbf{t}, W_{T_n}) \partial^{-k}$. The function $\Psi_{W_{T_n}}[g](z)$ becomes the eigenfunction of the operator Λ^m, namely $\Lambda^m w = z^m w$, for $m \in \mathbb{Z}$. Besides, $\partial_m w = \Lambda^m_{\geq 0} w$. In the view of (8.23) we can write this function as previously,

$$\Psi_{W_{T_n}}[g](z) = \left(1 + \sum_{k=1}^{n} \omega_k(\mathbf{t}, W_{T_n}) z^{-k} \right) e^{\xi(\mathbf{t}, z)}.$$

Proposition 8.8.1. *Let $n = 1$, and let the Baker–Akhiezer function be of the form*

$$\Psi_{W_{T_n}}[g](z) = e^{\xi(t,z)}\left(1 + \frac{\omega}{z}\right),$$

where $\omega = \omega_1$ is given by the formula (8.22). Then

$$\partial\omega = \frac{\partial^2 A}{1 - A} + \left(\frac{\partial A}{1 - A}\right)^2$$

is a solution to the KP equation with the Lax operator $L = \partial^2 - 2(\partial\omega)$.

Proof. First of all, we observe that

$$\Lambda_{\geq 0}^2 = \partial^2 + 2\lambda_1, \quad \Lambda_{\geq 0}^3 = \partial^3 + 3\lambda_1\partial + 3(\partial\lambda_1) + 3\lambda_2.$$

Given $\phi = 1 + \omega\partial^{-1}$, we are looking for the coefficient λ_1 checking the equality $\partial_{t_2}\Psi_{W_{T_n}}[g] = L\,\Psi_{W_{T_n}}[g]$, for the Lax operator $L = \Lambda_{\geq 0}^2 = \partial^2 + 2\lambda_1$.

First of all, we need some auxiliary calculations

$$\partial_{t_k} A = \partial^k A, \quad k = 1, 2, \ldots; \qquad \partial_{t_2}\omega = \partial^2\omega - 2\omega\partial\omega;$$

$$\partial_{t_2}\Psi = z^2\Psi + g\frac{\partial_{t_2}\omega}{z};$$

$$\partial^2\Psi = z^2\Psi + \frac{g}{z}(2z\partial\omega + 2\omega\partial\omega + \partial_{t_2}\omega).$$

Then, comparing the latter two equalities we conclude that $\partial_{t_2}\Psi = \partial^2\Psi - (2\partial\omega)\Psi$, and $\lambda_1 = -\partial\omega$. Now we use the formula for the KP hierarchy (8.24) and write the time evolutions

$$\partial_{t_2}\lambda_1 = \partial^2\lambda_1 + 2\partial\lambda_2,$$
$$\partial_{t_2}\lambda_2 = \partial^2\lambda_2 + 2\partial\lambda_3 + 2\lambda_1\partial\lambda_1,$$
$$\partial_{t_3}\lambda_1 = \partial^3\lambda_1 + 3\partial^2\lambda_2 + 3\partial\lambda_3 + 6\lambda_1\partial\lambda_1.$$

Finally, eliminating λ_2 and λ_3 we arrive at the first equation (KP equation) in the KP hierarchy for $\partial\omega$,

$$3\partial_{t_2}^2\lambda_1 = \partial(4\partial_{t_3}\lambda_1 - 12\lambda_1\partial\lambda_1 - \partial^3\lambda_1).$$

The latter is a standard procedure, see, e.g., [28]. □

Of course, one can express the Baker–Akhiezer function directly from the τ-function by the Sato formula

$$\Psi_{W_{T_n}}[g](z) = e^{\xi(t,z)}\frac{\tau_{W_{T_n}}\left(t_1 - \frac{1}{z}, t_2 - \frac{1}{2z^2}, t_3 - \frac{1}{3z^3}, \ldots\right)}{\tau_{W_{T_n}}(t_1, t_2, t_3 \ldots)},$$

or applying the *vertex operator* V acting on the Fock space $\mathbb{C}[\mathbf{t}]$ of homogeneous polynomials

$$\Psi_{W_{T_n}}[g](z) = \frac{1}{\tau_{W_{T_n}}} V \tau_{W_{T_n}},$$

where

$$V = \exp\left(\sum_{k=1}^{\infty} t_k z^k\right) \exp\left(-\sum_{k=1}^{\infty} \frac{1}{k} \frac{\partial}{\partial t_k} z^{-k}\right).$$

In the latter expression exp denotes the formal exponential series and z is another formal variable that commutes with all Heisenberg operators t_k and $\frac{\partial}{\partial t_k}$. Observe that the exponents in V do not commute and the product of exponentials is calculated by the Baker–Campbell–Hausdorff formula. The operator V is a vertex operator in which the coefficient V_k in the expansion of V is a well-defined linear operator on the space $\mathbb{C}[\mathbf{t}]$. The Lie algebra of operators spanned by $1, t_k, \frac{\partial}{\partial t_k}$, and V_k, is isomorphic to the affine Lie algebra $\hat{\mathfrak{sl}}(2)$. The vertex operator V plays a central role in the highest weight representation of affine Kac–Moody algebras [291], [394], and can be interpreted as the infinitesimal Bäcklund transformation for the Korteweg–de Vries equation [127].

The vertex operator V recovers the Virasoro algebra in the following sense. Taken in two close points $z + \lambda/2$ and $z - \lambda/2$ the operator product can be expanded into the following formal Laurent–Fourier series

$$: V\left(z + \frac{\lambda}{2}\right) V\left(z - \frac{\lambda}{2}\right) := \sum_{k \in \mathbb{Z}} W_k(z) \lambda^k,$$

where $: ab :$ stands for the bosonic normal ordering. Then $W_2(z) = T(z)$ is the stress-energy tensor which we expand again as

$$T(z) = \sum_{n \in \mathbb{Z}} L_n(\mathbf{t}) z^{n-2},$$

where the operators L_n are the Virasoro generators in the highest weight representation over $\mathbb{C}[\mathbf{t}]$. Observe that the generators L_n span the full Virasoro algebra with central extension and with the central charge 1.

8.9   Chordal Löwner equation and Benney moments

Benney [47] investigated long non-linear waves propagating on a free surface showing that the governing equations have an infinite number of conservation laws. Gibbons and Tsarëv [206] were the first to notice is that the chordal Löwner equation plays an essential role in the classification of reductions of the Benney equations.

The idea of this section is to revisit the Gibbons and Tsarëv observation and to show that the chordal Löwner evolution also possesses an infinite number of conservation laws, moments. We show that the Löwner PDE is exactly the Vlasov equation under an appropriate change of variables and the Löwner ODE implies the hydrodynamical type conservation equation.

Let us consider a Löwner chain of receding domains $\mathbb{H}_t = \mathbb{H} \setminus \gamma_t$ in the upper half-plane $\mathbb{H} = \{z \colon \mathrm{Im}\, z > 0\}$ and let $f \colon \mathbb{H} \to \mathbb{H}_t$ be normalized near infinity as

$$f(z,t) = z + \frac{A^0}{z} + O\left(\frac{1}{z^2}\right), \tag{8.25}$$

where $(-A^0(t))$ is the half-plane capacity of γ_t. Let γ_t be a Jordan curve in $\mathbb{H}$ except for an end point on the real axis $\mathbb{R}$, γ_t is parameterized by t. Then f satisfies the Löwner PDE

$$(z - \xi_t)\frac{\partial f(z,t)}{\partial t} - \frac{dA^0}{dt}\frac{\partial f(z,t)}{\partial z} = 0, \tag{8.26}$$

with a real-valued continuous driving function ξ_t and an initial condition $f(z,0) = f_0(z)$.

Remark 8.9.1. Actually, equation (8.26) appeared first in 1946, by Kufarev [329, Introduction].

For every $t \geq 0$, the function $f(z,t)$ has a continuous extension on the closure of $\mathbb{H}$, and the extended function denoted also by $f(z,t)$ satisfies equation (8.26) at least almost everywhere. The driving function ξ_t generates the growing slit γ_t.

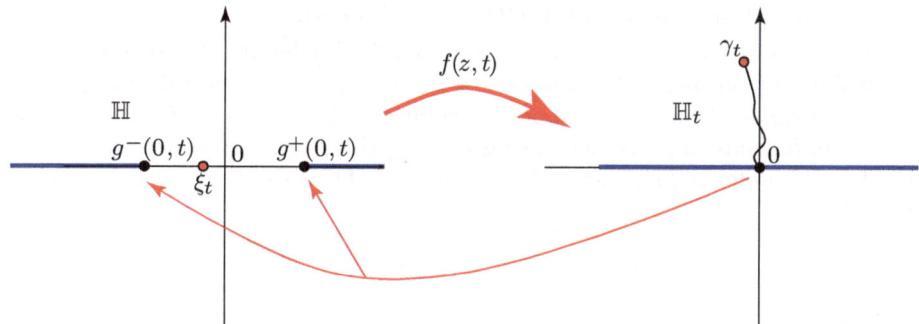

Fig. 8.1: Chordal decreasing Löwner chain

We will also use the two-parametric family of conformal maps

$$g(z,t,\tau) := f^{-1}(f(z,t),\tau),$$

where $0 \leq \tau \leq t < \infty$. We also write $g(z,t,0) =: g(z,t)$. The function g maps the half-plane $\mathbb{H}$ onto a subset of $\mathbb{H}$. It satisfies the Löwner ODE for the half-plane

$$\frac{\partial g(z,t,\tau)}{\partial t} = -\frac{dA^0/dt}{g(z,t,\tau) - \xi_t}, \quad 0 \leq \tau \leq t < \infty, \quad g(z,\tau,\tau) = z. \tag{8.27}$$

Moreover, $\lim_{t\to\infty} g(z,t,\tau) = \lim_{t\to\infty} f^{-1}(f(z,t),\tau) = f(z,\tau)$.

Define the real-valued functions $t = t(x, s)$, as a solution to the quasi-linear differential equation

$$\xi_t \frac{\partial t}{\partial x} + \frac{\partial t}{\partial s} = 0, \qquad (8.28)$$

satisfying the asymptotic behaviour $\lim_{x \to \infty} t(x, s) = \lim_{x \to -\infty} t(x, s) < \infty$. Assume that ξ_t is a function which admits a cone of solutions to (8.28) with the needed asymptotic behaviour.

Now, let us consider the superposition $f(z, t(x, s))$ and multiply both sides of (8.26) by $\frac{\partial t}{\partial x}$. By abuse of notation, we continue to write f for the function $f(z, t(x, s)) = f(z, x, s)$. Then

$$z \frac{\partial f}{\partial x} - \xi_t \frac{\partial f(z, t)}{\partial t} \frac{\partial t}{\partial x} - \frac{\partial A^0}{\partial x} \frac{\partial f}{\partial z} = 0.$$

If we use equation (8.28), then

$$z \frac{\partial f}{\partial x} + \frac{\partial f}{\partial s} - \frac{\partial A^0}{\partial x} \frac{\partial f}{\partial z} = 0, \qquad (8.29)$$

which is the Vlasov equation, see [206, 573], describing time evolution of the distribution function of plasma consisting of charged particles with long-range interaction. In fluid descriptions of plasmas one does not consider the velocity distribution but rather the plasma moments $A^n(t(x, s)) \equiv A^n(x, s)$.

Among solutions to the Vlasov equation (8.29) let us choose those which provide finite integrals for the moments A^n. Namely, for a given solution $f(z, x, s)$ with the normalization (8.25), choose a solution $\phi(z, x, s) = \varphi(f(z, x, s))$ where φ is an appropriate rapidly decreasing at infinities $z \to \pm\infty$ function, see, e.g., [427]. For example, $\varphi(f) = \exp(-f^2)$ is suitable. Then the moments $A^n(x, s)$ are defined by

$$A^n(x, s) = \int_{-\infty}^{\infty} w^n \, \phi(w, x, s) \, dw, \quad n \geq 1.$$

Direct computations imply that

$$A_s^n = \int_{-\infty}^{\infty} w^n \frac{\partial \phi}{\partial s} \, dw, \quad A_x^{n+1} = \int_{-\infty}^{\infty} w^{n+1} \frac{\partial \phi}{\partial x} \, dw.$$

Integrating by parts yields

$$A^{n-1} = -\int_{-\infty}^{\infty} \frac{w^n}{n} \frac{\partial \phi}{\partial w} \, dw.$$

Now we can use the Vlasov equation (8.29) and arrive at the equation for the moments

$$A_s^n + A_x^{n+1} + nA^{n-1}A_x^0 = 0, \qquad (8.30)$$

which is an infinite autonomous system, known as Benney's moment equations, see [47], which appear in long wavelength hydrodynamics of an ideal incompressible fluid of a finite depth in a gravitational field.

Following [206, 332] let us define a function $\lambda(z, x, s)$ by the Cauchy principal value of a singular integral

$$\lambda(z, x, s) = z + \int_{-\infty}^{\infty} \frac{\phi(w, x, s)}{z - w} dw = z + \sum_{n=0}^{\infty} \frac{A^n}{z^{n+1}}, \quad z \to \infty \text{ in } \mathbb{H},$$

where $z = g(w, t(x, s))$ and the coefficient A^0 is the same as for f. Then,

$$\lambda_s = \frac{\partial \lambda}{\partial s} = z_s + \sum_{n=0}^{\infty} \left(\frac{A_s^n}{z^{n+1}} - \frac{(n+1)A^n z_s}{z^{n+2}} \right),$$

$$\lambda_x = \frac{\partial \lambda}{\partial x} = z_x + \sum_{n=0}^{\infty} \left(\frac{A_x^n}{z^{n+1}} - \frac{(n+1)A^n z_x}{z^{n+2}} \right),$$

and

$$\lambda_s + z\lambda_x = z_s + z \cdot z_x + A_x^0 + \sum_{n=0}^{\infty} \frac{A_s^n + A_x^{n+1} - nA^{n-1}z_s - (n+1)A^n z_x}{z^{n+1}}.$$

Making use of the moment equations we come to

$$\lambda_s + z\lambda_x = z_s + z \cdot z_x + A_x^0 + \sum_{n=0}^{\infty} \frac{-nA^{n-1}A_x^0}{z^{n+1}} - \sum_{n=1}^{\infty} \frac{nA^{n-1}(z_s + z \cdot z_x)}{z^{n+1}}$$

$$= A_x^0 \lambda_z + (z_s + z \cdot z_x) \left(1 - \sum_{n=1}^{\infty} \frac{nA^{n-1}}{z^{n+1}} \right) = \lambda_z \left(A_x^0 + z_s + z \cdot z_x \right).$$

The Löwner ODE (8.27) implies that $A_x^0 = z_x(\xi_t - z)$ and the definition of the function $t(x, s)$ yields that

$$A_x^0 + z_s + z \cdot z_x = 0, \tag{8.31}$$

and therefore, the equality $\lambda_s + z\lambda_x = 0$ holds along the trajectories of the Löwner ODE (8.27). Equation (8.31) received the name "Gibbons' equation" in [428] following the original paper by Gibbons [205].

Let us consider the map $z(\lambda, x, s)$ which is the inverse to $\lambda(z, x, s)$ with respect to $\lambda \leftrightarrow z$,

$$\lambda(z, x, s) = z + \sum_{n=0}^{\infty} \frac{A^n}{z^{n+1}}, \quad z \to \infty \text{ in } \mathbb{H},$$

$$z(\lambda, x, s) = \lambda - \sum_{n=0}^{\infty} \frac{H^n}{\lambda^{n+1}}.$$

Then,

$$\sum_{n=0}^{\infty} \frac{A^n}{z^{n+1}} = \sum_{n=0}^{\infty} \frac{H^n}{\lambda^{n+1}},$$

and

$$\frac{\lambda}{z}\left(A^0 + \frac{A^1}{z} + \cdots\right) = H^0 + \frac{H^1}{\lambda} + \cdots$$

So $H^0 = A^0$. We continue by

$$\lambda\left(\frac{\lambda}{z} - 1\right)A^0 + \frac{\lambda^2}{z^2}\left(A^1 + \frac{A^2}{z} + \cdots\right) = H^1 + \frac{H^2}{\lambda} + \cdots,$$

and conclude $H^1 = A^1$. In the same fashion we come to

$$\lambda^2\left(\frac{\lambda}{z} - 1\right)A^0 + \lambda\left(\frac{\lambda^2}{z^2} - 1\right)A^1 + \frac{\lambda^3}{z^3}\left(A^2 + \frac{A^3}{z} + \cdots\right) = H^2 + \frac{H^3}{\lambda} + \cdots,$$

and $H^2 = A^2 + (A^0)^2$. Finally, we have

$$\sum_{k=0}^{n} \lambda^{n-k}\left(\frac{\lambda^{k+1}}{z^{k+1}}A^k - H^k\right) + \frac{\lambda^{n+1}}{z^{n+2}}\left(A^{n+1} + \frac{A^{n+2}}{z} + \cdots\right)$$
$$= \frac{1}{\lambda}\left(H^{n+1} + \frac{H^{n+2}}{\lambda} + \cdots\right),$$

and the coefficient H^n is calculated as $H^n = A^n + P(A^0, \ldots, A^{n-1})$, where $P(A^0, \ldots, A^{n-1})$ is a polynomial of $A^0, \ldots, A^{n-1}$, $n \geq 2$. The first coefficients are

$$H^0 = A^0, \quad H^1 = A^1, \quad H^2 = A^2 + (A^0)^2, \quad H^3 = A^3 + 3A^0 A^1,$$
$$H^4 = A^4 + 4A^0 A^2 + 2(A^1)^2 + 2(A^0)^3.$$

Analogous coefficients were calculated in, e.g., [332, 546].

This way the Löwner ODE (8.27) becomes the conservation equation in the following sense. According to (8.31)

$$\frac{d}{ds}\int_{-\infty}^{\infty} z(\lambda, x, s)\, dx = -\int_{-\infty}^{\infty} (A_x^0 + z \cdot z_x)\, dx,$$

where we integrate with respect to $x \in \mathbb{R}$ in the Cauchy principal value sense. The requirements on the asymptotic behaviour of $t(x, s)$ as $x \to \pm\infty$ imply that

$$\int_{-\infty}^{\infty} (A_x^0 + z \cdot z_x)\, dx = 0,$$

Therefore

$$\frac{d}{ds} \int_{-\infty}^{\infty} z(\lambda, x, s) \, dx = 0,$$

which corresponds to the momentum conservation law. So the conserved quantities of the evolution are the moments

$$I^n = \int_{-\infty}^{\infty} H^n(x, s) \, dx, \quad n \geq 0.$$

Analogous integrals of motion were studied in the original work by Benney [47] as well as in [332, 426, 594].

The Poisson structure allows us to reformulate the Benney moment equation (8.30) as an evolution equation with a Hamiltonian function. The Kupershmidt–Manin Poisson structure [332, 333] starts with the operators of differentiation and multiplication to the right for the moments $A^n \frac{\partial}{\partial x}$ as skew-symmetric operators with respect to the $L^2(\mathbb{R})$-paring, acting to the right by

$$\{A^m, A^n\}(\cdot) = -mA^{n+m-1}\frac{\partial}{\partial x}(\cdot) - n\frac{\partial}{\partial x}\left(A^{n+m-1}(\cdot)\right).$$

Then for any two observables $F(A)$ and $G(A)$, the Poisson bracket can be written as

$$\{F, G\}(A) = \sum_{m,n=0}^{\infty} \int_{-\infty}^{\infty} \frac{\delta F}{\delta A^m}\{A^m, A^n\}\frac{\delta G}{\delta A^n} \, dx.$$

Writing $\bar{H}^k = \frac{1}{k}\int_{-\infty}^{\infty} H^k dx = \frac{1}{k}I^k$, we have the hierarchy of commuting flows with the Hamiltonians $\bar{H}^k$: $\{\bar{H}^k, \bar{H}^j\} = 0$, in the form of evolution equations

$$\frac{\partial A^m}{\partial s} = \sum_{n=0}^{\infty} \{A^m, A^n\}\frac{\delta \bar{H}}{\delta A^n},$$

so that equation (8.30) becomes the equation in this hierarchy with the Hamiltonian $\bar{H}^2$.

Chapter 9

Stochastic Löwner and Löwner–Kufarev Evolution

The modern period of the development of the Löwner theory has been marked by a burst of interest in Stochastic (Schramm)–Löwner Evolution (SLE) which has implied an elegant description of several 2D conformally invariant statistical physical systems at criticality by means of a solution to the Cauchy problem for the Löwner equation with a random driving term $\sqrt{\kappa}B_t$ given by the one-dimensional Brownian motion B_t defined on the standard filtered probability space, see [351, 518]. Different values of κ correspond to different universality classes of critical behaviour. At the same time, several connections with mathematical physics were discovered, in particular, relations with a singular representation of the Virasoro algebra using Löwner martingales in [35, 195, 294] and with a Hamiltonian formulation and a construction of the KP integrable hierarchies in [366, 366, 537].

This chapter we dedicate to the stochastic counterpart of the Löwner–Kufarev theory first recalling that one of the last (but definitely not least) contributions to this growing theory was the description by Oded Schramm in 1999–2000 [518], of the stochastic Löwner evolution (SLE), also known as the Schramm–Löwner evolution. The SLE is a conformally invariant stochastic process; more precisely, it is a family of random planar curves generated by solving Löwner's differential equation with the Brownian motion as a driving term. This equation was studied and developed by Oded Schramm together with Greg Lawler and Wendelin Werner in a series of joint papers that led, among other things, to a proof of Mandelbrot's conjecture about the Hausdorff dimension of the Brownian frontier [351]. This achievement was one of the reasons Werner was awarded the Fields Medal in 2006. Sadly, Oded Schramm, born 10 December 1961 in Jerusalem, died in a tragic hiking accident on 01 September 2008 while climbing Guye Peak, north of Snoqualmie Pass in Washington.

We are not going to present a complete exposition of SLE, which has been done in several research and exposition texts, such as, e.g., [70, 349, 350, 479]. Instead we focus on other possible stochastic counterparts of the Löwner–Kufarev theory.

9.1   SLE

9.1.1   Half-plane version of the Löwner equation

In 1946, Kufarev [329, Introduction] first mentioned an evolution equation in the upper half-plane $\mathbb{H}$ analogous to the one introduced by Löwner in the unit disk, and this was first studied by Popova [449] in 1954. In 1968, Kufarev, Sobolev and Sporysheva [335] introduced a combination of Goluzin–Schiffer's variational and parametric methods for this equation for the class of univalent functions in the upper half-plane, which is known to be related to physical problems in hydrodynamics. They showed its application to the extremal problem of finding the range of $\{\mathrm{Re}\, e^{i\alpha} f(z),\ \mathrm{Im}\, f(z)\}$, $\mathrm{Im}\, z > 0$. Moreover, during the second half of the past century, the Soviet school intensively studied Kufarev's equations for $\mathbb{H}$. We ought to cite here at least contributions by Aleksandrov [17], Aleksandrov and Sobolev [18], Goryainov and Ba [221, 222]. However, this work was mostly unknown to many Western mathematicians, in particular, because some of it appeared in journals not easily accessible from outside of the Soviet Union. In fact, some of Kufarev's papers were not even reviewed by Mathematical Reviews. Anyhow, we refer the reader to [19], which contains a complete bibliography of his papers.

In order to introduce Kufarev's equation properly, let us fix some notation. Let γ be a Jordan arc in the upper half-plane $\mathbb{H}$ with starting point $\gamma(0) = 0$. Then there exists a unique conformal map $g_t : \mathbb{H} \setminus \gamma[0, t] \to \mathbb{H}$ with the normalisation

$$g_t(z) = z + \frac{c(t)}{z} + O\left(\frac{1}{z^2}\right), \quad z \sim \infty.$$

After a reparametrisation of the curve γ, one can assume that $c(t) = 2t$. Under this normalisation, one can show that g_t satisfies the following differential equation:

$$\frac{dg_t(z)}{dt} = \frac{2}{g_t(z) - \xi(t)}, \qquad g_0(z) = z. \tag{9.1}$$

The equation is valid up to a time $T_z \in (0, +\infty]$ that can be characterised as the first time t such that $g_t(z) \in \mathbb{R}$ and where $\xi(t)$ is a continuous real-valued function. Conversely, given a continuous function $\xi : [0, +\infty) \to \mathbb{R}$, one can consider the following initial value problem for each $z \in \mathbb{H}$:

$$\frac{dw}{dt} = \frac{2}{w - \xi(t)}, \qquad w(0) = z. \tag{9.2}$$

Let $t \mapsto w^z(t)$ denote the unique solution to this Cauchy problem and let $g_t(z) := w^z(t)$. Then g_t maps holomorphically a (not necessarily slit) subdomain of the upper half-plane $\mathbb{H}$ onto $\mathbb{H}$. Equation (9.2) is nowadays known as the *chordal Löwner differential equation* with the function ξ as the driving term. The name is due to the fact that the curve $\gamma[0, t]$ evolves in time as t tends to infinity into a sort of chord joining two boundary points. This kind of construction can be used to model evolutionary aspects of decreasing families of domains in the complex plane. The equation (8.5) with the function p given by (8.4) in this context is called the *radial Löwner equation*, because in the slit case, the tip of the slit tends to the origin in the unit disk.

Quite often the half-plane version is presented, considering the inverse of the functions g_t (see, i.e., [335]). Namely, the conformal mappings $f_t = g_t^{-1}$ from $\mathbb{H}$ onto $\mathbb{H} \setminus \gamma([0,t))$ satisfy the PDE

$$\frac{\partial f_t(z)}{\partial t} = -f_t'(z) \frac{2}{z - \xi(t)}. \tag{9.3}$$

We remark that using the Cayley transform $T(z) = \frac{z+1}{1-z}$ we obtain that the chordal Löwner equation in the unit disk takes the form

$$\frac{\partial h_t(z)}{\partial t} = -h_t'(z)(1 - z)^2 p(z, t), \tag{9.4}$$

where $\operatorname{Re} p(z, t) \geq 0$ for all $t \geq 0$ and $z \in \mathbb{D}$. From a geometric point of view, the difference between this family of parametric functions and those described by the Löwner–Kufarev equation (8.3) is clear: the ranges of the solutions of (9.4) decrease with t while in the former equation (8.3) they increase. This duality 'decreasing' versus 'increasing' has been recently analyzed in [102] and, roughly speaking, we can say that the 'decreasing' setting can be deduced from the 'increasing' one.

9.1.2   SLE

The *(chordal) stochastic Löwner evolution* with parameter $k \geq 0$ (SLE$_k$) starting at a point $x \in \mathbb{R}$ is the random family of maps (g_t) obtained from the chordal Löwner equation (9.1) by letting $\xi(t) = \sqrt{k}B_t$, where B_t is a standard one-dimensional Brownian motion such that $\sqrt{k}B_0 = x$. Namely, let us consider the equation

$$\frac{dg_t(z)}{dt} = \frac{2}{g_t(z) - \xi(t)}, \qquad g_0(z) = z, \tag{9.5}$$

where $\xi(t) = \sqrt{k}B_t = \sqrt{k}B_t(\omega)$, where $B_t(\omega)$ is the standard one-dimensional Brownian motion defined on the standard filtered probability space $(\Omega, \mathcal{G}, (\mathcal{G}_t), P)$ of Brownian motion with the sample space $\omega \in \Omega$, and $t \in [0, \infty)$, $B_0 = 0$. The solution to (9.5) exists as long as $g_t(z) - \xi(t)$ remains away from zero and we denote by T_z the first time that $\lim_{t \to T_z - 0}(g_t(z) - \xi(t)) = 0$. The function g_t satisfies the hydrodynamic normalization at infinity $g_t(z) = z + \frac{2t}{z} + \cdots$. Let $K_t = \{z \in \hat{\mathbb{H}} : T_z \leq t\}$, and let $\mathbb{H}_t$ be its complement $\mathbb{H} \setminus K_t = \{z \in \mathbb{H} : T_z > t\}$. The set K_t is called *SLE hull*. It is compact, $\mathbb{H}_t$ is a simply connected domain and g_t maps $\mathbb{H}_t$ onto $\mathbb{H}$. SLE hulls grow in time. The *trace* γ_t is defined as $\lim_{z \to \xi(t)} g_t^{-1}(z)$, where the limit is taken in $\mathbb{H}$. The unbounded component of $\mathbb{H} \setminus \gamma_t$ is $\mathbb{H}_t$. The Hausdorff dimension of the SLE$_k$ trace is $\min(1 + \frac{8}{k}, 2)$, see [39]. Similarly, one can define the radial stochastic Löwner evolution. The terminology comes from the fact that the Löwner trace tip tends almost surely to a boundary point in the chordal case (∞ in the half-plane version) or to the origin in the disk version of the radial case.

Chordal SLE enjoys two important properties: scaling invariance and the Markovian property. Namely,

- $g_t(z)$ and $\frac{1}{\lambda}g_{\lambda^2 t}(\lambda z)$ are identically distributed;
- $h_t(z) = g_t(z) - \xi(t)$ possesses the Markov property. Furthermore, $h_s \circ h_t^{-1}$ is distributed as h_{s-t} for $s > t$.

The SLE_k depends on the choice of ω and it comes in several flavors depending on the type of Brownian motion exploited. For example, it might start at a fixed point or start at a uniformly distributed point, or might have a built-in drift and so on. The parameter k controls the rate of diffusion of the Brownian motion and the behaviour of the SLE_k critically depends on the value of k.

The SLE_2 corresponds to the loop-erased random walk and the uniform spanning tree. The $\mathrm{SLE}_{8/3}$ is conjectured to be the scaling limit of self-avoiding random walks. The SLE_3 is proved [98] to be the limit of interfaces for the Ising model (another Fields Medal 2010 awarded to Stanislav Smirnov), while the SLE_4 corresponds to the harmonic explorer and the Gaussian free field. For all $0 \leq k \leq 4$, SLE gives slit maps. The SLE_6 was used by Lawler, Schramm and Werner in 2001 [351] to prove the conjecture of Mandelbrot (1982) that the boundary of planar Brownian motion has fractal dimension $4/3$. Moreover, Smirnov [527] proved that SLE_6 is the scaling limit of critical site percolation on the triangular lattice. This result follows from his celebrated proof of Cardy's formula. SLE_8 corresponds to the uniform spanning tree. For $4 < k < 8$ the curve intersects itself and every point is contained in a loop but the curve is not space-filling almost surely. For $k \geq 8$ the curve is almost sure space-filling. This phase change is due to the Bessel process interpretation of SLE, see (9.6).

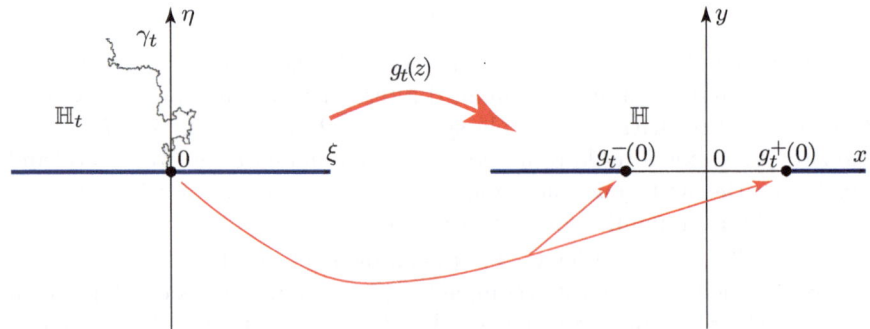

Fig. 9.1: SLE_k for $0 < k < 4$

An invariant approach to SLE starts with probability measures on non-self-crossing random curves in a domain Ω connecting two given points $a, b \in \partial\Omega$ and satisfying the properties of

- *Conformal invariance.* Consider a triple (Ω, a, b) and a conformal map ϕ. If γ is a trace $\mathrm{SLE}_k(\Omega, a, b)$, then $\phi(\gamma)$ is a trace $\mathrm{SLE}_k(\phi(\Omega), \phi(a), \phi(b))$.;
- *Domain Markov property.* Let $\{\mathcal{F}_t\}_{t \geq 0}$ be the filtration in $\mathcal{F}$ by $\{B_t\}_{t \geq 0}$ and let g_t be a Löwner flow generated by $\xi(t) = \sqrt{k}B_t$. Then the hulls $(g_t(K_{s+t} \cap \mathbb{H}_t) - \xi(t))_{s \geq 0}$ are also generated by SLE_k and independent of the sigma-algebra $\mathcal{F}$.

The expository paper [349] is perhaps the best option to start an exploration of this fascinating branch of mathematics. Nice papers [479, 480] give up-to-date exposition of developments of SLE so we do not intend to survey SLE in detail here.

Equation (9.5) is deterministic with a random entry and we solve it for every fixed ω. The corresponding stochastic differential equation (SDE) in the Itô form for the function $h_t(z) = g_t(z) - \xi(t)$ is

$$dh_t(z) = \frac{2}{h_t(z)}dt - \sqrt{k}dB_t, \tag{9.6}$$

where $-h_t/\sqrt{k}$ represents a Bessel process (of order $(4+k)/k$). For any holomorphic function $M(z)$ we have the Itô formula

$$(dM)(h_t) = -d\xi \mathcal{L}_{-1}M(h_t) + dt\left(\frac{k}{2}\mathcal{L}_{-1}^2 - 2\mathcal{L}_{-2}\right)M(h_t), \tag{9.7}$$

where $\mathcal{L}_n = -z^{n+1}\partial$. From the form of equation (9.6) one can see immediately that h_t is a (time-homogeneous) diffusion, i.e., a continuous strong Markov process. The infinitesimal generator of h_t is given by $A = (\frac{k}{2}\mathcal{L}_{-1}^2 - 2\mathcal{L}_{-2})$ and this operator appears here for the first time. This differential operator makes it possible to reformulate many probabilistic questions about h_t in the language of PDE theory. If we consider $h_t(z)$ with fixed z, then the equation (9.6) for h_t describes the motion of particles in the time-dependent field v with $dv = -d\xi \mathcal{L}_{-1} + dtA$. For instance, if we denote by $u_t(z)$ the mean function of $h_t(z)$,

$$u_t(z) = \mathbb{E}h_t(z),$$

then it follows from Kolmogorov's backward equation that u_t satisfies

$$\begin{cases} \dfrac{\partial u_t}{\partial t} = Au_t, \\ u_0(z) = z, \end{cases} \quad z \in \mathbb{H}.$$

The kernel of the operator A describes driftless observables with time-independent expectation known as local martingales or conservation (in mean) laws of the process.

9.2   Generalized Löwner–Kufarev evolution

The pioneering idea of Löwner [362] in 1923 contained two main ingredients: subordination chains and semigroups of conformal maps. In this section we concentrate our attention on a generalization of the Löwner–Kufarev approach that emerged recently in [3, 68, 69, 101]. One of the main results of the generalized Löwner–Kufarev theory is an essentially one-to-one correspondence between evolution families, Herglotz vector fields and (generalized) Löwner chains. Below we briefly describe this approach.

9.2.1   Evolution families and Herglotz vector fields

We start with the notion of an *evolution family*. Let us consider a semigroup $\mathcal{P}$ of conformal univalent maps from the unit disk $\mathbb{D}$ into itself with superposition as a semigroup operation. This makes $\mathcal{P}$ a topological semigroup with respect to the topology of local uniform convergence on $\mathbb{D}$.

Definition 9.2.1. An evolution family of order $d \in [1, +\infty]$ is a two-parameter family $(\phi_{s,t})_{0 \leq s \leq t < +\infty}$ of holomorphic endomorphisms of the unit disk from $\mathcal{P}$, such that the following three conditions are satisfied.

- $\phi_{s,s} = id_{\mathbb{D}}$;
- $\phi_{s,t} = \phi_{u,t} \circ \phi_{s,u}$ for all $0 \leq s \leq u \leq t < +\infty$;
- for any $z \in \mathbb{D}$ and $T > 0$ there is a function $k_{z,T} \in L^d([0,T], \mathbb{R})$ such that

$$|\phi_{s,u}(z) - \phi_{s,t}(z)| \leq \int_u^t k_{z,T}(\xi)d\xi,$$

  for all $0 \leq s \leq u \leq t \leq T$.

An infinitesimal description of an evolution family is given in terms of a *Herglotz vector field*.

Definition 9.2.2. A (generalized) Herglotz vector field of order d is a function $G : \mathbb{D} \times [0, +\infty) \to \mathbb{C}$ satisfying the following conditions:

- the function $[0, +\infty) \ni t \mapsto G(z, t)$ is measurable for all $z \in \mathbb{D}$;
- the function $z \mapsto G(z, t)$ is holomorphic in the unit disc for $t \in [0, +\infty)$;
- for any compact set $K \subset \mathbb{D}$ and for all $T > 0$ there exists a non-negative function $k_{K,T} \in L^d([0,T], \mathbb{R})$ such that

$$|G(z, t)| \leq k_{K,T}(t)$$

  for all $z \in K$ and almost every $t \in [0, T]$;
- for almost every $t \in [0, +\infty)$ the vector field $G(\cdot, t)$ is semicomplete.

By semicompleteness we mean that the solution to the problem

$$\begin{cases} \dfrac{dx(\tau)}{d\tau} = G(x(\tau), t), \\ x(s) = z \end{cases}$$

is defined for all times $\tau \in [s, +\infty)$, for any fixed $s \geq 0$, fixed $t \geq 0$ and fixed $z \in \mathbb{D}$.

An important result of general Löwner–Kufarev theory is the fact that the evolution families can be put into a one-to-one correspondence with the Herglotz vector fields by means of the so-called generalized Löwner–Kufarev ODE. This can be formulated as the following theorem.

Theorem 9.2.1 ([68, Theorem 1.1]). *For any evolution family $(\phi_{s,t})$ of order $d \geq 1$ in the unit disk there exists an essentially unique Herglotz vector field $G(z,t)$ of order d, such that for all $z \in \mathbb{D}$ and for almost all $t \in [0, +\infty)$*

$$\frac{\partial \phi_{s,t}(z)}{\partial t} = G(\phi_{s,t}(z), t).$$

Conversely, for any Herglotz vector field $G(z,t)$ of order $d \geq 1$ in the unit disk there exists a unique evolution family of order d, such that the equation above is satisfied.

Herglotz vector fields admit a convenient representation using so-called Herglotz functions.

Definition 9.2.3. A Herglotz function of order $d \in [1, +\infty)$ is a function

$$p : \mathbb{D} \times [0, +\infty) \to \mathbb{C}$$

such that

- the function $t \mapsto p(z,t)$ belongs to $L^d_{loc}([0, +\infty), \mathbb{C})$ for all $z \in \mathbb{D}$;
- the function $z \mapsto p(z,t)$ is holomorphic in $\mathbb{D}$ for each fixed $t \in [0, +\infty)$;
- $\operatorname{Re} p(z,t) \geq 0$ for all $z \in \mathbb{D}$ and for all $t \in [0, +\infty)$.

Now, the representation of Herglotz vector fields is given in the following theorem.

Theorem 9.2.2 ([68, Theorem 1.2]). *Given a Herglotz vector field of order $d \geq 1$ in the unit disk, there exists an essentially unique (i.e., defined uniquely for almost all t for which $G(\cdot, t) \neq 0$) measurable function $\tau : [0, +\infty) \to \overline{\mathbb{D}}$ and a Herglotz function $p(z,t)$ of order d, such that for all $z \in \mathbb{D}$ and almost all $t \in [0, +\infty)$*

$$G(z,t) = (z - \tau(t))(\overline{\tau(t)}z - 1)p(z,t). \qquad (9.8)$$

Conversely, given a measurable function $\tau : [0, +\infty) \to \overline{\mathbb{D}}$ and a Herglotz function $p(z,t)$ of order $d \geq 1$, the vector field defined by the formula above is a Herglotz vector field of order d.

According to Theorem 9.2.1, to every evolution family $(\phi_{s,t})$ one can associate an essentially unique Herglotz vector field $G(z,t)$. The pair of functions (p, τ) representing the vector field $G(z,t)$ is called the *Berkson–Porta data* of the evolution family $(\phi_{s,t})$.

To explain the geometrical meaning of the function $\tau(t)$ we need first to recall the notion of the Denjoy–Wolff point of a unit disk endomorphism.

A classical result by Denjoy and Wolff states that for a holomorphic self-map f of the unit disk $\mathbb{D}$ other than a (hyperbolic) rotation, there exists a unique fixed point τ in the closure of $\mathbb{D}$, such that the sequence of iterates $(f_n(z))$ converges locally uniformly on $\mathbb{D}$ to τ as $n \to \infty$. This point τ is called the Denjoy–Wolff point of f and it is also characterized as the only fixed point of f satisfying $f'(\tau) \in \mathbb{D}$. In other words, τ is the only attractive fixed point of f in the above multiplier sense. It follows from the hyperbolic metric principle that, if f is not the identity, there can be no other fixed points in $\mathbb{D}$ except the Denjoy–Wolff point but, nevertheless, f can have many other repulsive or non-regular boundary fixed points.

If $\tau \in \mathbb{D}$, then the endomorphism f is called *elliptic*. Otherwise, the angular limit $\angle \lim_{z \to \tau} f(z) = \tau$ exists as well as the andular derivative $\angle \lim_{z \to \tau} f'(z) = \alpha_f$. If the value $\alpha_f \in (0,1]$, then the map f in this case is said to be either *hyperbolic* (if $\alpha_f < 1$) or *parabolic* (if $\alpha_f = 1$) (for details and proofs see, e.g., [2]).

Now, let $(\phi_{s,t})$ be an evolution family with Berkson–Porta data (p, τ). In the simplest case when neither p, nor τ changes in time (i.e., the corresponding Herglotz vector field $G(z,t)$ is time-independent), τ turns out to be precisely the Denjoy–Wolff point of every endomorphism in the family $(\phi_{s,t})$. Moreover, for any $0 \le s < +\infty$, we have that $\phi_{s,t}(z) \to \tau$ uniformly on compacts subsets of $\mathbb{D}$, as $t \to +\infty$. For this reason, we call τ the *attractive point* of the evolution family $(\phi_{s,t})$.

In the case when the Herglotz field $G(z,t)$ is time-dependent, the meaning of τ is explained in the following theorem.

Theorem 9.2.3 ([68, Theorem 6.7]). *Let $(\phi_{s,t})$ be an evolution family of order $d \ge 1$ in the unit disk, and let $G(z,t) = (z - \tau(t))(\overline{\tau(t)}z - 1)p(z,t)$ be the corresponding Herglotz vector field. Then for almost every $s \in [0, +\infty)$, such that $G(z,s) \ne 0$, there exists a decreasing sequence $\{t_n(s)\}$ converging to s, such that $\phi_{s,t_n(s)} \ne id_{\mathbb{D}}$ and*

$$\tau(s) = \lim_{n \to \infty} \tau(s, n),$$

where $\tau(s, n)$ denotes the Denjoy–Wolff point of $\phi_{s,t_n(s)}$.

9.2.2   Generalized Löwner chains and Löwner–Kufarev PDE

We follow now the exposition [101] of the generalization of the classical notion of Löwner chains.

Definition 9.2.4. A family $(f_t)_{0 \le t < +\infty}$ of holomorphic maps of the unit disk is called a Löwner chain of order $d \in [1, +\infty]$ if

- each function $f_t : \mathbb{D} \to \mathbb{C}$ is univalent,
- $f_s(\mathbb{D}) \subset f_t(\mathbb{D})$ for $0 \le s < t < +\infty$,
- for any compact set $K \subset \mathbb{D}$ and all $T > 0$ there exists a non-negative function $k_{K,T} \in L^d([0,T],\mathbb{R})$ such that

$$|f_s(z) - f_t(z)| \le \int_s^t k_{K,T}(\xi)d\xi$$

for all $z \in K$ and all $0 \le s \le t \le T$.

Every Löwner chain $(f_t)_{0 \le t < +\infty}$ of order d generates an evolution family $\phi_{s,t}$ of the same order d defined by $\phi_{s,t} = f_t^{-1} \circ f_s$. This correspondence is, however, not one-to one; there may be many different Löwner chains associated to the given evolution family. Fortunately, they are unique up to normalization and composition with a univalent function, as the following theorem states.

Theorem 9.2.4 ([101, Theorems 1.6–1.7]). *For any evolution family* $(\phi_{s,t})$ *of order d, there exists a unique Löwner chain* (f_t) *of the same order d, such that*

(i) $\phi_{s,t} = f_t^{-1} \circ f_s$ *for any* $0 \le s \le t$;

(ii) $f(0) = 0$ *and* $f'(0) = 1$;

(iii) $\Omega := \cup_{t \ge 0} f_t(\mathbb{D}) = \{z : |z| < R\}$, *where* $R \in (0, +\infty]$.

Any other Löwner chain satisfying the condition (i) is of the form $(g_t) = (F \circ f_t)$, *where* $F : \Omega \to \mathbb{C}$ *is univalent.*

The number R is equal to $1/\beta_0$, *where*

$$\beta_0 = \lim_{t \to +\infty} \frac{|\phi_{0,t}'(0)|}{1 - |\phi_{0,t}(z)|^2}.$$

It was also shown [101, Theorem 4.1] that every Löwner chain (f_t) of order d satisfies the generalized Löwner PDE

$$\frac{\partial f_s(z)}{\partial s} = -G(z,s)f_s'(z) \quad \text{(for almost all } s \ge 0\text{)},$$

where $G(z,s)$ is the Herglotz vector field generating the associated evolution family $(\phi_{s,t})$.

9.3   Generalized Löwner–Kufarev stochastic evolution

The results of this part were obtained by G. Ivanov and A. Vasil'ev [286]. We considered a setup [286] in which the sample paths are represented by the trajectories of a point (e.g., the origin) in the unit disk $\mathbb{D}$ evolving randomly under the generalized Löwner equation. The driving mechanism differs from SLE. In the SLE case the Denjoy–Wolff attracting point (∞ in the chordal case or a boundary point of the unit disk in the radial case) is fixed. In our case, the attracting point is the driving mechanism and the Denjoy–Wolff point is different from it. 'The relationship between this model and CFT will be the subject of a forthcoming paper by A. Vasil'ev and I. Ivanov. Let us consider the generalized Löwner evolution driven by a Brownian particle on the unit circle. In other words, we study the following initial value problem.

$$\begin{cases} \dfrac{d}{dt}\phi_t(z,\omega) = \dfrac{(\tau(t,\omega) - \phi_t(z,\omega))^2}{\tau(t,\omega)} p(\phi_t(z,\omega),t,\omega), \\[2mm] \phi_0(z,\omega) = z, \end{cases} \quad t \geq 0,\ z \in \mathbb{D},\ \omega \in \Omega. \quad (9.9)$$

The function $p(z,t,\omega)$ is a Herglotz function for each fixed $\omega \in \Omega$. In order for $\phi_t(z,\omega)$ to be an Itô process adapted to the Brownian filtration, we require that the function $p(z,t,\omega)$ be adapted to the Brownian filtration for each $z \in \mathbb{D}$. Even though the driving mechanism in our case differs from that of SLE, the generated families of conformal maps still possess the important time-homogeneous Markov property.

For each fixed $\omega \in \Omega$, equation (9.9) similarly to SLE, may be considered as a deterministic generalized Löwner equation with the Berkson–Porta data $(\tau(\cdot,\omega), p(\cdot,\cdot,\omega))$. In particular, the solution $\phi_t(z,\omega)$ exists, is unique for each $t > 0$ and $\omega \in \Omega$, and moreover, is a family of holomorphic self-maps of the unit disk.

The equation in (9.9) is an example of a so-called *random differential equation* (see, for instance, [530]). Since for each fixed $\omega \in \Omega$ it may be regarded as an ordinary differential equation, the sample paths $t \mapsto \phi_t(z,\omega)$ have continuous first derivatives for almost all ω. See an example of a sample path of $\phi_t(0,\omega)$ for $p(z,t) = \frac{\tau(t)+z}{\tau(t)-z}$, $\tau(t) = e^{ikB_t}$, $k = 5$, $t \in [0,30]$ in the figure to the left.

In order to give an explicitly solvable example let

$$p(z,t,\omega) = \frac{\tau(t,\omega)}{\tau(t,\omega) - z} = \frac{e^{ikB_t(\omega)}}{e^{ikB_t(\omega)} - z}.$$

It makes equation (9.9) linear:

$$\frac{d}{dt}\phi_t(z,\omega) = e^{ikB_t(\omega)} - \phi_t(z,\omega),$$

and a well-known formula from the theory of ordinary differential equation yields

$$\phi_t(z,\omega) = e^{-t}\left(z + \int_0^t e^s e^{ikB_s(\omega)} ds\right).$$

Taking into account the fact that $\mathbb{E}e^{ikB_t(\omega)} = e^{-\frac{1}{2}tk^2}$, we can also write the expression for the mean function $\mathbb{E}\phi_t(z,\omega)$

$$\mathbb{E}\phi_t(z,\omega) = \begin{cases} e^{-t}(z+t), & k^2 = 2, \\ e^{-t}z + \dfrac{e^{-tk^2/2} - e^{-t}}{1 - k^2/2}, & \text{otherwise.} \end{cases} \qquad (9.10)$$

Thus, in this example all maps ϕ_t and $\mathbb{E}\phi_t$ are affine transformations (compositions of a scaling and a translation).

In general, solving the random differential equation (9.9) is much more complicated than solving its deterministic counterpart, mostly because of the fact that for almost all ω the function $t \mapsto \tau(t,\omega)$ is nowhere differentiable.

If we assume that the Herglotz function has the form $p(z,t,\omega) = \tilde{p}(z/\tau(t,\omega))$, then it turns out that the process $\phi_t(z,\omega)$ has an important invariance property, that was crucial in development of SLE.

Let $s > 0$ and introduce the notation

$$\tilde{\phi}_t(z) = \frac{\phi_{s+t}(z)}{\tau(s)}.$$

Then $\tilde{\phi}_t(z)$ is the solution to the initial-value problem

$$\begin{cases} \dfrac{d}{dt}\tilde{\phi}_t(z,\omega) = \dfrac{\left(\tilde{\tau}(t,\omega) - \tilde{\phi}_t(z,\omega)\right)^2}{\tilde{\tau}(t,\omega)} \tilde{p}\left(\tilde{\phi}_t(z,\omega)/\tilde{\tau}(t)\right), \\ \tilde{\phi}_0(z,\omega) = \phi_s(z,\omega)/\tilde{\tau}(s), \end{cases}$$

where $\tilde{\tau}(t) = \tau(s+t)/\tau(s) = e^{ik(B_{s+t}-B_s)}$ is again a Brownian motion on $\mathbb{T}$ (because $\tilde{B}_t = B_{s+t} - B_s$ is a standard Brownian motion). In other words, the conditional distribution of $\tilde{\phi}_t$ given ϕ_r, $r \in [0,s]$ is the same as the distribution of ϕ_t.

By the complex Itô formula, the process $\frac{1}{\tau(t,\omega)} = e^{-ikB_t}$ satisfies the equation

$$de^{-ikB_t} = -ike^{-ikB_t}dB_t - \frac{k^2}{2}e^{-ikB_t}dt.$$

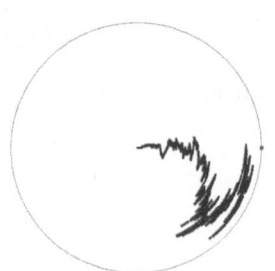

Let us write $\frac{\phi_t(z,\omega)}{\tau(t,\omega)}$ by $\Psi_t(z,\omega)$. Applying the integration by parts formula to Ψ_t, we arrive at the following initial value problem for the Itô stochastic differential equation:

$$
\begin{cases}
d\Psi_t = -ik\Psi_t dB_t + \left(-\frac{k^2}{2}\Psi_t + (\Psi_t - 1)^2 p(\Psi_t e^{ikB_t(\omega)}, t, \omega)\right) dt, \\
\Psi_0(z) = z.
\end{cases}
\tag{9.11}
$$

A numerical solution to this equation for a specific choice for $p(z,t) \equiv i$, $k = 1$, and $t \in [0,2]$, is shown in the figure to the right.

Analyzing the process $\frac{\phi_t(z,\omega)}{\tau(t,\omega)}$ instead of the original process $\phi_t(z,\omega)$ is in many ways similar to one of the approaches used in SLE theory.

The image domains $\Psi_t(\mathbb{D}, \omega)$ differ from $\phi_t(\mathbb{D}, \omega)$ only by rotation. Due to the fact that $|\Psi_t(z,\omega)| = |\phi_t(z,\omega)|$, if we compare the processes $\phi_t(0,\omega)$ and $\Psi_t(0,\omega)$, we note that their first hit times on the circle $\mathbb{T}_r$ with radius $r < 1$ coincide, i.e.,

$$
\inf\{t \geq 0, |\phi_t(0,\omega)| = r\} = \inf\{t \geq 0, |\Psi_t(0,\omega)| = r\}.
$$

In other words, the answers to probabilistic questions about the expected time of hitting the circle $\mathbb{T}_r$, the probability of exit from the disk $\mathbb{D}_r = \{z : |z| < r\}$, etc. are the same for $\phi_t(0,\omega)$ and $\Psi_t(0,\omega)$.

If the Herglotz function has the form $p(z,t,\omega) = \tilde{p}(z/\tau(t,\omega))$, then the equation (9.11) becomes

$$
\begin{cases}
d\Psi_t = -ik\Psi_t dB_t + \left(-\frac{k^2}{2}\Psi_t + (\Psi_t - 1)^2 \tilde{p}(\Psi_t)\right) dt, \\
\Psi_0(z) = z,
\end{cases}
\tag{9.12}
$$

and may be regarded as an equation of a two-dimensional time-homogeneous real diffusion written in complex form. This implies, in particular, that Ψ_t is a time-homogeneous strong Markov process. By construction, $\Psi_t(z)$ always stays in the unit disk.

Analogously to SLE we can consider random conformal Fock space fields defined on $\mathbb{D}$, changing correspondingly the definition using the Green function for $\mathbb{D}$ instead of $\mathbb{H}$. Coupling of equation (9.9) and the Gaussian free field in $\mathbb{D}$ we define the correlators $f_t(z_1, \ldots, z_n) = f(z_1, \ldots, z_n)\big|_{\mathbb{D}_t}$ as martingale-observables.

For a smooth function $f(z)$ defined in $\mathbb{D}$ we derive the Itô differential in the complex form

$$
df(\Psi_t) = -ikdB_t(L_{-1} - \bar{L}_{-1})f(\Psi_t) + dtAf(\Psi_t),
$$

where A is the infinitesimal generator of Ψ_t,

$$
\begin{aligned}
A = &\left(-\frac{k^2}{2}z + (z-1)^2\tilde{p}(z)\right)\frac{\partial}{\partial z} - \frac{1}{2}k^2 z^2 \frac{\partial^2}{\partial z^2} \\
&+ \left(-\frac{k^2}{2}\bar{z} + (\bar{z}-1)^2\overline{\tilde{p}(z)}\right)\frac{\partial}{\partial \bar{z}} - \frac{1}{2}k^2 \bar{z}^2 \frac{\partial^2}{\partial \bar{z}^2} + k^2|z|^2 \frac{\partial^2}{\partial z \partial \bar{z}}.
\end{aligned}
$$

In particular, if f is holomorphic, then

$$A = \left(-\frac{k^2}{2}z + (z-1)^2\tilde{p}(z)\right)\frac{\partial}{\partial z} - \frac{1}{2}k^2z^2\frac{\partial^2}{\partial z^2}. \tag{9.13}$$

If $f(z)$ is a martingale-observable, then $Af = 0$.

In [286] the authors proved the existence of a unique stationary point of Ψ_t in terms of the stochastic vector field

$$\frac{d}{dt}\Psi_t(z,\omega) = G_0(\Psi_t(z,\omega)),$$

where the Herglotz vector field $G_0(z,\omega)$ is given by

$$G_0(z,\omega) = -ikzW_t(\omega) - \frac{k^2}{2}z + (z-1)^2\tilde{p}(z).$$

Here, $W_t(\omega)$ denotes a generalized stochastic process known as *white noise*. Also nth moments were calculated and the boundary diffusion on the unit circle was considered, which corresponds, in particular, to North-South flow, see, e.g., [92].

References

[1] A. Abanov, M. Mineev-Weinstein, and A. Zabrodin, *Multi-cut solutions of Laplacian growth*, Phys. D **238** (2009), no. 17, 1787–1796.

[2] M. Abate, *Iteration theory of holomorphic maps on taut manifolds*, Mediterranean Press, Rende, Cosenza, 1989.

[3] M. Abate, F. Bracci, M.D. Contreras, and S. Díaz-Madrigal, *The evolution of Löwner's differential equations*. Eur. Math. Soc. Newsl. (2010), no. 78, 31–38.

[4] F. Abergel and J. Mossino, *Caractérisation du problème de Muskat multidimensionnel et existence de solutions régulières*, C. R. Acad. Sci. Paris Sér. I Math. **319** (1994), no. 1, 35–40.

[5] B. Ablett, *An investigation of the angle of contact between paraffin wax and water*, Phil. Mag. **46** (1923), 244–256.

[6] O. Agam, E. Bettelheim, P. Wiegmann, and A. Zabrodin, *Viscous fingering and a shape of an electronic droplet in the Quantum Hall regime*, Physical review letters **88** (2003), 236801.

[7] D. Aharonov and H.S. Shapiro, *Domains on which analytic functions satisfy quadrature identities*, J. Analyse Math. **30** (1976), 39–73.

[8] L.V. Ahlfors, *Lectures on quasiconformal mappings*, Van Nostrand Math. Stud., Princeton, N.J., 1966.

[9] L.V. Ahlfors, *Conformal invariants. Topics in geometric function theory*, McGraw-Hill, New York, 1973.

[10] L.V. Ahlfors and A. Beurling, *Conformal invariants and function-theoretic null-sets*, Acta Math. **83** (1950), no. 1-2, 101–129.

[11] L.V. Ahlfors and L. Bers, *Spaces of Riemann surfaces and quasiconformal mappings*, Collection of papers. Moscow: Inostrannaya Literatura., 1961.

[12] H. Airault and P. Malliavin, *Unitarizing probability measures for representations of Virasoro algebra*, J. Math. Pure Appl. **80** (2001), no. 6, 627–667.

[13] H. Airault, P. Malliavin, and A. Thalmaier, *Support of Virasoro unitarizing measures*, C. R. Acad. Sci. Paris, Ser. I **335** (2002), 621–626.

[14] H. Airault and Yu. Neretin, *On the action of Virasoro algebra on the space of univalent functions.* Bull. Sci. Math. **132** (2008), no. 1, 27–39.

[15] J.M. Aitchison and S.D. Howison, *Computation of Hele-Shaw flows with free boundaries*, J. Comput. Phys. **60** (1985), no. 3, 376–390.

[16] G. Akemann, Y.V. Fyodorov, and G. Vernizzi, *On matrix model partition functions for QCD with chemical potential*, Nuclear Phys. B **694** (2004), no. 1-2, 59–98.

[17] I.A. Aleksandrov, *Parametric continuations in the theory of univalent functions*, Nauka, Moscow, 1976 (in Russian).

[18] S.T. Aleksandrov, and V.V. Sobolev, *Extremal problems in some classes of functions, univalent in the half plane, having a finite angular residue at infinity*, Siberian Math. J. **27** (1986) 145–154. Translation from Sibirsk. Mat. Zh. **27** (1986) 3–13.

[19] I.A. Aleksandrov and V.V. Chernikov, *Pavel Parfen'evich Kufarev. Obituary* (in Russian), Uspehi Mat. Nauk **24** (1969) 181–184.

[20] M. Alimov, K. Kornev, and G. Mukhamadullina, *Hysteretic effects in the problems of artificial freezing*, SIAM J. Appl. Math. **59** (1998), no. 2, 387–410.

[21] B.L. Altshuler, V.E. Kravtsov, and I.V. Lerner, *Statistics of mesoscopic fluctuations and scaling theory*, in *Localization in disordered systems (Bad Schandau, 1986), Teubner-Texte Phys.*, vol. 16, Teubner, Leipzig, 1988, pp. 7–17

[22] S.N. Antontsev, C.R. Gonçalves, and A.M. Meirmanov, *Local existence of classical solutions to the well-posed Hele-Shaw problem*, Port. Math. (N.S.) **59** (2002), no. 4, 435–452.

[23] H. Aratyn, *Integrable Lax hierarchies, their symmetry reductions and multi-matrix models*, VIII J.A. Swieca Summer School on Particles and Fields (Rio de Janeiro, 1995), World Sci. Publ., River Edge, NJ, 1996, 419–472.

[24] D. Armitage and S. Gardiner, *Classical potential theory*, Springer-Verlag, London, 2001.

[25] A. Arnéodo, Y. Couder, G. Grasseau, V. Hakim, and M. Rabaud, *Uncovering the analytical Saffman–Taylor finger in unstable viscous fingering and diffusion-limited aggregation*, Phys. Rev. Lett. **63** (1989), no. 9. 984–987.

[26] V.I. Arnold, *Mathematical methods of classical mechanics*, Springer-Verlag, New York, 1989.

[27] F.G. Avkhadiev, L.A. Aksent'ev, and M.A. Elizarov, *Sufficient conditions for the finite-valence of analytic functions, and their applications*, Itogi Nauki i Tekhniki, Mathematical analysis, Vol. 25, Akad. Nauk SSSR, Vsesoyuz. Inst. Nauchn. i Tekhn. Inform., Moscow, 1987, 3–121; English Transl.: J. Soviet Math. **49** (1990), no. 1, 715–799.

[28] O. Babelon, D. Bernard, and M. Talon, *Introduction to classical integrable systems*, Cambridge Univ. Press, 2003.

[29] C. Baiocchi, *Su alcuni problemi a frontiera libera connessi a questioni di idraulica*, Rend. Sem. Mat. Univ. e Politec. Torino **31** (1971/73), 69–80 (in Italian).

[30] J.-H. Bailly, *Local existence of classical solutions to first-order parabolic equations describing free boundaries*, Nonlinear Anal. **32** (1998), no. 5, 583–599.

[31] G. Baker, M. Siegel, and S. Tanveer, *A well-posed numerical method to track isolated conformal map singularities in Hele-Shaw flow*, J. Comput. Phys. **120** (1995), no. 2, 348–364.

[32] F. Balogh and J. Harnad, *Superharmonic perturbations of a Gaussian measure, equilibrium measures and orthogonal polynomials*, Complex Anal. Oper. Theory **3** (2009), no. 2, 333–360.

[33] F. Balogh and R. Teodorescu, *Optimal approximation of harmonic growth clusters by orthogonal polynomials*, ArXiv preprint arXiv:0807.1700, 2008.

[34] F. Barra, B. Davidovitch, A. Levermann, and I. Procaccia, *Laplacian Growth and Diffusion Limited Aggregation: Different Universality Classes*, Phys. Rev. Lett. **87** (2001), no. 13, 134501.

[35] M. Bauer and D. Bernard, *SLE_k growth processes and conformal field theories*, Phys. Lett. B **543** (2002), no. 1-2, 135–138.

[36] C.M. Beaufoy, *Nautical and hydraulic experiments, with numerous scientific miscellanies*, London, at the private press of Henry Beaufoy, 1834.

[37] J. Bear, *Dynamics of fluids in porous media*, Elsevier (New York), 1972.

[38] B. Beckman, *Codebreakers. Arne Beurling and the Swedish Crypto program during World War II*, AMS, 2002.

[39] V. Beffara, *The dimension of the SLE curves*, Ann. Probab. **36** (2008), no. 4, 1421–1452.

[40] H. Begehr and R.P. Gilbert, *Hele-Shaw type flows in R^n*, Nonlinear Analysis **10** (1986), 65–85.

[41] H. Begehr and R.P. Gilbert, *Non-Newtonian Hele-Shaw flows in $n \geq 2$ dimensions*, Nonlinear Analysis **11** (1987), 17–47.

[42] P.P. Belinskii, *General properties of quasiconformal mappings*, Novosibirsk: Nauka., 1974 (Russian).

[43] M. Ben Amar, *Exact self-similar shapes in viscous fingering*, Phys. Review A **43** (1991), no. 10, 5724–5727.

[44] M. Ben Amar, *Viscous fingering in a wedge*, Phys. Review A **44** (1991), no. 6, 3673–3685.

[45] M. Ben Amar, V. Hakim, M. Mashaal, and Y. Couder, *Self-dilating viscous fingers in wedge-shaped Hele-Shaw cells*, Phys. Fluids A **3** (1991), no. 9, 2039–2042.

[46] E. Ben-Jacob, O. Shochet, I. Cohen, A. Tenenbaum, A. Czirok, and T. Vicsek, *Cooperative strategies in formation of complex bacterial patterns*, Fractals **4** (1995), 849–868.

[47] D.J. Benney, *Some properties of long nonlinear waves*, Stud. Appl. Math. **52** (1973), 45–50.

[48] D. Bensimon, *Stability of viscous fingering*, Phys. Rev. A bf 33 (1986), no. 2, 1302–1308.

[49] D. Bensimon and P. Pelcé, *Tip-splitting solutions to a Stefan problem*, Phys. Rev. A **33** (1986), no. 6, 4477–4478.

[50] D. Bensimon, L.P. Kadanoff, S. Liang, B.I. Shraiman, and C. Tang, *Viscows flows in two dimensions*, Rev. Modern. Phys. **58** (1986), no. 4, 977–999.

[51] S. Bergman and M. Schiffer, *Kernel functions and elliptic differential equations in mathematical physics*, Academic Press, New York, 1953.

[52] D. Berman, B.G. Orr, H.M. Jaeger, and A.M. Goldman, *Conductances of filled two-dimensional networks*, Phys. Rev. B **33** (1986), 4301–4302.

[53] D. Bernard, G. Boffetta, A. Celani, and G. Falkovich, *Inverse turbulent cascades and conformally invariant curves*, Phys. Rev. Letters **98** (2007), 024501.

[54] M. Bertola, B. Eynard, and J. Harnad, *Partition functions for matrix models and isomonodromic tau functions*, J. Phys. A. (2003), 36:3067.

[55] M. Bertola, B. Eynard, and J. Harnad, *Differential systems for biorthogonal polynomials appearing in 2-matrix models and the associated Riemann–Hilbert problem*, Communications in Mathematical Physics (2003), 243:193.

[56] M. Bertola, B. Eynard, and J. Harnad, *Duality of spectral curves arising in two-matrix models*, Theor. Math. Phys. (2003), 134:32.

[57] P. Bleher and A. Its, *Double scaling limit in the random matrix model: the Riemann–Hilbert approach*, Comm. Pure Appl. Math. **56** (2003), no. 4, 433–516.

[58] A. Beurling and L.V. Ahlfors, *The boundary correspondence under quasiconformal mappings*, Acta Math. **96** (1956), 125–142.

[59] L. Bieberbach, *Über die Koeffizienten derjenigen Potenzreihen, welche eine schlichte Abbildung des Einheitskreises vermitteln*, S.-B. Preuss. Akad. Wiss. (1916), 940–955.

[60] J. Billingham, *Surface-tension-driven flow in fat fluit wedges and cones*, J. Fluid Mech. **397** (1999), 45–71.

[61] R. Bott, *On the characteristics classes of groups of diffeomorphisms*, Enseignment Math. (2) **23** (1977), 209–220.

[62] A. Bourgeat, E. Marušić-Paloka, and A. Mikelić, *Weak nonlinear corrections for Darcy's law*, Math. Models Methods Appl. Sci. **6** (1996), no. 8, 1143–1155.

[63] M. Bocquet, D. Serban, and M.R. Zirnbauer, *Disordered 2d quasiparticles in class D: Dirac fermions with random mass, and dirty superconductors*, Nuclear Phys. B **578** (2000), no. 3, 628–680.

[64] A. Boggess, *CR manifolds and the tangential Cauchy–Riemann complex.* Studies in Advanced Mathematics. CRC Press, Boca Raton, FL, 1991. 364 pp.

[65] M. Boukrouche, *The quasiconformal mapping methods to solve a free boundary problem for generalized Hele-Shaw flows*, Complex Variables **28** (1996), 227–242.

[66] M.J. Boussinesq, *Calcul du pouvoir refroidissant des fluides*, J. Math. Pures et Appl., Sér. 6, **1** (1905), 285–332.

[67] P. Boutroux, *Recherches sur les transcendantes de M. Painlevé et l'étude asymptotique des équations différentielles du second ordre*, Ann. Sci. École Norm. Sup. (3) **30** (1913), 255–375.

[68] F. Bracci, M.D. Contreras, and S. Díaz-Madrigal, *Evolution Families and the Löwner Equation I: the unit disk*, J. Reine Angew. Math. **672** (2012), 1–37.

[69] F. Bracci, M.D. Contreras, and S. Díaz-Madrigal, *Evolution Families and the Löwner Equation II: complex hyperbolic manifolds*, Math. Ann. **344** (2009), 947–962.

[70] F. Bracci, M.D. Contreras, S. Díaz-Madrigal, and A. Vasil'ev, *Classical and stochastic Löwner–Kufarev equations*, Harmonic and Complex Analysis and its Applications. Trends Math., Birkhäuser, Basel (2014), 39–134.

[71] L. de Branges, *A proof of the Bieberbach conjecture*, Acta Math. **154** (1985) no. 1–2, 137–152.

[72] D.A. Brannan and W.E. Kirwan, *On some classes of bounded univalent functions*, J. London Math. Soc. (2) **1** (1969), 431–443.

[73] É. Brézin and V.A. Kazakov, *Exactly solvable field theories of closed strings*, Phys. Lett. B **236** (1990), no. 2, 144–150.

[74] E. Brézin, C. Itzykson, G. Parisi, and J.B. Zuber, *Planar diagrams*, Comm. Math. Phys. **59** (1978), no. 1, 35–51.

[75] G.O. Brown, Henry Darcy and the making of a law. Water Resour Res **38** (2002), no. 7, 11-1–11-12.

[76] L.A. Caffarelli, *The regularity of free boundaries in higher dimension*, Acta. Math. **139** (1977), 155–184.

[77] L.A. Caffarelli, *Compactness methods in free boundary problems*, Comm. Partial Diff. Equations **5** (1980), 427–448.

[78] L.A. Caffarelli, *A remark on the Hausdorff measure of a free boundary, and the convergence of coincedence sets*, Boll. Un. Mat. Ital. A(5) **18** (1981), no. 1, 109–113.

[79] L.A. Caffarelli, *The obstacle problem revisited, III. Existence theory, compactness and dependence on X*, J. Fourier Anal. Appl. **4** (1998), 383–402.

[80] L.A. Caffarelli, L. Karp, and H. Shahgholian, *Regularity of the free boundary with application to the Pompeiu problem*, Ann. Math. **151** (2000), 269–292.

[81] L.A. Caffarelli and J.L. Vázquez, *Viscosity solutions for the porous medium equation*, Differential equations: La Pietra 1996 (Florence), Proc. Sympos. Pure Math., 65, Amer. Math. Soc., Providence, RI, 1999, 13–26.

[82] G. Caginalp, *Stefan and Hele-Shaw type models as asymptotic limits of the phase-field equations*, Phys. Rev. A (3) **39** (1989), no. 11, 5887–5896.

[83] G. Caginalp, *The dynamics of a conserved phase field system: Stefan-like, Hele-Shaw, and Cahn–Hilliard models as asymptotic limits*, IMA J. Appl. Math. **44** (1990), no. 1, 77–94.

[84] G. Caginalp, *An analysis of a phase field model of a free boundary*, Arch. Rational Mech. Anal. **92** (1986), no. 3, 205–245.

[85] G. Caginalp and X. Chen, *Phase field equations in the singular limit of sharp interface problems*, Institute for Mathematics and Its Applications, 43, 1992.

[86] J.W. Cahn and J.E. Hilliard, *Free energy of a nonuniform system. i. Interfacial free energy*, The Journal of Chemical Physics **28** (1958), no. 2, 258–267.

[87] C. Carathéodory, *Untersuchungen über die konformen Abbildungen von festen und veränderlichen Gebieten*, Math. Ann. **72** (1912), 107–144.

[88] R.W. Carey and J.D. Pincus, *An exponential formula for determining functions*, Indiana Univ. Math. J. **23** (1973/74), 1031–1042.

[89] L. Carleson and N. Makarov, *Aggregation in the plane and Löwner's equation*, Comm. Math. Phys. **216** (2001), no. 3, 583–607.

[90] G. Carlet, B. Dubrovin, and Y. Zhang, *The extended Toda hierarchy*, Mosc. Math. J. **4** (2004), no. 2, 313–332.

[91] J. Carruthers, *Henry Selby Hele-Shaw: LLD, DSc, EngD, FRS, WhSch (1854–1941): Engineer, inventor and educationist*, South African J. Sci. **106** (2010), no. 1/2, art. 119.

[92] A. Carverhill, *Flows of stochastic dynamical systems: ergodic theory*, Stochastics **14** (1985), no. 4, 273–317.

[93] H.D. Ceniceros, T.Y. Hou, and H. Si, *Numerical study of Hele-Shaw flow with suction*, Physics of Fluids **11** (1999), no. 9, 2471–2486.

[94] H.D. Ceniceros and T.Y. Hou, *The singular perturbation of surface tension in Hele-Shaw flows*, J. Fluid Mech. **409** (2000), 251–272.

[95] H.D. Ceniceros and J.M. Villalobos, *Topological reconfiguration in expanding Hele-Shaw flow*, J. Turbulence **3** (2002), no. 37, 1–8.

[96] D.Y.C. Chan, B.D. Hughes, L. Paterson, and Ch. Sirakoff, *Simulating flow in porous media.* Phys. Rev. A (3) **38** (1988), no. 8, 4106–4120.

[97] L-L. Chau and O. Zaboronsky, *On the structure of normal matrix model*, Comm. Math. Phys. (1998), 196–203

[98] D. Chelkak and S. Smirnov, *Universality in the 2D Ising model and conformal invariance of fermionic observables*. Invent. Math. **189** (2012), no. 3, 515–580.

[99] A.J. Chorin and J. Marsden, *A mathematical introduction to fluid mechanics*, Third edition. Texts in Applied Mathematics, 4. Springer-Verlag, New York, 1993.

[100] P. Constantin and M. Pugh, *Global solution for small data to the Hele-Shaw problem*, Nonlinearity **6** (1993), 393–415.

[101] M.D. Contreras, S. Díaz-Madrigal, and P. Gumenyuk, *Löwner chains in the unit disk*. Rev. Mat. Iberoam. **26** (2010), no. 3, 975–1012.

[102] M. D. Contreras, S. Díaz-Madrigal, and P. Gumenyuk, *Local duality in Löwner equations*. J. Nonlinear Convex Anal. (to appear). Preprint available in http://arxiv.org/pdf/1202.2334.pdf

[103] J. Conway, *Functions of one complex variable II*, Springer-Verlag, New York, 1995.

[104] R. Courant and H. Robbins, *What Is Mathematics?* Oxford University Press, New York, 1941.

[105] R. Courant and D. Hilbert, *Methoden der mathematischen Physik. II*, Dritte Auflage. Heidelberger Taschenbücher, Band 30. Springer-Verlag, Berlin-New York, 1968.

[106] V.F. Cowling and W.C. Royster, *Domains of variability for univalent polynomials*, Proc. Amer. Math. Soc. **19** (1968), 767–772.

[107] R.G. Cox, *Inertial and viscous effects on dynamic contact angles*, J. Fluid Mech. **357** (1998), 249–278.

[108] M.G. Crandall and P.L. Lions, *Viscosity solutions of Hamilton–Jacoby equations*, Trans. Amer. Math. Soc. **277** (1983), 1–42.

[109] M.G. Crandall, H. Ishii, and P.-L. Lions, *User's guide to viscosity solutions of second order partial differential equations*, Bull. Amer. Math. Soc. (N.S.) **27** (1992), no. 1, 1–67.

[110] J. Crank, *Free and moving boundary problems*, Oxford University Press, New York, 1987.

[111] D. Crowdy, *Multiple steady bubbles in a Hele-Shaw cell*, Proc. R. Soc. Lond. Ser. A Math. Phys. Eng. Sci. **465** (2009), no. 2102, 421–435.

[112] D. Crowdy and J. Marshall, *Constructing multiply connected quadrature domains*, SIAM J. Appl. Math. **64** (2004), no. 4, 1334–1359.

[113] D. Crowdy and H. Kang, *Squeeze flow of multiply-connected fluid domains in a Hele-Shaw cell*, J. Nonlinear Sci. **11** (2001), no. 4, 279–304.

[114] D. Crowdy, *Theory of exact solutions for the evolution of a fluid annulus in a rotating Hele-Shaw cell*, Quart. Appl. Math. **60** (2002), no. 1, 11–36.

[115] D. Crowdy and J. Marshall, *Constructing multiply-connected quadrature domains*, SIAM J. Appl. Math. **64** (2004), 1334–1359.

[116] D. Crowdy, *Quadrature domains and fluid dynamics*, Quadrature Domains and Applications, a Harold S. Shapiro Anniversary Volume. Birkhäuser, 2004.

[117] A.B. Crowley, *On the weak solution of moving boundary problems*, J. Inst. Math. Appl. **24** (1979), 43–57.

[118] L.J. Cummings, *Flow around a wedge of arbitrary angle in a Hele-Shaw cell*, European J. Appl. Math. **10** (1999), 547–560.

[119] L.J. Cummings, S.D. Howison, and J.R. King, *Two-dimensional Stokes and Hele-Shaw flows with free surfaces*, European J. Appl. Math. **10** (1999), 635–680.

[120] L.J. Cummings, G. Richardson, and M. Ben Amar, *Models of void electromigration*, European J. Appl. Math. **12** (2001), 97–134.

[121] L.M. Cummings, Yu.E. Hohlov, S.D. Howison, and K. Kornev, *Two-dimensional solidification and melting in potential flows*, J. Fluid Mech. **378** (1999), 1–18.

[122] W.-S. Dai and M. Shelley, *A numerical study of the effect of surface tension and noise on an expanding Hele-Shaw bubble*, Phys. Fluids A **5** (1993), no. 9, 2131–2146.

[123] M. Dallaston, *Mathematical models of bubble contraction in a Hele-Shaw cell*, Doctoral thesis, October 7, 2013, Queensland University of Technology.

[124] M. Dallaston, and S. McCue, *New exact solutions for Hele-Shaw flow in doubly connected regions*, Physics of Fluids **24** (2012), 052101.

[125] M. Dallaston, and S. McCue, *Bubble extinction in Hele-Shaw flow with surface tension and kinetic undercooling regularisation*. Nonlinearity **26**, (2013), 1639–1665.

[126] M. Dallaston, and S. McCue, *Corner formation in Hele-Shaw flow with kinetic undercooling regularisation*, preprint 2013.

[127] E. Date, M. Kashiwara, and T. Miwa, *Vertex operators and functions: transformation groups for soliton equations II*, Proc. Japan Acad. Ser. A Math. Sci. **57** (1981), 387–392.

[128] E. Date, M. Kashiwara, M. Jimbo, and T. Miwa, *Transformation groups for soliton equations*, Nonlinear integrable systems-classical theory and quantum theory (Kyoto, 1981), 39–119, World Sci. Publishing, Singapore, 1983.

[129] B. Davidovitch, H.G.E. Hentschel, Z. Olami, I. Procaccia, L.M. Sander, and E. Somfai, *Diffusion limited aggregation and iterated conformal maps*, Phys. Rev. E (3) **59** (1999), no. 2, part A, 1368–1378.

[130] B. Davidovitch, M.J. Feigenbaum, H.G.E. Hentschel, and I. Procaccia, *Conformal dynamics of fractal growth patterns without randomness*, Phys. Rev. E **62** (2000), no. 2, 1706–1715.

[131] S.H. Davis, *Theory of solidification*, Cambridge Univ. Press, 2001.

[132] P. Davis, *The Schwarz function and its applications*, Carus Math. Monographs 17, Math. Assoc. of America, 1974.

[133] A.J. DeGregoria and L.W. Schwartz, *A boundary-integral method for two-phase displacement in Hele-Shaw cells*, J. Fluid Mech. **164** (1986), 383–400.

[134] P.A. Deift, *Orthogonal polynomials and random matrices: a Riemann–Hilbert approach*, volume 3 of *Courant Lecture Notes in Mathematics*, New York University Courant Institute of Mathematical Sciences, New York, 1999.

[135] P. Deift, J. Baik, A. Borodin, and T. Suidan, *A model for the bus system in Cuernavaca (Mexico)*, Journal of Physics A: Mathematical and General **39** (2006), no. 28, 8965–8975.

[136] A.S. Demidov and J.-P. Loheác, *The Stokes–Leibenson problem for Hele-Shaw flows*, Patterns and waves (Saint Petersburg, 2002), 103–124.

[137] A.S. Demidov, *Evolution of the perturbation of a circle in the Stokes–Leibenson problem for a Hele-Shaw flow*, Sovrem. Mat. Prilozh. no. 2, Differ. Uravn. Chast. Proizvod. (2003), 3–24; translation in J. Math. Sci. (N.Y.) **123** (2004), no. 5, 4381–4403.

[138] B. Derrida and V. Hakim, *Needle models of Laplacian growth*, Phys. Rev. A **45** (1992), no. 12, 8759–8765.

[139] E. Di Benedetto and A. Friedman, *The ill-posed Hele-Shaw model and the Stefan problem for supercooled water*, Trans. Amer. Math. Soc. **282** (1984), no. 1, 183–204.

[140] E. Di Benedetto and A. Friedman, *Bubble growth in porous media*, Indiana Univ. Math. J. **35** (1986), no. 3, 573–606.

[141] H. Darcy, *Les fontaines publiques de la ville de Dijon, Exposition et application des principes à suivre et des formules à employer dans les questions de distribution d'eau*, Paris, Victor Dalmont, 1856.

[142] P. Diaconis and W. Fulton, *A growth model, a game, an algebra, Lagrange inversion, and characteristic classes*, Rend. Sem. Mat. Univ. Politec. Torino **49** (1991), no. 1, 95–119 (1993), Commutative algebra and algebraic geometry, II (Italian) (Turin, 1990).

[143] L.A. Dickey, *Soliton equations and Hamiltonian systems*, volume 26 of *Advanced Series in Mathematical Physics*, World Scientific Publishing Co. Inc., River Edge, NJ, second edition, 2003.

[144] P.A. Dirac, *Selected papers on quantum electrodynamics*, Dover Publications, New York, 1958, 312–320.

[145] J. Doob, *Classical potential theory and its probabilistic counterpart*, Springer-
       Verlag, Berlin, 1983.

[146] M.R. Douglas and S.H. Shenker, *Strings in less than one dimension*, Nuclear
       Phys. B **335** (1990), no. 3, 635–654.

[147] R.G. Douglas, *Banach algebra techniques in operator theory*, second edi-
       tion. Graduate Texts in Mathematics **179**, Springer-Verlag, New York, 1998.
       194 pp.

[148] P.D. Dragnev and E.B. Saff, *Constrained energy problems with applications
       to orthogonal polynomials of a discrete variable*, J. Anal. Math. **72** (1997),
       223–259.

[149] V.N. Dubinin, *Symmetrization in the geometric theory of functions of a
       complex variable*, Uspekhi Mat. Nauk **49** (1994), no. 1(295), 3–76 (Russian).

[150] B.A. Dubrovin, S.P. Novikov, and A.T. Fomenko, *Modern geometry. Methods
       and applications*, Second edition. "Nauka", Moscow, 1986. (in Russian)

[151] B. Dubrovin and Y. Zhang, *Virasoro symmetries of the extended Toda hier-
       archy*, Comm. Math. Phys. **250** (2004), no. 1, 161–193.

[152] B.A. Dubrovin and S.P. Novikov, *Hydrodynamics of weakly deformed soliton
       lattices. Differential geometry and Hamiltonian theory*, Uspekhi Mat. Nauk
       44 (1989), no. 6, 29–98.

[153] W. Dunham, *The mathematical universe*, John Wiley & Sons, 1994.

[154] J. Duchon and R. Robert, *Evolution d'une interface par capillarité et diffu-
       sion de volume, 1: existence locale en temps*, Ann. Inst. H. Poincaré **1** (1984),
       361–378.

[155] I. Dumitriu and A. Edelman, *Matrix models for beta ensembles*, J. Math.
       Phys. **43** (2002), no. 11, 5830–5847.

[156] P. Duren, *Univalent functions*, Springer, New York, 1983.

[157] P. Duren and M.M. Schiffer, *Robin functions and energy functionals of mul-
       tiply connected domains*, Pacific J. Math. **148** (1991), 251–273.

[158] P. Duren and J. Pfaltzgraff, *Robin capacity and extremal length*, J. Math.
       Anal. Appl. **179** (1993), 110–119.

[159] P. Duren, M.M. Schiffer, *Robin functions and distortion of capacity under
       conformal mapping*, Complex Variables Theory Appl. **21** (1993), 189–196.

[160] P. Duren, J. Pfaltzgraff, and R.E. Thurman, *Physical interpretation and
       further properties of Robin capacity*, Algebra i Analiz **9** (1997), no. 3, 211–
       219; English vers.: St. Petersburg Math. J. **9** (1998), no. 3, 607–614.

[161] P. Duren and J. Pfaltzgraff, *Hyperbolic capacity and its distortion under
       conformal mapping*, J. Anal. Math. **78** (1999), 205–218.

[162] P. Duren, *Robin capacity*, Computational Methods and Function Theory,
       N. Papamichael, St. Ruscheweyh, E.B. Saff (eds.), World Sci. Publ. (1999),
       177–190.

[163] E.B. Dussan V., *On the spreading of liquids on solid surfaces: static and dynamic contact lines*, Ann. Rev. Fluid Mechanics **11** (1979), 371–400.

[164] A.M. Dykhne, *Conductivity of a two-dimensional two-phase system*, Sov. Phys. JETP **32** (1970), no. 1, 63–65.

[165] F.J. Dyson, *Statistical theory of the energy levels of complex systems. I.*, J. Mathematical Phys. **3** (1962), 140–156.

[166] F.J. Dyson, *Statistical theory of the energy levels of complex systems. II.*, J. Mathematical Phys. **3** (1962), 157–165.

[167] F.J. Dyson, *Statistical theory of the energy levels of complex systems. III.*, J. Mathematical Phys. **3** (1962), 166–175.

[168] F.J. Dyson, *The threefold way. Algebraic structure of symmetry groups and ensembles in quantum mechanics*, J. Mathematical Phys. **3** (1962), 1199–1215.

[169] P. Ebenfelt, B. Gustafsson, D. Khavinson, and M. Putinar (editors): *Quadrature Domains and Applications*, a Harold S. Shapiro Anniversary Volume. Birkhäuser, 2005.

[170] K.B. Efetov, *Supersymmetry and theory of disordered metals*, Adv. in Phys. **32** (1983), no. 1, 53–127.

[171] T. Eguchi and H. Ooguri, *Conformal and current algebras on a general Riemann surface*, Nucl. Phys. B **282** (1987), no. 2, 308–328.

[172] A. Einstein, *Zür Elektrodynamik der bewegten Körper*, Annalen der Physik **17** (1905), 891–921.

[173] C.M. Elliott and V. Janovský, *A variational inequality approach to Hele-Shaw flow with a moving boundary*, Proc. Roy. Soc. Edin. **A88** (1981), 93–107.

[174] C.M. Elliott and J.R. Ockendon, *Weak and variational methods for moving boundary problem*, Pitman, London, 1982.

[175] V.M. Entov and P.I. Ètingof, *Bubble contraction in Hele-Shaw cells*, Quart. J. Mech. Appl. Math. **44** (1991), no. 4, 507–535.

[176] V.M. Entov and P.I. Ètingof, and D.Ya. Kleinbock, *On nonlinear inteface dynamics in Hele-Shaw flows*, Euro. J. Appl. Math. **6** (1995), no. 5, 399–420.

[177] V.M. Entov and P.I. Ètingof, *On the breakup of air bubbles in a Hele-Shaw cell*, Euro. J. Appl. Math. **22** (2011), 125–149.

[178] J. Escher and G. Simonett, *On Hele-Shaw models with surface tension*, Math. Res. Lett. **3** (1996), no. 4, 467–474.

[179] J. Escher and G. Simonett, *Classical solutions for Hele-Shaw models with surface tension*, Adv. Differential Equations **2** (1997), no. 4, 619–642.

[180] J. Escher and G. Simonett, *Classical solutions of multidimensional Hele-Shaw models*, SIAM J. Math. Anal. **28** (1997), no. 5, 1028–1047.

[181] J. Escher and G. Simonett, *A center manifold analysis for the Mullins–Sekerka model*, J. Differential Equations **143** (1998), no. 2, 267–292.

[182] J. Escher and G. Simonett, *Moving surfaces and abstract parabolic evolution equations*, Topics in nonlinear analysis, 183–212, Progr. Nonlinear Differential Equations Appl. 35, Birkhäuser, Basel, 1999.

[183] H.M. Farkas and I. Kra, *Riemann surfaces*, second ed., Graduate Texts in Mathematics, vol. 71, Springer-Verlag, New York, 1992.

[184] J.D. Fay, *Theta functions on Riemann surfaces*, Lecture Notes in Mathematics, Vol. 352, Springer-Verlag, Berlin, 1973.

[185] L. Fejér, *Mechanische Quadraturen mit positiven Cotesschen Zahlen*, Math. Zeitschrift **37** (1933), no. 2, 287–309.

[186] L. Fejér and G. Szegő, *Special conformal mappings*, Duke Math. J. **18** (1951), 535–548.

[187] L.G. Fel and K.M. Khanin, *On Effective Conductivity on $\mathbb{Z}^d$ Lattice*, ArXiv Mathematical Physics e-prints, 2001.

[188] R.P. Feynman, *The principle of least action in quantum mechanics*, Ph.D. Thesis, Princeton Univ., Princeton, N.J., 1942.

[189] M. Firdaouss and J.-L. Guermond, *Sur l'homogénéisation des équations de Navier-Stokes à faible nombre de Reynolds*, C. R. Acad. Sci. Paris Sér. I Math. **320** (1995), no. 2, 245–251.

[190] J.M. Fitz-Gerald and J.A. McGeough, *Mathematical theory of electrochemical machining: Anodic smoothing*, J. Inst. Math. Appl. **5** (1969), 387–408; *Anodic shaping*, J. Inst. Math. Appl. **5** (1969), 409–421; *Deburring and cavity forming*, J. Inst. Math. Appl. **6** (1970), 102–110.

[191] H. Flaschka, M.G. Forest, and D.W. McLaughlin, *Multiphase averaging and the inverse spectral solution of the Korteweg–de Vries equation*, Comm. Pure Appl. Math. **33** (1980), no. 6, 739–784.

[192] T. Frankel, *The geometry of physics. An introduction.* Cambridge University Press, Cambridge, 3rd ed., 2012.

[193] A. Friedman, *Variational principles and free boundary problems*, Wiley-Interscience, New York, 1982.

[194] A. Friedman and M. Sakai, *A characterization of null quadrature domains in $\mathbf{R}^N$*, Indiana Univ. Math. J. **35** (1986), no. 3, 607–610.

[195] R. Friedrich and W. Werner, *Conformal fields, restriction properties, degenerate representations and SLE*, C. R. Math. Acad. Sci. Paris **335** (2002), no. 11, 947–952.

[196] R. Friedrich, *The global geometry of stochastic Loewner evolutions*, Probabilistic approach to geometry, Adv. Stud. Pure Math., 57, Math. Soc. Japan, Tokyo, 2010, 79–117.

[197] Y.V. Fyodorov, *Complexity of random energy landscapes, glass transition, and absolute value of the spectral determinant of random matrices*, Phys. Rev. Lett. **92** (2004), no. 24, 240601.

[198] F.D. Gakhov, *Boundary value problems*, Addison-Wesley, 1966.

[199] L.A. Galin, *Unsteady filtration with a free surface*, Dokl. Akad. Nauk USSR **47** (1945), 246–249 (in Russian).

[200] F. Fucito, A. Gamba, M. Martellini, and O. Ragnisco, *Non-linear WKB analysis of the string equation*, International Journal of Modern Physics B **6** (1992), 2123.

[201] S.J. Gardiner and T. Sjödin, *Partial balayage and the exterior inverse problem of potential theory*, Potential theory and stochastics in Albac, Theta Ser. Adv. Math., vol. 11, Theta, Bucharest, 2009, pp. 111–123.

[202] S.J. Gardiner and T. Sjödin, *Stationary boundary points for a Laplacian growth problem in higher dimensions*, (2013).

[203] C.F. Gauss, *Allgemeine Lehrsätze in Beziehung auf die im verkehrten Verhältnisse des Quadrats der Entfernung wirkenden Anziehungs- und Abstossungs-Kräfte*. Gauss Werke **5**, pp. 197–242, 1840, Göttingen, 1867.

[204] I.M. Gel'fand and D.B. Fuchs, *Cohomology of the Lie algebra of vector fields on the circle*, Func. Anal. Appl. **2** (1968), no. 4, 342–343.

[205] J. Gibbons, *Linearisation of Benney's equations* Phys. Lett. A **90** (1982), no. 1-2, 7–8.

[206] J. Gibbons and S.P. Tsarev, *Conformal maps and reductions of the Benney equations*, Phys. Lett. A **258** (1999), no. 4–6, 263–271.

[207] R.P. Gilbert and P. Shi, *Anisotropic Hele-Shaw flows*, Math. Methods Appl. Sci. **11** (1989), no. 4, 417–429.

[208] K.A. Gillow and S.D. Howison, *A bibliography of free and moving boundary problems for Hele-Shaw and Stokes flow*, Published electronically at URL>http://www.maths.ox.ac.uk/~howison/Hele-Shaw.

[209] J. Ginibre, *Statistical ensembles of complex, quaternion, and real matrices*, J. Mathematical Phys. **6** (1965), 440–449.

[210] V.L. Girko, *The elliptic law*, Teor. Veroyatnost. i Primenen. **30** (1965), no. 4, 640–651.

[211] M.E. Goldstein and R.I. Reid, *Effect of flow on freezing and thawing of saturated porous media*, Proc. Royal Soc. London, Ser. A **364** (1978), 45–73.

[212] J.P. Gollub and J.S. Langer, *Pattern formation in nonequilibrium physics*, Rev. Mod. Phys. **71** (1999), no. 2, S396–S403.

[213] G.M. Goluzin, *Geometric theory of functions of a complex variable*, Transl. Math. Monogr., Vol. 26, American Mathematical Society, Providence, R.I. 1969.

[214] D. Grier, E. Ben-Jacob, R. Clarke, and L.M. Sander, *Morphology and Microstructure in Electrochemical Deposition of Zinc*, Phys. Rev. Lett. **56** (1986), 1264–1267.

[215] A.W. Goodman and E.B. Saff, *On univalent functions convex in one direction*, Proc. Amer. Math. Soc. **73** (1979), no. 2, 183–187.

[216] A.W. Goodman, *Univalent functions, Vol. I, II*, Mariner Publishing Company, Inc. U. South Florida, 1983.

[217] Y.N. Gordeev and K.G. Kornev, *Crystallization in forced flow: the Saffman–Taylor problem*, Euro. J. Appl. Math. **10** (1999), 535–545.

[218] V.V. Goryainov, *Fractional iterates of functions that are analytic in the unit disk with given fixed points*, Mat. Sb. **182** (1991), no. 9, 1281–1299; Engl. Transl. in Math. USSR-Sb. **74** (1993), no. 1, 29–46.

[219] V.V. Goryainov, *One-parameter semigroups of analytic functions*, Geometric function theory and applications of complex analysis to mechanics: studies in complex analysis and its applications to partial differential equations, 2 (Halle, 1988), Pitman Res. Notes Math. Ser., 257, Longman Sci. Tech., Harlow, 1991, 160–164.

[220] V.V. Goryainov, *One-parameter semigroups of analytic functions and a compositional analogue of infinite divisibility*. Proceedings of the Institute of Applied Mathematics and Mechanics, Vol. 5, Tr. Inst. Prikl. Mat. Mekh., 5, Nats. Akad. Nauk Ukrainy Inst. Prikl. Mat. Mekh., Donetsk, 2000, 44–57 (in Russian).

[221] V.V. Goryainov, *Semigroups of conformal mappings*, Mat. Sb. (N.S.) **129(171)** (1986), no. 4, 451–472 (Russian); translation in Math. USSR Sbornik **57** (1987), 463–483.

[222] V.V. Goryainov and I. Ba, *Semigroups of conformal mappings of the upper half-plane into itself with hydrodynamic normalization at infinity*, Ukrainian Math. J. **44** (1992), 1209–1217.

[223] I. Graham and G. Kohr, *Geometric function theory in one and higher dimensions*, Marcel Dekker, Inc., New York, 2003.

[224] D.-A. Grave, *Sur le problème de Dirichlet*, Assoc. Française pour l' Avancement des Sciences, Comptes-Rendus (Bordeaux), 111–136.

[225] Great Soviet Encyclopedia, Vol. 14, Part 2, (1969–1978), p. 380

[226] P.G. Grinevich and S.P. Novikov, *String equation. II. Physical solution*, Algebra i Analiz **6** (1994), no. 3, 118–140.

[227] L. Gromova and A. Vasil'ev, *On the estimate of the fourth-order homogeneous coefficient functional for univalent functions*, Ann. Polon. Math. **63** (1996), 7–12.

[228] D.J. Gross and A.A. Migdal, *Nonperturbative two-dimensional quantum gravity*, Phys. Rev. Lett. **64** (1990), no. 2, 127–130.

[229] H. Grötzsch, *Über einige Extremalprobleme der konformen Abbildungen. I–II*, Ber. Verh. – Sächs. Akad. Wiss. Leipzig, Math.–Phys. Kl. **80** (1928), 367–376.

[230] H. Grötzsch, *Über die Verzerrung bei nichtkonformen schlichten Abbildungen mehrfach zusammenhängender schlichten Bereiche*, Ber. Verh. – Sächs. Akad. Wiss. Leipzig **82** (1930), 69–80.

[231] K. Gustafson and T. Abe, *The third boundary condition– was it Robin's?* Math. Intelligencer **20** (1998), no. 1, 63–71.

[232] B. Gustafsson, *Quadrature identities and the Schottky double*, Acta Appl. Math. **1** (1983), no. 3, 209–240.

[233] B. Gustafsson, *On a differential equation arising in a Hele-Shaw flow moving boundary problem*, Arkiv för Mat. **22** (1984), no. 1, 251–268.

[234] B. Gustafsson, *Applications of variational inequalities to a moving boundary problem for Hele-Shaw flows*, SIAM J. Math. Analysis **16** (1985), no. 2, 279–300.

[235] B. Gustafsson: *An ill-posed moving boundary problem for doubly-connected domains*, Ark. Mat. **25** (1987), 331–353.

[236] B. Gustafsson, *Existence of weak backward solutions to a generalized Hele-Shaw flow moving boundary problem*, Nonlinear Analysis **9** (1985), 203–215.

[237] B. Gustafsson, *On mother bodies of convex polyhedra*, SIAM J. Math. Anal. **29** (1998), 1106–1117.

[238] B. Gustafsson, *Lectures on balayage*, in: Sirkka-Liisa Eriksson (ed.) Clifford Algebras and Potential Theory, Univ. Joensuu Dept. Math. Rep. Ser. **7**, pp. 17–63, University of Joensuu, Joensuu, 2004.

[239] B. Gustafsson and Y.-L. Lin, *On the dynamics of roots and poles for solutions of the Polubarinova–Galin equation*, Ann. Acad. Sci. Fenn. Math. **38** (2013), no. 1, 259–286.

[240] B. Gustafsson and M. Putinar, *An exponential transform and regularity of free boundaries in two dimensions*, Ann. Scuola Norm. Sup. Pisa Cl. Sci. (4) **26** (1998), 507–543.

[241] B. Gustafsson and M. Putinar, *Linear analysis of quadrature domains II*, Israel J.Math. **119** (2000), 187–216.

[242] B. Gustafsson and M. Putinar, *Linear analysis of quadrature domains. IV*, in [169].

[243] B. Gustafsson and M. Sakai *Properties of some balayage operators, with applications to quadrature domains and moving boundary problems*, Nonlinear Analysis **22** (1994), no. 10, 1221–1245.

[244] B. Gustafsson and M. Sakai, *Sharp estimates of the curvature of some free boundaries in two dimensions*, Ann. Acad. Sci. Fenn. Math. **28** (2003), no. 1, 123–142.

[245] B. Gustafsson and M. Sakai, *On the curvature of the free boundary for the obstacle problem in two dimensions*, Monatsh. Math. **142** (2004), no. 1-2, 1–5.

[246] B. Gustafsson and H. S Shapiro, *What is a quadrature domain?*, in [169].

[247] B. Gustafsson, D. Prokhorov, and A. Vasil'ev, *Infinite lifetime for the starlike dynamics in Hele-Shaw cells*, Proc. Amer. Math. Soc. **132** (2004), no. 9, 2661–2669.

[248] B. Gustafsson and A. Vasil'ev, *Nonbranching weak and starshaped strong solutions for Hele-Shaw dynamics*, Arch. Math. (Basel) **84** (2005), no. 6, 551–558.

[249] B. Gustafsson and A. Vasil'ev, *Conformal and potential analysis in Hele-Shaw cells*, Birkhäuser Verlag, 2006.

[250] V.Ya. Gutlyanskiĭ, *Parametric representation of univalent functions*, Dokl. Akad. Nauk SSSR **194** (1970), 750–753; Engl. Transl. in Soviet Math. Dokl. **11** (1970), 1273–1276.

[251] V.Ya. Gutlyanskiĭ, *On some classes of univalent functions*. Theory of functions and mappings, "Naukova Dumka", Kiev, 1979, 85–97 (in Russian).

[252] V.Ya. Gutljanskiĭ, *The method of variations for univalent analytic functions with a quasiconformal extension*, Sibirsk. Mat. Zh. **21** (1980), no. 2, 61–78; Engl. Transl. in Siberian Math. J. **21** (1980), no. 2, 190–204.

[253] H.L. Guy, *H.S. Hele-Shaw 1854–1941*, Obituary Notices of Fellows of The Royal Society **3**(1941), no. 10. 790–811.

[254] H.E. Haber and G.L. Kane, *Is nature supersymmetric?*, Scientific American (1986), June, page 42.

[255] W.A. Hall, *An analytical derivation of the Darcy equation*, EOS Trans. Am. Geophys. Union **37** (1956), 185–188.

[256] T.C. Halsey, *Diffusion-Limited Aggregation: A Model for Pattern Formation*, Physics Today **53** (2000), no. 11, 36.

[257] M.B. Hastings, *Renormalization theory of stochastic growth*, Physical Review E **55** (1987), no. 1, 135–152.

[258] M.B. Hastings and L.S. Levitov, *Laplacian growth as one-dimensional turbulence*, Physica D **116** (1998), no. 1-2, 244–252.

[259] W.K. Hayman, *Multivalent functions*, Cambridge University Press, Cambridge, 1958.

[260] C.Q. He, *A parametric representation of quasiconformal extensions*, Kexue Tongbao **25** (1980), no. 9, 721–724.

[261] H. Hedenmalm and S. Shimorin, *Hele-Shaw flow on hyperbolic surfaces*, J. Math. Pures et Appl. **81** (2002), 187–222.

[262] H. Hedenmalm and N. Makarov, *Coulomb gas ensembles and Laplacian growth*, Proc. Lond. Math. Soc. (3) **106** (2013), no. 4, 859–907.

[263] H. Hedenmalm and A. Olofsson, *Hele-Shaw flow on weakly hyperbolic surfaces*, Indiana Univ. Math. J. **54** (2005), no. 4, 1161–1180.

[264] P. Heinzner, A. Huckleberry, and M.R. Zirnbauer, *Symmetry classes of disordered fermions*, Comm. Math. Phys. **257** (2005), no. 3, 725–771.

[265] H.S. Hele-Shaw, *The flow of water*, Nature **58** (1898), no. 1489, 34–36, 520.

[266] H.S. Hele-Shaw, *On the motion of a viscous fluid between two parallel plates*, Trans. Royal Inst. Nav. Archit., London **40** (1898), 21.

[267] W. Hengartner and G. Schober, *A remark on level curves for domains convex in one direction*, Collection of articles dedicated to Eberhard Hopf on the occasion of his 70th birthday, Applicable Anal. **3** (1973), 101–106.

[268] G. Herglotz, *Über die analytische Fortsetzung des Potentials ins Innere der anziehenden Massen*, Gekrönte Preisschr. der Jablonowskischen Gesellsch. zu Leipzig, 56 pp.

[269] Yu.E. Hohlov and S.D. Howison, *On the classification of solutions to the zero surface tension model for Hele-Shaw free boundary flows*, Quart. Appl. Math. **54** (1994), no. 4, 777–789.

[270] Yu.E. Hohlov, S.D. Howison, C. Huntingford, J.R. Ockendon, and A.A. Lacey, *A model for non-smooth free boundaries in Hele-Shaw flows*, Quart. J. Mech. Appl. Math. **47** (1994), 107–128.

[271] Yu.E. Hohlov and M. Reissig, *On classical solvability for the Hele-Shaw moving boundary problem with kinetic undercooling regularization*, Euro. J. Applied Math. **6** (1995), 421–439.

[272] Yu.E. Hohlov, D.V. Prokhorov, and A. Vasil'ev, *On geometrical properties of free boundaries in the Hele-Shaw flow moving boundary problem*, Lobachevskiĭ J. Math. **1** (1998), 3–13 (electronic).

[273] G. 't Hooft, *Planar diagram theory for strong interactions*, Nucl. Phys. B **72** (1974), no. 3, 461.

[274] R.W. Hopper, *Capillarity-driven plane stokes flow exterior to a parabola*, Q. J. Mechanics Appl. Math. **46** (1993), no. 2, 193–210.

[275] T.Y. Hou, L. Lowengrub, and M. Shelley, *Removing the stiffness from interfacial flow with surface tension*, J. Comput. Phys. **114** (1994), no. 2, 312–338.

[276] T.Y. Hou, *Numerical solutions to free boundary problems*, Acta Numerica, 1995, 335–415.

[277] S.D. Howison, *Fingering in Hele-Shaw cells*, J. Fluid Mech. **167** (1986), 439–453.

[278] S.D. Howison, *Cusp development in Hele-Shaw flow with a free surface*, SIAM J. Appl. Math. **46** (1986), no. 1, 20–26.

[279] S.D. Howison, *Bubble growth in porous media and Hele-Shaw cells*, Proc. Roy. Soc. Edinburgh, Ser. A **102** (1986), no. 1-2, 141–148.

[280] S.D. Howison, *Complex Variable methods in Hele-Shaw moving Boundary Problems*, European J. Appl. Math. **3** (1992), no. 3, 209–224.

[281] S.D. Howison, *A note on the two-phase Hele-Shaw problem*, J. Fluid Mech. **409** (2000), 243–249.

[282] S.D. Howison and J. King, *Explicit solutions to six free-boundary problems in fluid flow and diffusion*, IMA J. Appl. Math. **42** (1989), 155–175.

[283] M.K. Hubbert, *Darcy's law and the field equations of the flow of underground fluids*, Amer. Inst. Mining Engin., Petr. Trans. **207** (1956), 222–239.

[284] C. Huntingford, *An exact solution to the one-phase zero-surface-tension Hele-Shaw free-boundary problem*, Comput. Math. Appl. **29** (1995), no. 10, 45–50.

[285] G.P. Ivantzov, *The temperature field around a spherical, cylindrical, or point crystal growing in a cooling solution*, Dokl. Acad. Nauk USSR **58** (1947), 567–569 (in Russian).

[286] G. Ivanov and A. Vasil'ev, *Löwner evolution driven by a stochastic boundary point*, Anal. Math. Phys. **1** (2011), no. 4, 387–412.

[287] J. Jenkins, *On the existence of certain general extremal metrics*, Ann. of Math. **66** (1957), no. 3, 440–453.

[288] J. Jenkins, *Univalent functions and conformal mapping*, Springer-Verlag, 1958.

[289] Ø. Johnsen, C. Chevalier, A. Lindner, R. Toussaint, E. Clément, K.J. Måløy, E.G. Flekkøy, and J. Schmittbuhl, *Decompaction and fluidization of a saturated and confined granular medium by injection of a viscous liquid or gas*. Phys. Rev. E 78 (2008), 051302

[290] G.W. Johnson and M.L. Lapidus, *The Feynman integral and Feynman's operational calculus*, Oxford Mathematical Monographs. Oxford Science Publications. The Clarendon Press, Oxford University Press, New York, 2000.

[291] V. Kac, *Vertex algebras for beginners*, Amer. Math. Soc., Providence, RI, 1998.

[292] L.P. Kadanoff, *Exact soutions for the Saffman–Taylor problem with surface tension*, Phys. Review Letters **65** (1990), no. 24, 2986–2988.

[293] M. Kaku, *Introduction to superstrings*, Springer-Verlag, Berlin-New York, 1988.

[294] N.-G. Kang and N.G. Makarov, *Gaussian free field and conformal field theory*, Astérisque **353** (2013), viii+136 pp.

[295] W. Kaplan, *Close-to-convex schlicht functions*, Michigan Math. J. **1** (1952), 169–185.

[296] L. Karp, *Global solutions to bubble growth in porous media*, J. Math. Anal. Appl. **382** (2011), no. 1, 132–139.

[297] L. Karp and A.S. Margulis, *Newtonian potential theory for unbounded sources and applications to free boundary problems*, J. Analyse Math. J. **70** (1996), 653–669.

[298] L. Karp and A.S. Margulis, *Null quadrature domains and a free boundary problem for the Laplacian*, Indiana Univ. Math. J. **61** (2012), no. 2, 859–882. MR 3043599

[299] V.A. Kazakov, *Ising model on a dynamical planar random lattice: exact solution*, Phys. Lett. A **119** (1986), no. 3, 140–144.

[300] J.B. Keller and M. Miksis, *Surface tension driven flows*, SIAM J. Appl. Math. **34** (1983), no. 2, 268–277.

[301] J.B. Keller, P.A. Milewski, and J.-M. Vanden-Broeck, *Merging and wetting driven by surface tension*, European J. Mech. B-Fluids **19** (2000), 491–502.

[302] J.B. Keller, *A Theorem on the Conductivity of a Composite Medium*, J. Math. Phys. **5** (1964), 548–549.

[303] D.A. Kessler, J. Koplik, and H. Levine, *Pattern selection in fingering growth phenomena*, Adv. Physics. **37** (1988), no. 3, 255–339.

[304] H. Kesten, *Hitting probabilities of random walks on* $\mathbb{Z}^d$, Stoch. Proc. Appl. **25** (1987), 165–184.

[305] D. Khavinson, M. Mineev-Weinstein, and M. Putinar, *Planar elliptic growth*, Complex Anal. Oper. Theory **3** (2009), no. 2, 425–451.

[306] D. Khavinson, M. Mineev-Weinstein, R. Teodorescu, and M. Putinar, *Lemniscates do not survive Laplacian growth*, Math. Res. Lett. **17** (2010), 335–341.

[307] D. Khavinson and G. Neumann, *On the number of zeros of certain rational harmonic functions*, Proc. Amer. Math. Soc. **134** (2006), no. 4, 1077–1085.

[308] I.C. Kim, *Uniqueness and existence results on the Hele-Shaw and the Stefan problems*, Arch. Ration. Mech. Anal. **168** (2003), no. 4, 299–328.

[309] M. Kimura, *Time local existence of a moving boundary of the Hele-Shaw flow with suction*, Europ. J. Appl. Math. **10** (1999), no. 6, 581–605.

[310] D. Kinderlehrer and G. Stampacchia, *An introduction to variational inequalities and their applications*, Academic Press, New York-London, 1980.

[311] J.R. King, A.A. Lacey, and J.L. Vázquez, *Persistence of corners in free boundaries in Hele-Shaw flows*, Euro. J. Appl. Math. **6** (1995), no. 5, 455–490.

[312] J.R. King, *Development of singularities in some moving boundary problems*, Euro. J. Appl. Math. **6** (1995), no. 5, 491–507.

[313] A.A. Kirillov and D.V. Yuriev, *Kähler geometry of the infinite-dimensional homogeneous space* $M = \mathrm{Diff}_+(S^1)/\mathrm{Rot}(S^1)$, Funktsional. Anal. i Prilozhen. **21** (1987), no. 4, 35–46 (in Russian).

[314] A.A. Kirillov and D.V. Yuriev, *Representations of the Virasoro algebra by the orbit method*, J. Geom. Phys. **5** (1988), no. 3, 351–363.

[315] A.A. Kirillov, *Kähler structure on the K-orbits of a group of diffeomorphisms of the circle*, Funktsional. Anal. i Prilozhen. **21** (1987), no. 2, 42–45.

[316] A.A. Kirillov, *Geometric approach to discrete series of unirreps for Vir*, J. Math. Pures Appl. **77** (1998), 735–746.

[317] A.N. Kolmogorov, *Foundations of probability theory*, Chelsea, New York, 1950.

[318] K. Kornev and G. Mukhamadullina, *Mathematical theory of freezing in porous media*, Proc. Royal Soc. London, Ser. A **447** (1994), 281–297.

[319] K. Kornev and A. Vasil'ev, *Geometric properties of the solutions of a Hele-Shaw type equation*, Proc. Amer. Math. Soc. **128** (2000), no. 9, 2683–2685.

[320] D.J. Korteweg, G. de Vries, *On the change of form of long waves advancing in a rectangular channel, and a new type of long stationary waves*, Phil. Mag. **38** (1895), 422–443.

[321] I.K. Kostov, I. Krichever, M. Mineev-Weinstein, P.B. Wiegmann, and A. Zabrodin, *The τ-function for analytic curves*, Random matrix models and their applications, Math. Sci. Res. Inst. Publ., 40, Cambridge Univ. Press, Cambridge, 2001, 285–299.

[322] I.K. Kostov, *Matrix models as conformal field theories*, In *Applications of random matrices in physics*, volume 221 of *NATO Sci. Ser. II Math. Phys. Chem.*, pages 459–487. Springer, Dordrecht, 2006.

[323] I.M. Krichever, *On the rational solutions of the Zaharov-Shabat equations and completely integrable systems of N particles on the line*, Zap. Nauchn. Sem. Leningrad. Otdel. Mat. Inst. Steklov. (LOMI) **84** (1979), 117–130.

[324] I.M. Krichever, *The dispersionless Lax equations and topological minimal models*, Comm. Math. Phys. **143** (1992), no. 2, 415–429.

[325] I.M. Krichever, *The τ-function of the universal Whitham hierarchy, matrix models and topological field theories*, Comm. Pure Appl. Math. **47** (1994), no. 4, 437–475.

[326] I. Krichever, M. Mineev-Weinstein, P. Wiegmann, and A. Zabrodin, *Laplacian growth and Whitham equations of soliton theory*, Phys. D **198** (2004), no. 1-2, 1–28.

[327] I. Krichever, A. Marshakov, and A. Zabrodin, *Integrable structure of the Dirichlet boundary problem in multiply-connected domains*, Comm. Math. Phys. **259** (2005), no. 1, 1–44.

[328] P.P. Kufarev, *On one-parameter families of analytic functions*, Rec. Math. [Mat. Sbornik] N.S. **13** (1943), no. 55, 87–118.

[329] P.P. Kufarev, *On integrals of a simplest differential equation with movable polar singularity in the right-hand side*, Tomsk. Gos. Univ. Uchen. Zap. **1** (1946), 35–48.

[330] P.P. Kufarev, *A solution of the boundary problem for an oil well in a circle*, Doklady Akad. Nauk SSSR (N. S.) **60** (1948), 1333–1334 (in Russian).

[331] P.P. Kufarev, *The problem of the contour of the oil-bearing region for a circle with an arbitrary number of gaps*, Doklady Akad. Nauk SSSR (N.S.) **75** (1950), 507–510 (in Russian).

[332] B.A. Kupershmidt and Yu.I. Manin, *Long wave equations with a free surface. I. Conservation laws and solutions.* Funktsional. Anal. i Prilozhen. **11** (1977), no. 3, 31–42.

[333] B.A. Kupershmidt and Yu.I. Manin, *Long wave equations with a free surface. II. The Hamiltonian structure and the higher equations*, Funktsional. Anal. i Prilozhen. **12** (1978), no. 1, 25–37.

[334] O. Kuznetsova, *On polynomial solutions of the Hele-Shaw problem*, Sibirsk. Mat. Zh. **42** (2001), no. 5, 1084–1093; Engl. transl.: Siberian Math. J. **42** (2001), no. 5, 907–915.

[335] P.P. Kufarev, V.V. Sobolev, and L.V. Sporysheva, *A certain method of investigation of extremal problems for functions that are univalent in the half-plane*, Trudy Tomsk. Gos. Univ. Ser. Meh.-Mat. **200** (1968), 142–164.

[336] O.S. Kuznetsova, *On polynomial solutions of the Hele-Shaw problem*, Sibirsk. Mat. Zh. **42** (2001), no. 5, 1084–1093; translation in Siberian Math. J. **42** (2001), no. 5, 907–915.

[337] O.S. Kuznetsova, *Invariant families in the Hele-Shaw problem*, Preprint TRITA-MAT-2003-07, Royal Institute of Technology, Stokholm, Sweden, 2003.

[338] O.S. Kuznetsova and V.G. Tkachev, *Ullemar's formula for the Jacobian of the complex moment mapping*, Complex Var. Theory Appl. **49** (2004), no. 1, 55–72.

[339] A.A. Lacey and J.R. Ockendon, *Ill-posed free boundary problems*, Control Cybernet. **14** (1985), no. 1-3, 275–296.

[340] H. Lamb, *Treatise on the Motion of Fluids*, 1st Edition, Cambridge University Press, 1879.

[341] H. Lamb, *Hydrodynamics*, 3rd Edition, Cambridge University Press, 1906.

[342] H. Lamb, *Hydrodynamics*, 6th Edition, Cambridge University Press, 1932.

[343] L.D. Landau and E.M. Lifshits, *Teoreticheskaya fizika. Tom VI.* "Nauka", Moscow, third edition, 1986. Gidrodinamika. [Fluid dynamics].

[344] J.S. Langer, *Eutectic solidification and marginal stability*, Phys. Rev. Lett. **44** (1980), no. 15, 1023–1026.

[345] J.S. Langer, *Models of pattern formation in first-order phase transitions*, In *Directions in condensed matter physics*, volume 1 of *World Sci. Ser. Dir. Condensed Matter Phys.*, pages 165–186. World Sci. Publishing, Singapore, 1986.

[346] J.S. Langer and H. Müller-Krumbhaar, *Mode selection in a dendritelike nonlinear system*, Phys. Rev. A **27** (1983), no. 1, 499–514.

[347] B. Launder, Horace Lamb and the circumstances of his appointment at Owens College, Notes Rec. R. Soc. **67** (2012) , 139–158.

[348] M.A. Lavrentiev, *A general problem of the theory of quasi-conformal representation of plane regions*, Mat. Sbornik (N.S) **21** (1947), no. 63, 85–320.

[349] G.F. Lawler, *An introduction to the stochastic Loewner evolution*, in Random Walks and Geometry, V. Kaimonovich, ed., de Gruyter (2004), pp. 261–293.

[350] G.F. Lawler, *Conformally invariant processes in the plane*, Mathematical Surveys and Monographs, 114. American Mathematical Society, Providence, RI, 2005.

[351] G. Lawler, O. Schramm, and W. Werner, *Values of Brownian intersection exponents. I. Half-plane exponents.* Acta Math. **187** (2001), no. 2, 237–273; *Values of Brownian intersection exponents. II. Plane exponents.* Acta Math. **187** (2001), no. 2, 275–308; *Values of Brownian intersection exponents. III. Two-sided exponents.* Ann. Inst. H. Poincaré. Probab. Statist. **38** (2002), no. 1, 109–123.

[352] S.-Y. Lee, R. Teodorescu, and P. Wiegmann, *Shocks and finite-time singularities in Hele-Shaw flow*, Phys. D **238** (2009), no. 14, 1113–1128.

[353] S.-Y. Lee and N. Makarov, *Topology of quadrature domains*, arXiv:1307.0487 [math.CV].

[354] P. Lehto, *On fourth-order homogeneous functionals in the class of bounded univalent functions*, Ann. Acad. Sci. Fenn. Ser. A I Math. Dissertationes no. 48, (1984), 1–46.

[355] L.S. Leibenzon, *Ob odnom sluchae izoterminicheskogo techeniya vyazkogo gaza v grunte i o debete gazovykh skvazhin*, Azerb. neft. khoz-vo (1923) no. 4, pp. 109–113; no. 5, pp. 74–79; no. 6-7, pp. 52–59; no. 11, pp. 77–79.

[356] L.S. Leibenzon, *The motion of a gas in a porous medium*, Complete works, vol. 2, Acad. Sciences USSR, Moscow, 1953; First published in Neftyanoe and Slantsevoe Hozyaistvo **10** (1929) and Neftyanoe Hozyaistvo **8-9**, 1930 (in Russian).

[357] L.S. Leibenzon, *Oil producing mechanics, Part II*, Neftizdat, Moscow, 1934.

[358] L. Lempert, *The Virasoro group as a complex manifold*, Math. Res. Lett. **2** (1995), 479–495.

[359] L. Levine and Y. Peres, *Scaling limits for internal aggregation models with multiple sources*, J. Anal. Math. **111** (2010), 151–219.

[360] Yu-L. Lin, *Large-time rescaling behaviours of Stokes and Hele-Shaw flows driven by injection*, European J. Appl. Math. **22** (2011), no. 1, 7–19.

[361] Yu-L. Lin, *Perturbation theorems for Hele-Shaw flows and their applications*, Ark. Mat. **49** (2011), no. 2, 357–382.

[362] K. Löwner, *Untersuchungen über schlichte konforme Abbildungen des Einheitskreises*, Math. Ann. **89** (1923), 103–121.

[363] D. Lüst and S. Theisen, *Lectures on string theory*, Lecture Notes in Physics, Springer-Verlag, 1989.

[364] I. Loutsenko, *The variable coefficient Hele-Shaw problem, integrability and quadrature identities*, Comm. Math. Phys. **268** (2006), no. 2, 465–479.

[365] A.S. Margulis, *The moving boundary problem of potential theory*, Adv. Math. Sci. Appl. **5** (1995), no. 2, 603–629. MR 1361007 (96k:35197)

[366] I. Markina and A. Vasil'ev, *Virasoro algebra and dynamics in the space of univalent functions*, Contemporary Math. **525** (2010), 85–116.

[367] I. Markina and A. Vasil'ev, *Löwner–Kufarev evolution in the Segal–Wilson Grassmannian*, Geometric Methods in Physics. XXX Workshop 2011, Trends in Mathematics, Birkhäuser, Basel (2012), 367–376.

[368] I. Markina and A. Vasil'ev, *Long-pin perturbations of the trivial solution for Hele-Shaw corner flows*, Scientia. Ser. A, Math. Sci. **9** (2003), 33–43.

[369] I. Markina and A. Vasil'ev, *Explicit solutions for Hele-Shaw corner flows*, European J. Appl. Math. **15** (2004), 1–9.

[370] A. Marshakov, P. Wiegmann, and A. Zabrodin, *Integrable structure of the Dirichlet boundary problem in two dimensions*, Comm. Math. Phys. **227** (2002), no. 1, 131–153.

[371] A. Marshakov, *On nonabelian theories and abelian differentials*, Differential equations: geometry, symmetries and integrability, Abel Symp., vol. 5, Springer, Berlin, 2009, pp. 257–274.

[372] A. Marshakov and A. Zabrodin, *On the Dirichlet boundary problem and Hirota equations*, Bilinear integrable systems: from classical to quantum, continuous to discrete, NATO Sci. Ser. II Math. Phys. Chem., vol. 201, Springer, Dordrecht, 2006, pp. 175–190.

[373] A. Marshakov, *Matrix models, complex geometry, and integrable systems. I*, Teoret. Mat. Fiz. **147** (2006), no. 2, 163–228.

[374] A. Marshakov, P. Wiegmann, and A. Zabrodin, *Integrable structure of the Dirichlet boundary problem in two dimensions.* Comm. Math. Phys. **227** (2002), no. 1, 131–153.

[375] M. Martin and M. Putinar, *Lectures on hyponormal operators*, Operator Theory: Advances and Applications, vol. 39, Birkhäuser Verlag, Basel, 1989.

[376] V.A. Maksimov, *On the stable shape of bodies solidified around a could sourse in a stream fluid*, Izv. Acad. Nauk USSR, Mekhanica **4** (1965), no. 41, 41–45 (in Russian).

[377] V.A. Maksimov, *On the determination of the shape of bodies formed by solidification of the fluid phase of the stream*, Prikl. Mat. Mekh. (J. Appl. Math. Mech.) **40** (1976), no. 264, 264–272.

[378] O. Martio and B. Øksendal, *Fluid flow in a medium distorted by a quasiconformal map can produce fractal boundaries*, European J. Appl. Math. **7** (1996), no. 1, 1–10.

[379] A. Marx, *Untersuchungen über schlichte Abbildungen*, Math. Ann. **107** (1932/33), 40–67.

[380] M. Matsushita and H. Fujikawa, *Diffusion-limited growth in bacterial colony formation*, Physica A Statistical Mechanics and its Applications **168** (1990), 498–506.

[381] V.B. Matveev, *Some comments on the rational solutions of the Zakharov-Schabat equations*, Letters Math. Phys. **3** (1979), no. 6, 503–512.

[382] J.C. Maxwell, *A dynamical theory of the electromagnetic field*, Royal Society Transactions **155** (1865), 459–512.

[383] J.W. McLean and P.G. Saffman, *The effect of surface tension on the shape of fingers in a Hele Shaw cell*, J. Fluid Mech. **102** (1981), 455–469.

[384] E.B. McLeod, *The explicit solution of a free boundary problem involving surface tension*, J. Rat. Mech. Analysis **4** (1955), 557–567.

[385] M.L. Mehta, *Random matrices*, volume 142 of *Pure and Applied Mathematics (Amsterdam)*, Elsevier/Academic Press, Amsterdam, third edition, 2004.

[386] E. Meiburg and G.M. Homsy, *Nonlinear unstable viscous fingers in Hele-Shaw flows. 2. Numerical simulation*, Phys. Fluids **31** (1988), no. 3, 429–439.

[387] A.M. Meirmanov and B. Zaltzman, *Global in time solution to the Hele-Shaw problem with change of topology*, Eur. J. Appl. Math. **13** (2002), 431–447.

[388] R. Merks, A. Hoekstra, J. Kaandorp, and P. Sloot, *Models of coral growth, sponatneous branching, compactification and Laplacian growth applications*, J. Theoret. Biol. **224** (2003), 153–166.

[389] J. Milnor, *Remarks on infinite-dimensional Lie groups*, pp. 1007–1057 in: 'Relativité, Groupes et Topologie II', B. DeWitt and R. Stora (eds.), North-Holland, Amsterdam, 1984.

[390] M. Mineev-Weinstein and S.P. Dawson, *Class of nonsingular exact solutions for Laplacian pattern formation*, Physical Rev. E **50** (1994), no. 1, R24.

[391] M. Mineev-Weinstein, *Selection of the Saffman–Taylor finger width in the absence of surface tension: an exact result*, Physical Rev. Lett. **80** (1998), no. 10, 2113–2116.

[392] M. Mineev-Weinstein, P.B. Wiegmann, and A. Zabrodin, *Integrable structure of interface dynamics*, Physical Review Letters (2000), 84:5106.

[393] M. Mineev-Weinstein and A. Zabrodin, *Whitham–Toda hierarchy in the Laplacian growth problem*, J. Nonlinear Math. Phys. **8** (2001), 212–218.

[394] R V. Moody, *A new class of Lie algebras*, J. Algebra **10** (1968), 211–230.

[395] J. Mossino, *Inégalités isopérimétriques et applications en physique*, Travaux en Cours. Hermann, Paris, 1984 (in French).

[396] W.W. Mullins, *Grain boundary grooving by volume diffusion*, Trans. Metallurgical Society of AIME **218** (1960), 354–361.

[397] D. Mumford, *Pattern theory: the mathematics of perception*, Proceedings ICM 2002, vol. 1, 401–422.

[398] J. Muñoz Porras and F. Pablos Romo, *Generalized reciprocity laws*. Trans. Amer. Math. Soc. 360 (2008), no. **7**, 3473–3492.

[399] N.I. Muskhelishvili, *Singular integral equations*, P. Noordhoff, Groningen, the Netherlands, 1953.

[400] M. Muskat, *Two fluid systems in porous media. The encroachment of water into an oil sand*, Physics **5** (1934), 250–264.

[401] M. Muskat, *The flow of homogeneous fluids through porous media*, McGraw-Hill, New York, 1937.

[402] U. Nakaya, *Snow Crystals*, Harvard University Press, Cambridge, 1954.

[403] Y. Nambu, *Duality and hydrodynamics*, Lectures at the Copenhagen symposium, 1970.

[404] Z. Nehari, *Conformal mapping*, McGraw-Hill Book Co., Inc., New York, Toronto, London, 1952.

[405] Z. Nehari, *Schwarzian derivatives and schlicht functions*, Bull. Amer. Math. Soc. **55** (1949), no. 6, 545–551.

[406] Yu.A. Neretin, *Representations of Virasoro and affine Lie algebras*, Encyclopedia of Mathematical Sciences, vol. 22, Springer-Verlag, 1994, pp. 157–225.

[407] S.P. Neuman, *Theoretical derivation of Darcy's law*, Acta Mech. **25** (1977), 153–170.

[408] Q. Nie and F.R. Tian, *Singularities in Hele-Shaw flows*, SIAM J. Appl. Math. **58** (1998), no. 1, 34–54.

[409] L. Niemeyer, L. Pietronero, and H.J. Wiessmann, *Fractal dimension of dielectric breakdown*, Phys. Rev. Lett. **52** (1984), no. 12, 1033–1036.

[410] T. Nishida, *A note on a theorem of Nirenberg*, J. Differential Geom. **12** (1977), no. 4, 629–633.

[411] E. Noether, *Invariante Varlationsprobleme*, Nachr. d. König. Gesellsch. d. Wiss. zu Göttingen, Math-phys. Klasse (1918), 235–257.

[412] P.S. Novikov, *On the uniqueness of the inverse problem of potential theory*, Dokl. Akad. Nauk SSSR **18** (1938), 165–168 (in Russian).

[413] H. Ockendon and J.R. Ockendon, *Viscous Flow*, Cambridge U.P., 1995.

[414] J.R. Ockendon and S.D. Howison, *P. Ya. Kochina and Hele-Shaw in modern mathematics, natural sciences, and technology*. Prikl. Mat. Mekh. **66** (2002), no. 3, 515–524; Engl. transl.: J. Appl. Math. Mech. **66** (2002), no. 3, 505–512.

[415] J.R. Ockendon, S.D. Howison, and A.A. Lacey, *Mushy regions in negative squeeze films*, Quart. J. Mech. Appl. Math. **56** (2003), no. 3, 361–379.

[416] M. Onodera, *Geometric flows for quadrature identities*. Mathematische Annalen (2014), DOI 10.1007/s00208-014-1062-2.

[417] M. Onodera, *Asymptotics of Hele-Shaw flows with multiple point sources*, Proc. Roy. Soc. Edinburgh Sect. A **140** (2010), no. 6, 1217–1247.

[418] M. Onodera, *Stability of the interface of a Hele-Shaw flow with two injection points*, SIAM J. Math. Anal. **43** (2011), no. 4, 1810–1834.

[419] A. Onuki, *Phase Transition Dynamics*, Cambridge University Press, 2002.

[420] M. Ohtsuka, *Dirichlet problem, extremal length, and prime ends*, Van Nostrand, New York, 1970.

[421] N.C. Overgaard, *Application of variational inequalities to the moving-boundary problem in a fluid model for biofilm growth*, Nonlinear Anal. **70** (2009), 3658–3664.

[422] E. Ozugurlu and J.-M. Vanden-Broeck, *The distortion of a bubble in a corner flow*, Euro. J. Appl. Math. **11** (2000), 171–179.

[423] C.-W. Park and G.M. Homsy, *Two-phase displacement in Hele-Shaw cells: theory*, J. Fluid Mech. **139** (1984), 291–308.

[424] L.A. Pastur, *The spectrum of random matrices*, Teoret. Mat. Fiz. **10** (1972), no. 1, 102–112.

[425] L.A. Pastur, *The distribution of eigenvalues of the Schrödinger equation with a random potential*, Funkcional. Anal. i Priložen. **6** (1972), no. 2, 93–94.

[426] M.V. Pavlov and S.P. Tsarev, *Conservation laws for the Benney equations*, Uspekhi Mat. Nauk **46** (1991), no. 4(280), 169–170; translation in Russian Math. Surveys **46** (1991), no. 4, 196–197.

[427] M.V. Pavlov and S.P. Tsarev, *Classical mechanical systems with one-and-a-half degrees of freedom and Vlasov kinetic equation*, arXiv 1306.3737, 2013, 37 pp.

[428] M.V. Pavlov, *Algebro-geometric approach in the theory of integrable hydrodynamic type systems*, Comm. Math. Phys. **272** (2007), no. 2, 469–505.

[429] (ed) P. Pelce, *Dynamics of Curved Fronts*, Academic (Boston), 1988.

[430] M.A. Peterson and J. Ferry, *Spontaneous symmetry breaking in needle crystal growth*, Phys. Rev. A **39** (1989), no. 5, 2740–2741.

[431] A. Petrosyan, H. Shahgholian, and N. Uraltseva, *Regularity of free bound-aries in obstacle-type problems*, Graduate Studies in Mathematics, 136. American Mathematical Society, Providence, RI, 2012, 221 pp.

[432] N.B. Pleshchinskii and M. Reissig, *Hele-Shaw flows with nonlinear kinetic undercooling regularization*, Nonlinear Anal. **50** (2002), no. 2, 191–203.

[433] H. Poincaré, *Sur les équations aux dérivées partielles de la physique mathé-matique*, Amer. J. Math. **12** (1890), 211–294.

[434] H. Poincaré, *Théorie du potentiel Newtonien*, Paris, Gauthier-Villars, 1899.

[435] H. Poincaré, *L'état actuel et l'avenir de la physique mathématique*, Bulletin des sciences mathématiques **28** (1904), no. 2, 302–324.

[436] H. Poincaré, *Sur la dynamique de l'électron*, Compte rendus de l'Académie des Sciences de Paris **140** (1905), 1504–1508.

[437] H. Poincaré, *Sur la dynamique de l'électron*, Rendiconti del circolo matem-atico di Palermo **21** (1906), 129–175.

[438] P.Ya. Polubarinova-Kochina, *On a problem of the motion of the contour of a petroleum shell*, Dokl. Akad. Nauk USSR **47** (1945), no. 4, 254–257 (in Russian).

[439] P.Ya. Polubarinova-Kochina, *Concerning unsteady motions in the theory of filtration*, Prikl. Matem. Mech. **9** (1945), no. 1, 79–90 (in Russian).

[440] P.Ya. Polubarinova-Kochina, *Theory of motion of ground water*, Gosudarstv. Izdat. Tehn.-Teor. Lit., Moscow, 1952.

[441] G. Pólya and G. Szegő, *Isoperimetric Inequalities in Mathematical Physics*, Princeton, N.J.: Princeton University Press, 1951.

[442] G. Pólya and M. Schiffer, *Sur la représentation conforme de l'extérieur d'une courbe fermée convexe*, C. R. Acad. Sci. Paris **248** (1959), 2837–2839.

[443] A.M. Polyakov, *Quantum geometry of bosonic strings*, Phys. Lett. B **103** (1981), no. 3, 207–210.

[444] Ch. Pommerenke, *Über die Subordination analytischer Funktionen*, J. Reine Angew. Math. **218** (1965), 159–173.

[445] Ch. Pommerenke, *On the logarithmic capacity and conformal mapping*, Duke Math. J. **35** (1968), 321–325.

[446] Ch. Pommerenke, *Univalent functions, with a chapter on quadratic differen-tials by G. Jensen*, Vandenhoeck & Ruprecht, Göttingen, 1975.

[447] Ch. Pommerenke, *Boundary behaviour of conformal maps*, Springer, Berlin, 1992.

[448] Ch. Pommerenke and A. Vasil'ev, *Angular derivatives of bounded univa-lent functions and extremal partitions of the unit disk*, Pacific. J. Math. **206** (2002), no. 2, 425–450.

[449] N.V. Popova, *Connection between the Löwner equation and the equation* $\frac{dw}{dt} = \frac{1}{w - \lambda(t)}$, Izv. Acad. Sci. Belorussian SSR **6** (1954), 97–98.

[450] O Praud and H.L. Swinney, *Fractal dimension and unscreened angles measured for radial viscous fingering*, Phys. Rev.z E **72** (2005), no. 1, 011406.

[451] A. Pressley and G. Segal, *Loop groups*, Oxford Mathematical Monographs, The Clarendon Press, Oxford University Press, New York, 1986. 318 pp.

[452] G. Prokert, *Existence results for Hele-Shaw flow driven by surface tension*, European J. Appl. Math. **9** (1998), no. 2, 195–221.

[453] D. Prokhorov, *Level lines of functions that are convex in the direction of an axis.*, Mat. Zametki (Math. Notes) **44** (1988), no. 4, 523–527 (in Russian).

[454] D. Prokhorov and A. Vasil'ev, *Convex Dynamics in Hele-Shaw cells*, Intern. J. Math. and Math. Sci. **31** (2002), no. 11, 639–650.

[455] D. Prokhorov and A. Vasil'ev, *Univalent functions and integrable systems*, Comm. Math. Phys. **262** (2006), no. 2, 393–410.

[456] M. Putinar, *On a class of finitely determined planar domains*, Math. Res. Lett. **1** (1994), no. 3, 389–398.

[457] M. Putinar, *Extremal solutions of the two-dimensional L-problem of moments*, J. Funct. Anal. **136** (1996), no. 2, 331–364.

[458] M. Putinar, *Extremal solutions of the two-dimensional L-problem of moments. II*, J. Approx. Theory **92** (1998), no. 1, 38–58.

[459] Q. Nie, F.R. Tian, *Singularities in Hele-Shaw flows*, SIAM J. Appl. Math. **58** (1998), no. 1, 34–54.

[460] P. Raboin, *Le problème du $\bar{\partial}$ sur en espace de Hilbert*, Bull. Soc. Math. France **107** (1979), 225–240.

[461] A. Rákos and G.M. Schütz, *Current distribution and random matrix ensembles for an integrable asymmetric fragmentation process*, J. Stat. Phys. **118** (2005), no. 3-4, 511–530.

[462] A. Rákos and G.M. Schütz, *Bethe ansatz and current distribution for the TASEP with particle-dependent hopping rates*, Markov Process. Related Fields **12** (2006), no. 2, 323–334.

[463] T. Ransford, *Potential theory in the complex plane*, London Math. Soc. Student Texts, 28, Cambridge Univ. Press, 1995.

[464] M. Reed and B. Simon, *Methods of modern mathematical physics. I. Functional analysis.* Second edition. Academic Press, Inc. [Harcourt Brace Jovanovich, Publishers], New York, 1980. 400 pp.

[465] Y. Reichelt, *Moving boundary problems for degenrate elliptic equations*, Nonlinear Analysis **27** (1996), 1207–1227.

[466] O. Reynolds, *An experimental investigation of the circumstances which determine whether the motion of water shall be direct or sinuous, and of the law of resistance in parallel channels*, Royal Society, Phil. Trans. **174** (1883), 935–982.

[467] M. Reissig and L. von Wolfersdorf, *A simplified proof for a moving boundary problem for Hele-Shaw flows in the plane*, Ark. Mat. **31** (1993), no. 1, 101–116.

[468] M. Reissig and F. Hübner, *Analytical and numerical treatment of Hele-Shaw models with and without regularization*, Generalized analytic functions (Graz, 1997), Int. Soc. Anal. Appl. Comput., 1, Kluwer, 1998, 271–287.

[469] M. Reissig and S.V. Rogosin, *Analytical and numerical treatment of a complex model for Hele-Shaw moving boundary value problems with kinetic undercooling regularization*, European J. Appl. Math. **10** (1999), no. 6, 561–579.

[470] S. Richardson, *Hele-Shaw flows with a free boundary produced by the injecton of fluid into a narrow channel*, J. Fluid Mech. **56** (1972), no. 4, 609–618.

[471] S. Richardson, *On the classification of solutions to the zero surface tension model for Hele-Shaw free boundary flows*. Quart. Appl. Math. **55** (1997), no. 2, 313–319.

[472] S. Richardson, *Hele-Shaw flows with time-dependent free boundaries involving a multiply-connected fluid region*, Europ. J. Appl. Math. **12** (2001), 677–688.

[473] S. Richardson, *Hele-Shaw flows with free boundaries in a corner or around a wedge. Part I: Liquid at the vertex*, Europ. J. Appl. Math. **12** (2001), 665–676.

[474] S. Richardson, *Hele-Shaw flows with free boundaries in a corner or around a wedge. Part II: Air at the vertex*, Europ. J. Appl. Math. **12** (2001), 677–688.

[475] B. Rider, *Deviations from the circular law*, Probab. Theory Related Fields **130** (2004), no. 3, 337–367.

[476] L. Ristroph, M. Thrasher, M.B. Mineev-Weinstein, and H.L. Swinney, *Fjords in viscous fingering: Selection of width and opening angle*, Physical Review E (Statistical, Nonlinear, and Soft Matter Physics) **74** (2006), no. 1, 015201.

[477] M.S. Robertson, *Analytic functions starlike in one direction*, American J. Math. **58** (1936), 465–72.

[478] A. Robertson, Dr. Stanley Richardson 1943–2008. In memory of an outstanding mathematician (Obituary), The Journal. Scottland's Student Newspaper, March 2008.

[479] S. Rohde and O. Schramm, *Basic properties of SLE*, Ann. of Math. (2) **161** (2005), no. 2, 883–924.

[480] S. Rohde, *Oded Schramm: from circle packing to SLE*, Ann. Probab. **39** (2011), no. 5, 1621–1667.

[481] S. Rodhe and M. Zinsmeister, *Some remarks on Laplacian growth*, Topology Appl. **152** (2005), no. 1-2, 26–43.

[482] M. Rosenblum and J. Rovnyak, *Topics in Hardy classes and univalent functions*, Birkhäuser Verlag, Basel, 1994.

[483] J. Roos, *Equilibrium measures and partial balayage*, arXiv:1309.5252 [math. CV] and Compl. Anal. Oper. Theory (DOI: 10.1007/s11785-014-0372-4, to appear).

[484] J. Ross and D. Witt Nyström, *The Hele-Shaw flow and moduli of holomorphic discs* (2012), arXiv:1212.2337 [math.CV].

[485] J. Rubinstein, *Effective equations for flow in random porous media with a large number of scales*, J. Fluid Mech. **170**, (1986) 379–383.

[486] S. Ruscheweyh and L.C. Salinas, *On the preservation of direction-convexity and the Goodman-Saff conjecture*, Ann. Acad. Sci. Fenn. Ser. A-I Math. **14** (1989), no. 1, 63–73.

[487] E.B. Saff and V. Totik, *Logarithmic potentials with external fields* (with Appendix B by Thomas Bloom), Springer-Verlag, Berlin.

[488] P.G. Saffman and G.I. Taylor, *The penetration of a fluid into a porous medium or Hele-Shaw cell containing a more viscous liquid*, Proc. Royal Soc. London, Ser. A **245** (1958), no. 281, 312–329.

[489] P.G. Saffman and G.I. Taylor, *A note on the motion of bubbles in a Hele-Shaw cell and porous medium*, Quart. J. Mech. Appl. Math. **17** (1959), no. 3, 265–279.

[490] M. Sakai, *Quadrature domains*, Lecture Notes in Math. 934, Springer-Verlag, New York, 1982.

[491] M. Sakai, *Regularity of boundaries having a Schwarz function*, Acta Math. **166** (1991), 263–297.

[492] M. Sakai, *Regularity of boundaries in two dimensions*, Ann. Scuola Norm. Sup. Pisa Cl. Sci. (4) **20** (1993), 323–339.

[493] M. Sakai, *Application of variational inequalities to the existence theorem on quadrature domains*, Trans. Amer. Math. Soc. **276** (1993), 267–279.

[494] M. Sakai, *Regularity of boundaries of quadrature domains in two dimensions*, SIAM J. Math. Analysis **24** (1993), no. 2, 341–364.

[495] M. Sakai, *Sharp estimate of the distance from a fixed point to the frontier of a Hele-Shaw flow*, Potential Analysis **8** (1998), no. 3, 277–302.

[496] M. Sakai, *Linear combinations of harmonic measures and quadrature domains of signed measures with small support*, Proc. Edinburgh. Math. Soc. **42** (1999), 433–444.

[497] M. Sakai, *Restriction, localization, and microlocalization*, in [169].

[498] M. Sakai, *A moment problem on Jordan domains*, Proc. Amer. Math. Soc. **70** (1978), no. 1, 35–38.

[499] M. Sakai, *Null quadrature domains*, J. Analyse Math. **40** (1981), 144–154.

[500] M. Sakai, *Quadrature domains*, Lecture Notes in Mathematics, vol. 934, Springer-Verlag, Berlin, 1982.

[501] M. Sakai, *An index theorem on singular points and cusps of quadrature domains*, Holomorphic functions and moduli, Vol. I (Berkeley, CA, 1986), Math. Sci. Res. Inst. Publ., vol. 10, Springer, New York, 1988, pp. 119–131.

[502] M. Sakai, *Quadrature domains with infinite volume*, Compl. Anal. Oper. Theory **3** (2009), 525–549.

[503] M. Sakai, *Small modifications of quadrature domains*, Mem. Amer. Math. Soc. **206** (2010), no. 969, vi+269.

[504] M. Sato and Y. Sato, *Soliton equations as dynamical systems on infinite-dimensional Grassmann manifold*, Nonlinear Partial Differential Equations in Applied Science Tokyo, 1982, North-Holland Math. Stud. vol. 81, North-Holland, Amsterdam (1983), pp. 259–271.

[505] A. Savakis and S. Maggelakis, *Models of shrinking clusters with applications to epidermal wound healing*, Math. Comp. Modeling **25** (1997), no. 6, 1–6.

[506] T.V. Savina, B. Yu. Sternin, and V.E. Shatalov, *On a minimal element for a family of bodies producing the same external gravitational field*, Appl. Anal. **84** (2005), no. 7, 649–668.

[507] Y. Sawada, A. Dougherty, and J.P. Gollub, *Dendritic and fractal patterns in electrolytic metal deposits*, Phys. Rev. Lett. **56** (1986), no. 12, 1260–1263.

[508] A.C. Schaeffer and D.C. Spencer, *Coefficient Regions for Schlicht Functions (With a Chapter on the Region of the Derivative of a Schlicht Function by Arthur Grad)*, American Mathematical Society Colloquium Publications, Vol. 35. American Mathematical Society, New York, 1950.

[509] D.G. Schaeffer, *A stability theorem for the obstacle problem*, Advances in Math. **17** (1975), no. 1, 34–47.

[510] D.G. Schaeffer, *Some examples of singularities in a free boundary*, Ann. Scuola Norm. Sup. Pisa Cl. Sci. (4) **4** (1977), 133–144.

[511] M. Schiffer and N.S. Hawley, *Connections and conformal mapping*, Acta Math. **107** (1962), 175–274.

[512] G. Segal and G. Wilson, *Loop groups and equations of KdV type*, Publ. Math. IHES **61** (1985), no. 1, 5–65.

[513] S. Dao-Shing, *Parametric representation of quasiconformal mappings*, Science Record **3** (1959), 400–407.

[514] H. Shahgholian, *Unbounded quadrature domains in $\mathbb{R}^n$ ($n \geq 3$)*, J. Analyse Math. **56** (1991), 281–291.

[515] H. Shahgholian, *Quadrature surfaces as free boundaries*, Ark. Mat. **32** (1994), no. 2, 475–492.

[516] H.S. Shapiro, *Unbounded quadraure domains*, pp. 287–331 in Complex analysis, I (College Park, Md., 1985–86), Springer Lecture Notes 1275, 1987

[517] H.S. Shapiro, *The Schwarz function and its generalization to higher dimensions*, University of Arkansas Lecture Notes in the Mathematical Sciences, 9. A Wiley-Interscience Publication. John Wiley & Sons, Inc., New York, 1992.

[518] O. Schramm, *Scaling limits of loop-erased random walks and uniform spanninig trees*, Israel J. Math. **118** (2000), 221–288.

[519] M.J. Shelley, F.-R. Tian, and K. Wlodarski, *Hele-Shaw flow and pattern formation in a time-dependent gap*, Nonlinearity **10** (1997), no. 6, 1471–1495.

[520] D. Shoikhet, *Semigroups in geometrical function theory*, Kluwer Academic Publishers, Dordrecht, 2001.

[521] B. Shraiman and D. Bensimon, *Singularities in nonlocal interface dynamics*, Phys. Rev. A **30** (1984), no. 5, 2840–2842.

[522] M. Siegel, S. Tanveer, and W.S. Dai, *Singular effects of surface tension in evolving Hele-Shaw flows*, J. Fluid Mech. **323** (1996), 201–236.

[523] M. Siegel, R.E. Caflisch, and S.D. Howison, *Global existence, singular solutions and ill-posedness for the Muskat problem*, Comm. Pure Appl. Math. **57** (2004), 1374–1411.

[524] A.M.P. Silva and G.L. Vasconcelos, *Doubly periodic array of bubbles in a Hele-Shaw cell*, Proc. R. Soc. Lond. Ser. A Math. Phys. Eng. Sci. **467** (2011), no. 2126, 346–360.

[525] T. Sjödin, *Mother bodies of algebraic domains in the complex plane*, Complex Var. Elliptic Eq. **51** (2006), 357–369.

[526] T. Sjödin, *On the structure of partial balayage*, Nonlinear Anal. **67** (2007), no. 1, 94–102.

[527] S. Smirnov, *Critical percolation in the plane: conformal invariance, Cardy's formula, scaling limits*, C. R. Acad. Sci. Paris Sér. I Math. **333** (2001), 239–244.

[528] A. Yu. Solynin, *Modules and extremal metric problems*, Algebra i Analiz **11** (1999), no. 1, 3–86; English transl.: St. Petersburg Math. J. **11** (2000), no. 1, 1–70.

[529] A. Soshnikov and Y.V. Fyodorov, *On the largest singular values of random matrices with independent Cauchy entries*, J. Math. Phys. **46** (2005), no. 3, 033302, 15 pp.

[530] T.T. Soong, *Random differential equations in science and engineering*. Academic Press, New York, 1973.

[531] J. Stankiewicz, *Some remarks concerning starlike functions*, Bull. Acad. Polon. Sci., Sér. Sci. Math. Astronom. Phys. **18** (1970), 143–146.

[532] The correspondence between Sir George Gabriel Stokes and Sir William Thomson Baron Kelvin of Largs, Vol. 2, 1807–1901, letter 607.

[533] K. Strebel, *Quadratic differentials*, Springer-Verlag, 1984.

[534] G. Szegő, *Orthogonal polynomials*, American Mathematical Society, Providence, R.I., third edition, 1967, American Mathematical Society Colloquium Publications, Vol. 23.

[535] K. Takasaki and T. Takebe, *Integrable hierarchies and dispersionless limit*, Rev. Math. Phys. **7** (1995), no. 5, 743–808.

[536] K. Takasaki and T. Takebe, *Radial Löwner equation and dispersionless cmKP hierarcy*, arXiv: nlin.SI/0601063, (2006), 1–18.

[537] T. Takebe, L.-P. Teo, and A. Zabrodin, *Löwner equations and dispersionless hierarchies*, J. Phys. A: Math. Gen. **39** (2006), 11479–11501.

[538] L.A. Takhtajan, *Liouville theory: quantum geometry of Riemann surfaces*, Modern Phys. Lett. A **8** (1993), no. 37, 3529–3535.

[539] L.A. Takhtajan and L.-P. Teo, *Weil–Petersson metric on the universal Teichmüller space I. Curvature properties and Chern forms*, arXiv: math. CV/0312172, 2004.

[540] L.A. Takhtajan and L.-P. Teo, *Weil–Petersson metric on the universal Teichmüller space II. Kähler potential and period mapping*, arXiv: math. CV/0406408, 2004.

[541] L.A. Takhtajan and L.-P. Teo, *Weil–Petersson geometry of the universal Teichmuller space*, Progress in Math. **237** (2005), 225–233.

[542] L.A. Takhtajan and L.-P. Teo, *Weil–Petersson geometry on the universal Teichmuller space*, Mem. Amer. Math. Soc. **183** (2006), no. 861, 119 pp.

[543] C. Tam, *The drag on a cloud of spherical particles in low Reynolds number flow*, J. Fluid Mech. **38** (1969), no. 3, 537–546, .

[544] O. Tammi, *Extremum problems for bounded univalent functions*, Lecture Notes in Mathematics, 646. Springer-Verlag, Berlin – New York, 1978.

[545] O. Tammi, *Extremum problems for bounded univalent functions II*, Lecture Notes in Mathematics, 913. Springer-Verlag, Berlin – New York, 1982.

[546] O. Tammi, *Some coefficient estimations in the class Σ_b' of meromorphic univalent functions*, Ann. Acad. Sci. Fenn. Ser. A I Math. **13** (1988), no. 1, 125–136.

[547] S. Tanveer, *Evolution of Hele-Shaw interface for small surface tension*, Phil. Trans. R. Soc. Lond., A **343** (1993), no. 1668, 155–204.

[548] S. Tanveer and X. Xie, *Rigorous results in steady finger selection in viscous fingering*, Arch. Rational Mech. Anal. **166** (2003), 219–286.

[549] S. Tanveer and X. Xie, *Analyticity and nonexistence of classical steady Hele-Shaw fingers*, Comm. Pure Appl. Math. **56** (2003), no. 3, 353–402.

[550] O. Teichmüller, *Extremale quasikonforme Abbildungen und quadratische Differentiale*, Abhandl. Preuss. Akad. Wiss., Math.-Naturwiss. Kl. **22** (1940), 3–197.

[551] R. Teodorescu, E. Bettelheim, O. Agam, A. Zabrodin, and P. Wiegmann, *Normal random matrix ensemble as a growth problem*, Nuclear Phys. B **704** (2005), no. 3, 407–444.

[552] H. Thomé, M. Rabaud, V. Hakim, and Y. Couder, *The Saffman–Taylor Instability: From the Linear to the Circular Geometry* Phys. Fluids A **1** (1989), 224–240.

[553] F.R. Tian, *A Cauchy integral approach to Hele-Shaw problems with a free boundary*, Arch. Rational Mech. Anal. **135** (1996), no. 2, 175–196.

[554] F.R. Tian, *Hele-Shaw problems in multidimensional spaces*, J. Nonlinear Science **10** (2000), 275–290.

[555] Y. Tu, *Saffman–Taylor problem in sector geometry*. Asymptotics beyond all orders (La Jolla, CA, 1991), NATO Adv. Sci. Inst. Ser. B Phys., 284, Plenum, New York, 1991, 175–186.

[556] V.G. Tkachev, *Ullemar's formula for the moment map. II*, Linear Algebra Appl. **404** (2005), 380–388.

[557] D. Tse and P. Viswanath, *On the capacity of the multiple antenna broadcast channel*, in *Multiantenna channels: capacity, coding and signal processing* (*Piscataway, NJ, 2002*), volume 62 of *DIMACS Ser. Discrete Math. Theoret. Comput. Sci.*, pages 87–105. Amer. Math. Soc., Providence, RI, 2003.

[558] C. Ullemar, *Uniqueness theorem for domains satisfying a quadrature identity for analytic functions*, Research Bulletin TRITA-MAT-1980-37, Royal Institute of Technology, Department of Mathematics, Stockholm, 1980.

[559] J.-M. Vanden-Broeck and J.B. Keller, *Deformation of a bubble or drop in a uniform flow*, J. Fluid. Mech. **101** (1980), 673–686.

[560] A.N. Varchenko and P.I. Ètingof, *Why the boundary of a round drop becomes a curve of order four*, University Lecture Series, vol. 3, AMS, 1992.

[561] A. Vasil'ev, *Univalent functions in the dynamics of viscous flows*, Comp. Methods and Func. Theory **1** (2001), no. 2, 311–337.

[562] A. Vasil'ev, *Moduli of families of curves for conformal and quasiconformal mappings*. Lecture Notes in Mathematics, vol. 1788, Springer-Verlag, Berlin-New York, 2002.

[563] A. Vasil'ev, *Univalent functions in two-dimensional free boundary problems*, Acta Applic. Math **79** (2003), no. 3, 249–280.

[564] A. Vasil'ev and I. Markina, *On the geometry of Hele-Shaw flows with small surface tension*, Interfaces and Free Boundaries **5** (2003), no. 2, 183–192.

[565] A. Vasil'ev, *Evolution of conformal maps with quasiconformal extensions*, Bull. Sci. Math. **129** (2005), no. 10, 831–859.

[566] A. Vasil'ev, *From the Hele-Shaw experiment to integrable systems: a historical overview*, Complex Anal. Oper. Theory **3** (2009), no. 2, 551–585.

[567] J. Verbaarschot, *The spectrum of the Dirac operator near zero virtuality for $N_c = 2$ and chiral random matrix theory*, Nuclear Phys. B **426** (1994), no. 3, 559–574.

[568] Vestnik. Tomsk State University Journal of Mathematics and Mechanics 4 (2009)

[569] Yu.P. Vinogradov and P.P. Kufarev, *On a problem of filtration*, Akad. Nauk SSSR. Prikl. Mat. Meh. **12** (1948), 181–198 (in Russian).

[570] Yu.P. Vinogradov and P.P. Kufarev, *On some particular solutions of the problem of filtration*, Doklady Akad. Nauk SSSR (N.S.) **57** (1947), 335–338 (in Russian).

[571] T. Vicsek, *Fractal growth phenomena*, World Scientific Publishing Co, Singapore, 1989.

[572] M.A. Virasoro, *Subsitiary conditions and ghosts in dual resonance models*, Phys. Rev. **D1** (1970), 2933–2936.

[573] A.A. Vlasov, *Many-particle theory and its application to plasma*. Gordon and Breach Science Publishers, Inc., New York, 1961.

[574] M. Volpato, *Sulla derivabilità, rispetto a valori iniziali ed a parametri, delle soluzioni dei sistemi di equazioni differenziali ordinarie del primo ordine*, Rend. Sem. Mat. Univ. Padova **28** (1958), 71–106.

[575] E. Vondenhoff, *Long-time asymptotics of Hele-Shaw flow for perturbed balls with injection and suction*, Interfaces Free Bound. **10** (2008), no. 4, 483–502.

[576] Y.U. Wang, Y.M. Jin, A.M. Cuitiño, and A.G. Khachaturyan, *Phase field microelasticity theory and modeling of multiple dislocation dynamics*, Applied Physics Letters (2001), 78.

[577] W. Wasow, *Linear turning point theory*, volume 54 of *Applied Mathematical Sciences*, Springer-Verlag, New York, 1985.

[578] J. Weiner, Philip G. Saffman. 1931–2008 (Obituary), Engineering and Science, Caltech **71** (2008), no. 3, 44.

[579] N. Whitaker, *Numerical solution of the Hele-Shaw equations*, J. Comput. Phys. **90** (1990), no. 1, 176–199.

[580] P.B. Wiegmann and A. Zabrodin, *Conformal maps and integrable hierarchies*, Comm. Math. Phys. **213** (2000), no. 3, 523–538.

[581] P. Wiegmann and A. Zabrodin, *Large scale correlations in normal and general non-hermitian matrix ensembles*, J. Phys. A. **36** (2003), 3411.

[582] E. Witten, *Quantum field theory, Grassmannians, and algebraic curves.* Comm. Math. Phys. **113** (1988), no. 4, 529–600.

[583] T.A. Witten, Jr., and L.M. Sander, *Diffusion-Limited Aggregation, and kinetic critical phenomenon*, Phys. Rev. Letters **47** (1981), no. 2, 1400–1403.

[584] T.A. Witten and L.M. Sander, *Diffusion-Limited Aggregation*, Phys. Rev. B **27** (1983), no. 9, 5686–5697.

[585] E.P. Wigner, *On the statistical distribution of the widths and spacings of nuclear resonance levels*, Proc. Cambridge Philos. Soc. **47** (1950), 790.

[586] E.P. Wigner, *Characteristic vectors of bordered matrices with infinite dimensions*, Ann. of Math. (2) **62** (1955), 548–564.

[587] E.P. Wigner, *Characteristic vectors of bordered matrices with infinite dimensions. II.*, Ann. of Math. (2) **65** (1957), 203–207.

[588] G. Wilson, *Bispectral commutative ordinary differential operators*, J. Reine Angew. Math. **442** (1993), 177–204.

[589] G. Wilson, *Collision of Calogero–Moser particles and an adelic Grassmanian* (with an appendix by I.G. Macdonald), Invent. Math. **133** (1998), 1–41.

[590] G.B. Whitham, *Linear and nonlinear waves.* Pure and Applied Mathematics (New York). John Wiley & Sons Inc., New York, 1999.

[591] S. Wolpert, *Thurston's Riemannian metric for Teichmüller space*, J. Differential Geom. **23** (1986), no. 2, 143–174.

[592] D.V. Yakubovich, *Real separated algebraic curves, quadrature domains, Ahlfors type functions and operator theory*, J. Funct. Anal. **236** (2006), no. 1, 25–58.

[593] I. Zakharevich, *A generalization of Wigner's law*, Comm. Math. Phys. **268** (2006), no. 2, 403–414.

[594] V.E. Zakharov, *Benney equations and quasiclassical approximation in the inverse problem method*, Funktsional Anal. i Prilozhen. **14** (1980), no. 2, 15–24.

[595] D. Zidarov, *Inverse gravimetric problem in geoprospecting and geodesy*, Developments in Solid Earth Geophysics, no. 19, Elsevier, 1990.

[596] P.G. Zograf and L.A. Takhtajan, *On the Liouville equation, accessory parameters and the geometry of Teichmüller space for Riemann surfaces of genus 0*, Mat. Sb. (N.S.) **132(174)** (1987), no. 2, 147–166; translation in Math. USSR-Sb. **60** (1988), no. 1, 143–161.

[597] P. Zograf and L. Takhtajan, *Hyperbolic 2-spheres with conical singularities, accessory parameters and Kähler metrics on $M_{0,n}$*, Trans. Amer. Math. Soc. **355** (2003), no. 5, 1857–1867.

[598] N.M. Zubarev and O.V. Zubareva, *Exact solutions for equilibrium configurations of charged conducting liquid jets*, Physical Review E **71** (2005), no. 1, 016307.

List of Symbols

$\mathbb{C}$ complex plane

$\overline{\mathbb{C}}$ Riemann sphere

$\mathbb{D}$ open unit disk

$\partial \mathbb{D} = S^1$,   unit circle

$\mathbb{D}^*$ exterior part of unit disk

$\mathbb{D}_r(a) = \mathbb{D}(a, r)$,   open disk with center a and radius r

$\mathbb{D}_r = \mathbb{D}_r(0)$

$\mathbb{H}$ upper half-plane

$\mathbb{H}^+$ right half-plane

$\mathbb{H}^-$ left half-plane

$\mathbb{R}$ real line

$\mathbb{R}^+$ positive real axis

$\mathbb{R}^-$ negative real axis

$\overline{D}$ closure of D

int D,   interior of D

$|D|$ area of D

$f \circ g$,   superposition f and g

$\delta_a(z)$,   Dirac's distribution in $z \in \mathbb{C}$ supported at a

$G_\Omega(z, a)$,   Green's function of Ω with pole at a

$\mathbf{F} \equiv {}_2\mathbf{F}_1$,   Gauss hypergeometric function

$\mathbf{B}$ Euler's Beta-function

$\mathbf{K}$ complete elliptic integral

$S(z)$ Schwarz function

χ_Ω characteristic function of Ω

$d\sigma_z = dxdy = d^2z$,   area element in z-plane

$ds = d\ell = |dz|$,   arc length measure in z-plane

Bal   partial balayage

dist (Γ, a), distance from set Γ to point a

h^* holomorphic reflection in the unit circle: $h^*(\zeta) = \overline{h(1/\bar{\zeta})}$

$*u$ conjugate harmonic function (or Hodge star on differential form)

$\mathbb{E}$ expectation value

S class of univalent functions in $\mathbb{D}$ normalized by $f(\zeta) = \zeta + a_2\zeta^2 + \cdots$

$\mathcal{F}_0$ subclass of S of functions smooth at the boundary of $\mathbb{D}$

Σ class of univalent functions in $\mathbb{D}^*$

       normalized by $f(\zeta) = \zeta + a_0 + a_1/\zeta + \cdots$

Σ_0 subclass of Σ of functions with $a_0 = 0$

Hol $(\mathbb{D})$, holomorphic functions in $\mathbb{D}$

$\tilde{\Sigma}$ subclass of Σ of functions smooth on the boundary

$\tilde{\Sigma}_0$ subclass of Σ_0 of functions smooth on the boundary

S^* class of starlike functions in $\mathbb{D}$

S_α^* class of starlike functions of order α in $\mathbb{D}$

$S^*(\alpha)$, class of strongly starlike functions of order α in $\mathbb{D}$

C class of convex functions in $\mathbb{D}$

$C_\mathbb{R}$ class of convex functions in $\mathbb{D}$ in the direction of $\mathbb{R}$

$H_\mathbb{R}^-$ class of convex functions in $\mathbb{H}^+$ in the negative direction of $\mathbb{R}$

$H_\mathbb{R}^-(\alpha)$, class of convex functions in $\mathbb{H}^+$ of order α in the negative direction of $\mathbb{R}$

$m(D, \Gamma)$, modulus of a family of curves Γ in D

$M(D)$, conformal modulus of a doubly connected domain D

$R(D, a)$, conformal radius of D with respect to a

$m(D, a)$, reduced modulus of D with respect to a

cap C, capacity of a condenser C

cap $^{(h)}C$, hyperbolic capacity of a continuum C

$m_\Delta(D, a)$, reduced modulus of a triangle D with respect to its vertex a

$S_f(z)$, Schwarzian derivative

Diff S^1, Lie group of C^∞ sense preserving diffeomorphisms of the unit circle S^1

Rot S^1, group of rotations of S^1

Vect S^1, Lie algebra of smooth tangent vector fields to S^1

Vir Virasoro–Bott group

Gr (H), Sato–Segal–Wilson Grassmannian over a Hilbert space H

Gr $_\infty(H)$, infinite dimensional smooth Grassmannian

Gr $^0(H)$, the component of Gr $_\infty(H)$ of virtual dimension 0

Index